Pilgrims: Values and Identities

Pilgrims: Values and Identities

Edited by

Darius Liutikas

Lithuanian Social Research Centre, Vilnius, Lithuania

CABI

CABI is a trading name of CAB International

CABI
Nosworthy Way
Wallingford
Oxfordshire OX10 8DE
UK

Tel: +44 (0)1491 832111
Fax: +44 (0)1491 833508
E-mail: info@cabi.org
Website: www.cabi.org

CABI
WeWork
One Lincoln St
24th Floor
Boston, MA 02111
USA

T: +1 (617)682-9015
E-mail: cabi-nao@cabi.org

British Library Cataloguing-in-Publication Data
A catalogue record for this book is available from the British Library, London.

Library of Congress Cataloging-in-Publication Data

ISBN-13: 978 1 78924 565 3 (hardback)
978 1 78924 566 0 (ePDF)
978 1 78924 567 7 (ePub)

Commissioning Editor: Claire Parfitt
Editorial Assistant: Emma McCann
Production Editor: James Bishop

Typeset by Exeter Premedia Services Pvt Ltd, Chennai, India
Printed and bound in the UK by Severn, Gloucester

Contents

Contributors

María Ángeles Piñeiro Antelo has a PhD in Geography (2015) and Master's degrees in Territorial Planning and Development (2002) and Geographic Information Systems (2016). Since 2003, she has worked as a researcher at the Galician Studies and Development University Institute (IDEGA). Between 2007 and 2009 she was Director of the Institute of Tourist Studies of Galicia (IETG). She has some 15 years' professional experience in the private and public sector; in European, Spanish and regional projects; and in contracts with companies and public administrations. She has taught at the Universities of Santiago and Vigo, and is currently a lecturer in the Geography Department of the University of Santiago de Compostela.

University of Santiago, Faculty of Geography and History, Praza da Universidade, nº 1, 15782, Santiago de Compostela, Spain. E-mail: manxeles.pineiro@usc.es

Pedro Azevedo holds a degree in History and a Master's in Heritage and Cultural Tourism from the University of Minho. He is currently a researcher at CETRAD (Centre for Transdisciplinary Development Studies) of UTAD (Portugal). His current research interests focus on the Ways of Saint James in the region of Trás-os-Montes and Alto Douro (northern Portugal).

Department of Economics, Sociology and Management (DESG), School of Human and Social Sciences (ECHS), University of Trás-os-Montes and Alto Douro (UTAD), ECHS Pole II Building, Office E2.03 (2nd floor), Quinta de Prados 5000-103, Folhadela -Vila Real (Portugal). E-mail: pedrodosrc@gmail.com

Derek Dalton is an Associate Professor in the College of Business, Government and Law at Flinders University (Australia). Previously he lectured at Charles Sturt University, NSW. In 2015, he published *Dark Tourism and Crime*, a Routledge monograph exploring tourism at sites across the world connected to crime, mass murder and genocide. In 2019 he published *Encountering Nazi Tourism Sites*. This book explores how the terrible legacy of Nazi criminality is experienced by tourists, bridging the gap between cultural criminology and tourism studies to make a significant contribution to our understanding of how Nazi criminality is evoked and invoked in the landscape of modern Germany.

College of Business, Government & Law, Flinders University, Sturt Road, Bedford Park, South Australia 5042, GPO Box 2100, Adelaide SA 5001. E-mail: derek.dalton@flinders.edu.au

Luciana Thais Villa Gonzalez is Senior Lecturer at the Management and Tourism Department at the Federal Rural University of Rio de Janeiro, Brazil. She received her PhD in Social Sciences from Rio de Janeiro State University. Her research interests focus on religious or spiritual tourism, anthropology perspectives over hospitality and sharing economy.

Departamento de Administração e Turismo (UFRRJ), Av. Governador Roberto da Silveira, s/n. Nova Iguaçu, Rio de Janeiro, Brasil, CEP 26285-060. E-mail: ltvgonzalez@yahoo.com.br

Rubén Camilo Lois González is full professor of Geography, Department of Geography, University of Santiago de Compostela. He has been General Director for Tourism of the Autonomous Government of Galicia (2005–2009). Currently he is the co-ordinator of the research group ANTE (Territorial Analysis), Director of the Department of Geography and Vice-President of the IGU. His main research lines are Geography and Urban Planning, Regional Studies, Social Geography and Regional Planning. He has been an invited professor at the Universities of Le Mans, Caen and Toulouse-Le Mirail (France), Bergen and Finmark Universitary College (Norway), and visiting professor in Calgary (Canada). He is the author many articles in scientific reviews such as *Geographical Research, Tourism Management Perspective, Die Erde, Journal of Tourism and Cultural Change* and *Tourism and Hospitality Research.*

University of Santiago, Faculty of Geography and History, Praza da Universidade, nº 1, 15782, Santiago de Compostela, Spain. E-mail: rubencamilo.lois@usc.es

Luis Alfonso Escudero Gómez was awarded a PhD in Geography from the Universidad de Santiago de Compostela in 2000. He is a Senior Lecturer in Human Geography in the Department of Geography of the Universidad de Castilla-La Mancha. He has participated in national and international research projects and is the author of many books and chapters in books. He has published in scientific journals such as *BAGE, Sustainability, Urban Science* and *The Open Urban Studies Journal.*

University of Castilla-La Mancha, Faculty of Humanities, Department of Geography, Plaza de Padilla No. 4, 45001 Toledo, Spain. E-mail: luisalfonso.escudero@uclm.es

Rami K. Isaac was born in Palestine. He did his undergraduate studies in The Netherlands and his graduate studies in the UK, and earned his PhD in Spatial Sciences from the University of Groningen, The Netherlands. He is currently Senior Lecturer in Tourism at Breda University of Applied Sciences in The Netherlands. He is an assistant professor at the Faculty of Tourism and Hotel Management at Bethlehem University, Palestine. He is the Vice-President of the Research Committee 50 on International Tourism, International Sociologist Association ISA (2018–2022). His research interests lie in the areas of tourism development and management, critical theory and political aspects of tourism. He has published numerous articles and book chapters on tourism and political (in)stability, occupation, tourism and war, dark tourism, violence and transformational tourism.

Centre for Sustainability, Tourism and Transport, Breda University of Applied Sciences, Mgr. Hopmansstraat 2, 4817 JS Breda, PO Box 3917, 4800 DX Breda, The Netherlands. E-mail: Isaac.R@buas.nl

Elyor E. Karimov is an Uzbekistani orientalist and scholar of Islamic, Central Eurasian Studies and Diplomatics. He graduated from Tashkent State University (Faculty of Oriental Studies) in 1986. He has a PhD (1990) in Medieval History. He worked in the Academy of Sciences of Uzbekistan. He has been a member of the UNESCO Memory of the World Register Sub-Committee from 2014. He has been Professor of the Middle East and Central Asia at Hofstra University (USA) since 2017. He is an expert in the History of Civilization and Culture in Central Asia and the Middle East with numerous research publications in this field.

Hofstra University (USA), Hempstead, New York 11549-1000. E-mail: ekarimov@yahoo.com

Kumi Kato, PhD (Queensland), is a professor at the Faculty of Tourism of Wakayama University, Japan. Her research areas include sustainability, community resilience and ethics related to tourism, focusing on disaster recovery, slow tourism and dark tourism.

Wakayama University, Sakaedani 930, Wakayama City, Japan (640-8510). E-mail: kumikato@wakayama-u.ac.jp

Darius Liutikas, PhD, is a senior scientific researcher at the Lithuanian Social Research Centre (Vilnius). Research interests include: modern and traditional pilgrimage, sociology of tourism, anthropology of tourism, forms of mobility, sacred places and sacred spaces, and geography of religions. Darius Liutikas accomplished his PhD studies in 2008 at Vilnius University, his thesis being 'Traditional and modern pilgrimage as an expression of values and identity in Lithuanian society'. He has published around 40 articles and book chapters in Lithuanian and international academic journals. He has prepared and published 5 books (3 in the Lithuanian language and 2 in English). The author's monograph about manifestation of values and identity in the journeys was published in 2009 (in Lithuanian). He has participated in more than 40 international scientific conferences as guest speaker or speaker. He is a member of the editorial boards of the journals *Creativity Studies*, *International Journal of Religious Tourism and Pilgrimage*, and *International Journal of Research in Sociology and Anthropology*. He is Chairman of the Community of Lithuanian Pilgrims (www.piligrimai.lt). He initiated and co-ordinated different projects on pilgrimage and religious tourism.

Lithuanian Social Research Centre, Bitininku 1C-6, LT-08417, Vilnius, Lithuania. E-mail: darius.liutikas@gmail.com

Lucrezia Lopez holds a PhD in Geography and is Lecturer in Geography at the Department of Geography in the University of Santiago de Compostela. Her main research lines include: Human and Cultural Geography; Tourism Geography; Heritage and Heritage Management; Geography of Pilgrimages; Geography of Sacred Spaces; and Geo-Humanities. She has published in international scientific reviews (*Mobilities*, *Gender, Place and Culture*, *Emotion, Space and Society* and *Sustainability*). She participates in various research projects at international, national and regional level. She has been a visiting researcher at the University of Bari (Italy), the Université de Paris 1 (La Sorbonne) and Royal Holloway University of London (RHUL).

University of Santiago, Faculty of Geography and History, Praza da Universidade, nº 1, 15782. Santiago de Compostela, Spain. E-mail: lucrezia.lopez@usc.es

Dane Munro (1958, Melbourne) is a Resident Academic at the Institute for Tourism, Travel and Culture (ITTC) at the University of Malta. His academic educational trajectory is marked by an MA in the Classics (Neo-Latin) and a Ph.D., both at the University of Malta, in the cultural anthropology of faith-based tourism and pilgrimage. At the University of Groningen, the Netherlands, he read for a further Ph.D. in history, covering the culture of memoria and the identity of the knights of the Order of St John throughout the ages. Munro lectures in tourism while remaining an active researcher in the field of faith-based tourism and history. He has published widely on both topics.

Institute for Tourism, Travel and Culture (ITTC), University of Malta, Msida MSD 2080, Malta. E-mail: dane.munro@um.edu.mt

Daniel H. Olsen, PhD, is an Associate Professor in the Department of Geography at Brigham Young University. His research interests revolve around religious and spiritual tourism, heritage tourism, and the management of sacred sites, with secondary research interests in tourism in peripheral areas and tourism and disabilities. He is co-editor of *Religion, Tourism, and Spiritual Journeys* (Routledge, 2006), *Religious Pilgrimage Routes and Trails* (CABI, 2018), and *Dark Tourism and Pilgrimage* (CABI, 2020), and has published some 50 journal articles and book chapters on religious and spiritual tourism.

Department of Geography, Brigham Young University, Provo, UT, 84660, USA. E-mail: dholsen@byu.edu

Josephine Pryce is a scholar of organizational behaviour with a keen interest in the nature of work. She came to academia with a strong industry background having worked for *Australia Post* and in the hotel industry. During that time she developed a strong appreciation of the challenges faced by workers, employers and organizations in service industries. Her PhD extended understanding of service predispositions, human relations in the workplace, and organizational culture. She seeks to enrich her perspectives through research that leads to better understanding of the lives of people in communities and the world, especially in relation to their working lives.

James Cook University, PO Box 6811, Cairns, Queensland, Australia 4870. E-mail: josephine.pryce@jcu.edu.au

Ricardo Nicolas Progano PhD, is a lecturer at the Center for Tourism Research, Wakayama University, Japan. His research interests include pilgrimage and religious travel, cultural anthropology and heritage tourism. Has has carried fieldwork on the contemporary tourism development of Japanese pilgrimage sites, particularly Kumano Kodo.

Wakayama University, Sakaedani 930, Wakayama City, Japan (640-8510). E-mail: nprogano@wakayama-u.ac.jp

Xerardo Pereiro holds a PhD in Social Anthropology from the University of Santiago de Compostela and another international PhD in Tourism from the University of La Laguna. He is an Associate Professor of anthropology and cultural tourism in UTAD. He conducts research into the anthropology of tourism and cultural heritage in CETRAD. He is the head of the Tourism First Degree Course of UTAD, and he was a visiting teacher at the universities of Vigo, Coruña, Santiago de Compostela, Pablo Olavide (Seville), Salamanca, Panamá, 'Universidade Nova de Lisboa', Costa Rica, UNICAMP (Brazil) and others. He was awarded the 1994 Vicente Risco Award of Social Anthropology and Social Sciences. At present, he is doing research on the Portuguese Inner Way of Pilgrimage to Santiago de Compostela and about tourism in the Douro region.

University of Trás-os-Montes and Alto Douro (UTAD). E-mail: xperez@utad.pt

Kip Redick is Professor and Chair of the Department of Philosophy and Religion at Christopher Newport University. Professional interests include pilgrimage studies; spiritual journey; spirituality of place; aesthetics; media ecology; visual, religious and environmental rhetoric; and film studies. His interest centres on the study of wilderness trails as sites of spiritual journey. Among his publications are: Interpreting Contemporary Pilgrimage as Spiritual Journey or Aesthetic Tourism along the Appalachian Trail (*International Journal of Religious Tourism and Pilgrimage* 6 (2018)); Sensuous Encounter where Journey and Festival Meet: A Phenomenology of Pilgrimage (*Pilgrimage as Transformative Process: The Movement from Fractured to Integrated* (ed. Heather Warfield and Kate Hetherington, Brill Press, 2018)); Spiritual Rambling: Long Distance Wilderness Sojourning as Meaning-Making (*Journal of Ritual Studies* 30(2), 41–52); Profane Experience and Sacred Encounter: Journeys to Disney and the Camino De Santiago (*Environment, Space, Place* 5(1), 46– 72).

Christopher Newport University, Avenue of the Arts, Newport News, VA 23606. E-mail: kredick@cnu.edu

Larry Russell has been engaged since 1999 in a dialogue with pilgrims about the impact of their shared experience as they walk to an adobe shrine of healing at Chimayó, New Mexico. He walked the journey 11 times, wrote journal articles and contributed chapters on pilgrimage to books. He won the Autoethnography and Personal Narrative Research Award for the best ethnographic article in the National Communication Association. His undergraduate work was at Purdue University, and his graduate studies were at Harvard University and the University of Southern Illinois. Before retirement, he was an associate professor at Hofstra University, teaching performance studies.

(Retired from) Hofstra University, Hempstead, New York. 15 Birch Street, Asheville, North Carolina 28801, USA. E-mail: l.russell.pilgrim@gmail.com

Pravin S. Rana PhD, is Assistant Professor in Tourism Management, Faculty of Arts, Banaras Hindu University, Varanasi. Having experiences of two decades of teaching and research, he has published twenty papers and five books (e.g. *Banaras Region: A Spiritual and Cultural Guide* (2002/2020), co-authored with R.P.B. Singh), *Tourism Geography* (2006, with A.K. Singh), and *Pilgrimage Tourism: A Study of Varanasi Region* (2014). He has delivered seminars in Norway, Canada, Malaysia, Korea and Singapore. He is a member of Indian Tourism Congress, INTACH, and Forum UNESCO -- University and Heritage International Network (FUUH), and Secretary ACLA – Asian Cultural Landscape Association.

'Vaikunth Nilayam', N-3/75, KVR 1, R.K. Puram Colony, Karmajeetpur (Sunderpur), PO: B.H.U., Varanasi, UP 221005, India. E-mail: psranabhu@gmail.com; psrana@bhu.ac.in

Rana P.B. Singh, PhD, was Professor of Cultural Landscapes & Heritage Studies at Banaras Hindu University, and during the last four decades has been involved in studying and promoting the sacred landscapes and heritage planning in the Varanasi region, as well as field studies in Japan, Sweden, Italy and Korea. On these topics he has delivered lectures and seminars at various centres in all parts of the world. His publications include 320 papers and 41 books and anthologies, including *The Spirit and Power of Place* (1994), *Cosmic Order & Cultural Astronomy* (2009), *Sacredscapes and Pilgrimage Systems* (2011), and *Hindu Tradition of Pilgrimage* (2013).

'Usha Apartment', N 9/87 – D 41 (Flat 101-102), Lane 5 – Janaki Nagar Colony, Flyover DLW end – Patia Road, PO: Bajardiha, Varanasi, UP 221106, India. E-mail: ranapbs@gmail.com; https://banaras.academia.edu/RanaPBSINGH/Papers/

Xosé M. Santos is a full professor in Human Geography, at the University of Santiago de Compostela. He has specialized in tourism management of world heritage cities and the Way of Saint-James. He has carried out work on the recent changes in rural areas and other human geographic aspects. He has been Director of the Centre for Tourism Studies and Research (CETUR) at the University of Santiago de Compostela (2005–2014). His professional career includes research in Ireland (University College Cork), Norway (University of Trondheim), UK (University of Edinburgh) and Canada (Memorial University of Newfoundland and UQAM-Montreal).

Department of Geography, University of Santiago de Compostela, 1 University Square, 15782, Santiago de Compostela. E-mail: xosemanuel.santos@usc.es

Augusta X. Thomson earned her BA in Archaeology and Anthropology from Oxford University. She is currently a PhD student and Teaching Fellow in Anthropology at New York University, where her research meanders between mobility studies, ecology, place and space, memory, digital media, visual culture, materiality, religion, and pilgrimage. Trained in NYU's Graduate Certificate Program in Culture and Media, Augusta is currently undertaking visual and ethnographic research along the Camino de Santiago, her long-term field site. Committed to the trend towards 'open access' in academia, she has contributed to publications including *Al Jazeera English*, *New Internationalist*, *The Daily Beast*, and *Geographical Magazine*.

New York University, 855 Ocean Avenue, Apt. 1H, Brooklyn, NY 11226. Email: axt203@nyu.edu

Dallen J. Timothy, PhD, is Professor of Community Resources and Development, Arizona State University, and Senior Sustainability Scientist at the Wrigley Global Institute of Sustainability. He is also currently a visiting professor at several universities in Spain and China, and Senior Research Associate at the University of Johannesburg, South Africa. He is the editor of the *Journal of Heritage Tourism*, Associate Editor of *Annals of Tourism Research* and serves on the editorial boards of 24 scholarly journals. He is commissioning editor for Routledge's *Cultural Heritage and Tourism* book series, and co-commissioning editor for three book series at Channel View Publications.

School of Community Resources and Development, Arizona State University, 411 N. Central Avenue, Suite 550, Phoenix, AZ 85004, USA.

Slawoj Tanas, PhD, Associate Professor at the University of Lodz, Poland, specializing in geography of tourism. His research interests focus on cultural tourism, mainly on the issue of using the space of death in tourism – thanatourism and dark tourism. His research results have been published in dozens of articles in Polish, English and German. He has published two books: *Cemetery Tourism Space: Introduction to Thanatourism* (2008) and *Thanatourism: From Death Space to Tourism Space* (2013). He is the head of the Department of Geography of Recreation, University of Lodz.

University of Lodz, Faculty of Geographical Sciences, Institute of Urban Geography and Tourism Studies, ul. Kopcinskiego 31, 90-142 Lodz, Poland. E-mail: slawoj.tanas@geo.uni.lodz.pl

Shin Yasuda, PhD, is Associate Professor at Takasaki City University of Economics in Takasaki City, Gunma Prefecture in Japan. Yasuda received his BA degree in history from Sophia University in Tokyo. He obtained his Master's and PhD degrees in area study from the Graduate School of Asian and African Area Studies in Kyoto University, Japan. He is presently working at the Department of Tourism Policy, Faculty of Regional Policy, in Takasaki City University of Economics. His research interests include Islamic tourism, religious tourism in Islamic countries, history of tourism in the Middle Eastern countries, and theoretical studies in religious tourism, anthropology of tourism and sociology of tourism. He edited a book in CABI's *Religious Tourism and Pilgrimage* series with Razaq Raj and Kevin Griffin – *Religious Tourism in Asia: Tradition and Change through Case Studies and Narratives* (2018).

Department of Tourism Policy, Faculty of Regional Policy, Takasaki City University of Economics (Takasaki, Gunma, Japan), 1300 Kaminamie-Machi, Takasaki City, 3700801, Gunma, Japan. Email: syasuda@tcue.ac.jp; shinmarcie@gmail.com

1 Introduction: Motivation of the Pilgrimage Journey

Darius Liutikas*
Lithuanian Social Research Centre, Vilnius, Lithuania

Pilgrims travel to sacred or valuable places. They seek to manifest their identities and values, to get some spiritual or physical benefits, to express gratitude or just experience a new place and get new impressions. Pilgrimage has a great tradition in religions, but in recent years, as a phenomenon, it has expanded to the secular sphere as well. We each have a pilgrimage experience. Maybe you went to a religious centre, maybe to an artistic, cultural or sporting event that is important to your identity, or maybe you just went to a cemetery to honour the people who are important to you. There is a growing interest in this multi-layered social, cultural and economic phenomenon. Research on values and identities in pilgrimage could offer a theoretically and empirically grounded framework for understanding pilgrims' motivations, needs and behaviour. Travellers' identities and values indicate directions that frame and define the pilgrim experience and the meanings of their choices.

The phenomenon of pilgrimage can be studied in different academic disciplines, including sociology, geography, social psychology, anthropology, ethnology and history, as well as in economic, religious, literary and tourism studies. The question that shapes a pilgrim's motivation and actions has been the core subject of my own research on this topic. My personal experience from social geography (Bachelor's studies) to sociology (Master's studies and PhD thesis) allows me to look at this phenomenon in an interdisciplinary way. A phenomenological approach led me to discover previously unknown dimensions of pilgrimage. Various classifications (e.g. segmentation of pilgrims and rites of pilgrims) enable the structuring of this phenomenon.

This book deals with the significance of values and identity in contemporary pilgrim journeys. Identity and values help to explain the differences in the motivation and behaviour of pilgrims. Motivations for travel help to construct or affirm personal and/or social identity. A pilgrimage becomes a valuistic or values-rich journey and makes a strong impact on the self-concept of the traveller. There are many motivations for undertaking values-rich journeys. The most important include personal values, personal and social identity, life experience, lifestyle, and social and cultural influence. Accordingly, the manifestation of personal and social identity has different forms and rituals and constitutes different models of a specific behaviour. The journeys are often embraced as potential instruments for life-altering experiences. An analysis of the values and negotiated identities during

*darius.liutikas@gmail.com

a pilgrimage is commonly based on theoretical models (Liutikas, 2017), blended spiritual and touristic experiences (Norman, 2013), or a 'segmentation' approach (Mehmetoglu *et al.*, 2010). Consequently, this book gathers contributions that address pilgrim motivation, identity and values as they are shaped by the broader sociological, psychological, cultural and anthropological perspectives. The focus of the book is the travellers themselves and their inner world through the lens of their pilgrimage. In a postmodern consumer society, intentional journeys as a pilgrimage can foster self-development and new forms of self-expression. The case studies in the book provide interesting and challenging perspectives on the identity of pilgrims in the 21st century. In particular, the book presents the perceptions of values-rich journeys in the age of value transformations and identity challenges. The collection of case studies and models of pilgrims' identities reveal the complexity of the pilgrimage phenomenon and its ability to adapt to various changes.

What Is Pilgrimage?

Contemporary research on pilgrimage stretches over various disciplines and theoretical lenses. The most common definitions of pilgrimage include three elements: mobility, motivation and indication of a destination, e.g. 'a trip taken for religious and devotional reasons to a holy place or shrine' (Espin, 2007) or 'a journey to a holy place for religious reasons' (Beaver, 2012). Pilgrimage could also be defined as a spiritual magnetism (Preston, 1992), archetype (Clift and Clift, 1996), meaningful journey (Norman, 2020), ritual transformation (Turner and Turner, 1978), transformation journey (Morinis, 1984), searching of authenticity (MacCannell, 1973), arena for competing religious and secular discourses (Eade and Sallnow, 2000), an act of empowerment (Winkleman and Dubisch, 2005), or recreation illusion (Louw, 2007). Pilgrimage as whole travel sector could be related to the manifestation of social functions and social dynamics and the satisfaction of psychological functions and needs (Cohen, 1979; Pearce, 1988, 1993).

Religious pilgrims want to find access to God or the divine figures (Digance, 2003). Pilgrims also could be motivated to get some spiritual merit (indulgences in the Catholic tradition) or to seek healing from serious illness. Religious pilgrimage involves rituals, holy objects and sacred architecture (Coleman and Elsner, 1995). Earlier Christian pilgrimage had a striking penitential aspect and now it has become a complex of concrete motivations, practices and spatial relationships (Voye, 2002).

In recent years, substantial discussions have been launched regarding the relationship between pilgrimage and tourism (MacCannell (1999 [1976]), Graburn (1989), Cohen (1972, 1979, 1992, 2001), Smith (1992), Badone and Roseman (2004), Olsen (2010), see also Chapters 3 and 14). However, various studies have shown that it is not possible to draw clear boundaries between these phenomena (Badone and Roseman, 2004). From the research and case studies presented in this book, we see that the classical dichotomy, where there is a religious pilgrim at one point on the continuum and a secular tourist at another, needs to be reviewed and supplemented. Studies of travellers' motivations, identities and values show that most of them occupy intermediate positions between these points (see Chapter 2).

Pilgrimage is not restricted to religions any more (Dransart, 2004). Secular pilgrimage has been investigated by Reader and Walter (1993), Clift and Clift (1996) and Margry (2008). Places associated with national identity, with death and grief (see Chapters 4 and 17) and cultural places such as art exhibition centres are the destinations of the contemporary pilgrimage. Reader (2015) noticed that fan culture is the area where the image of pilgrimage has been widely adopted (see Chapter 2). We can identify journeys of sport and music fans, and fans of art, famous novels or TV shows and movies. During their holidays, representatives of various other subcultures (hippies, goths, windsurfers, snowboarders, climbers, nudists) travel to places that are sacred to them.

My personal research on pilgrimage is related to the conception of valuistic journeys (Liutikas, 2012, 2013, 2014). Valuistic journeys express values and the personal or social identity of the traveller. A similar pilgrimage could be defined as a journey to the particular

Fig. 1.1. Youth pilgrimage from Vilnius to Trakai, Lithuania. (Photo: the author)

places or events of searching and expressing a traveller's values and identities. The geographical destination of these journeys embodies a certain value. But the most important notion is that there exists a strongly expressed link between the journey and the identity of the traveller (Fig. 1.1). The motivation of such journeys is related to value orientations. Specific behaviour during the journey and at the pilgrimage site is characterized by typical rites and symbolic actions (Liutikas, 2012).

Identity related research in tourism occupies an important place (Cohen, 1979; Desforges, 2000; Burns and Novelli, 2006; Wearing *et al*., 2010; Smed, 2012; Bond and Falk, 2013). However, the majority of the identity-focused tourism literature is related to the national identity (Lepp and Harris, 2008; Frew and White, 2011; Bhandari, 2014). There are some works related to religious identity and tourism (Timothy and Olsen, 2006; Olsen, 2012; Norman, 2013). Pilgrimage is not the only religious-related travelling form. Another traditional form is the mission. Missionaries travel to share their religious identity, values and faith with the people who have a different religious identity or don't have a religious identity at all. These trips are often to foreign countries and travellers encounter different cultural traditions.

Motivation for Pilgrimage

There are different theories and perspectives to understand the potential and actual motivations of pilgrims. Each theory of tourist motivation may have some aspects that are suitable for pilgrimage as well. An analysis of these theories could help to reveal the complex nature of pilgrimage motivation and to identify the essential features of the motivation structure. Different definitions of pilgrimage could support one or another approach of travel motivation. However, it is important to recognize that there is a range of motivations among individuals visiting a religious site. Understanding different motivations could provide insights into the segmentation and typologization of religious travellers.

Page (2019) summarizes some of the main theoretical approaches developed in tourism motivation research. He identifies needs-based approaches (that assume that tourists select destinations to satisfy their needs), values-based approaches (that highlight the importance of human values on tourism motivation), benefits sought and realized approaches (that focus on causal factors in terms of what benefit a tourist expects to gain from travel and holidays), expectancy-based approaches (based upon the assumption that the concept of attractiveness of achieving an outcome is the prime motivator) and other approaches (push-and-pull factors) (Page, 2019, p. 84).

Needs-based approaches (Pearce, 1988, 1993) use a classical model, Maslow's (1954) hierarchy of needs (Fig. 2.1). The individual must satisfy lower-level needs (physiological or safety) before going on to meet higher-level needs (social or belonging, esteem and self-actualization). In this theory, pilgrimage could help to satisfy social needs, esteem and self-actualization needs. Pearce's (1988) Travel Career Ladder (TCL) and Pearce and Lee's (2005) Travel Career Patterns (TCP) model are based on Maslow's hierarchy of needs. The TCL describes tourist motivation as having five different levels: relaxation needs, safety and security needs, relationship needs, self-esteem and development needs, and self-actualization and fulfilment needs. Pearce (2005) explains that in the TCL model, travellers were considered to have more than one level of travel motivation, but one set of needs in the ladder levels might be dominant (Pearce, 2005, p. 53). Pearce and Lee (2005) argue that travel motivation is influenced by travel experience, stage in the life cycle and age. The idea of the TCP model is that tourists have different motivating patterns over their life cycle which impact their experience of travelling (Page, 2019).

Benefit-based approaches (Haley, 1968) are used for market segmentation and consumers' behaviour. Frochot and Morrison (2000) notice the lack of understanding of the benefit for travellers and tourists. They note that three possibilities exist: attribute-based (the attributes or features of the destination or service), psychologically based (associated with travellers' motivation and expectations), or a combination of both. Benefits sought (prior to the consumption experience) and benefits gained (after the consumption experience or based on past experiences) are two different constructs. Benefits sought is related to expectations and motivation, and benefits gained or realized is related to an improved psycho-physiological condition and satisfaction. Such psychological and psychophysiological benefits as personal development and growth, and mental health and health maintenance, could be related to pilgrimages. Brandon (1970, p. 501) provides a definition of pilgrimage which involves three factors – a holy place, the attraction of individuals or crowds to this place and a specific aim, i.e. to obtain some spiritual or material benefit. More particular benefits could be indulgence for sins, recovery from illness or the solution of a personal problem.

The expectancy theory of motivation (Vroom, 1964) is one of the process theories most relevant to the understanding of the motivations for pilgrimage and religious tourism (Blackwell, 2007). The idea of this theory is that people's motivations towards doing anything will be determined by the value they place on the outcome of their effort, multiplied by the confidence they have that their efforts will materially aid in achieving a goal (Koontz and Weihrich, 2008). The three main components of this theory are expectancy (the belief that one's effort will result in a desired performance), valence (the importance an individual places on the rewards of an outcome, which is based on their needs, goals or values) and instrumentality (the belief that a valued outcome will be received if you perform well).

Actually, Vroom's expectancy theory is related to perception. For example, an individual must believe that undertaking a pilgrimage will be rewarded. In Catholic traditions such rewards could be getting indulgence and the redemption of sins, expressing gratitude to Jesus or the Virgin Mary, receiving God's grace, accomplishing an act of repentance or obtaining spiritual renewal. Undertaking a pilgrimage (expectancy to get religious rewards) and experiencing hardship or self-restrictions on the journey, and participation in religious rituals (instrumentality), lead to intrinsic or extrinsic, religious, social or personal rewards (valence).

Fig. 1.2. Tourists and pilgrims at Gergeti Trinity Church, Georgia. (Photo: the author)

Pilgrims' motivations are determined by internal (push) and external (pull) factors. Internal factors could be pilgrims' attitudes, perceptions and beliefs, their preferences and expectations, knowledge and prior interaction. Cultural and communal traditions and destination images are external factors attracting pilgrims. In general, the push-and-pull framework (Crompton, 1979; Dann, 1981; Iso-Ahola, 1982; Mansfeld, 1992; Yoon and Uysal, 2005) is a useful approach for uncovering both tourists' and pilgrims' motivations. Push factors form a desire to travel and pull factors affect the choice of the destination (Fig. 1.2). Crompton (1979) analysed motivations for pleasure vacations. However, all seven socio-psychological motives (escape from a perceived mundane environment, exploration and evaluation of self, relaxation, prestige, regression, enhancement of kinship relationships and facilitation of social interaction) and two cultural motives (novelty and education) identified in his research could be analysed in the context of pilgrimage motivation.

However, pilgrimage as a form of travel is most often analysed on the basis of values-based approaches. Morinis (1992) defined pilgrimages as journeys undertaken to a place or state that embody a valued ideal. According to Morinis (1992), a pilgrimage is a movement towards aspirational ideals that cannot be realized at home, and a pilgrimage becomes a value-based endeavour mingling the sacrum and profanum spheres. Other authors (Timothy, 1997; Smith, 2003; Smith and Kelly, 2006; Norman, 2011) use different concepts of values-rich journeys, including spiritual travel, personal heritage tourism, and holistic tourism. However, in all of these concepts, the motivations for travelling are related to the construction or manifestation of personal and/or social identity and self-fulfilment (Liutikas, 2017).

The concept of valuistic journeys (Liutikas, 2012, 2013, 2014) includes an expression of valuistic ideals, as well as the confirmation and demonstration of identity during the journeys. Social and personal identity and different value

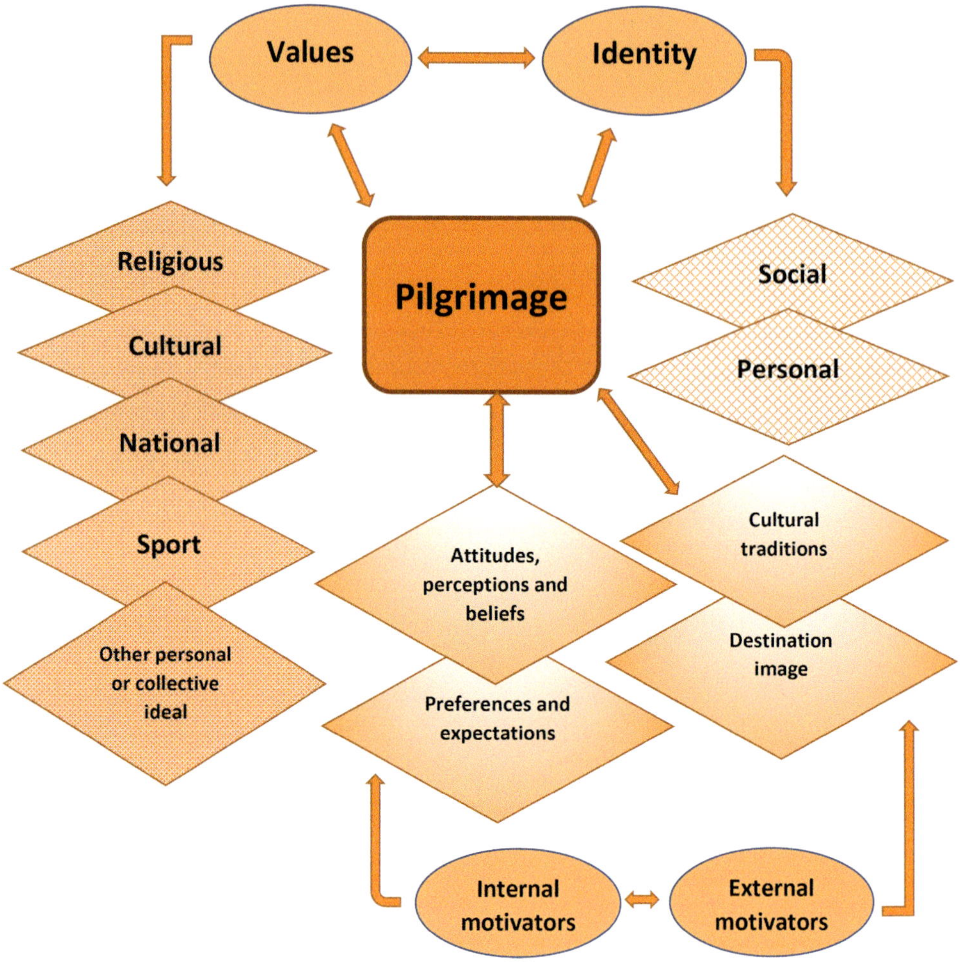

Fig. 1.3. The concept of pilgrims' motivation.

orientations related to religion, culture, sport, nation or personal experience become a principal driving force for choosing a pilgrim journey (Fig. 1.3). This notion has three main components: identity, values and their expression or demonstration.

Identity is one of the key categories of sociology. It plays a central role for pilgrims' motivations and behaviour. Identity is not a static category; it has dynamic influences that move pilgrims among different types of religious-motivated tourists. Values are essentially abstract beliefs that are culturally defined and serve as guidelines for worldviews, principles and behaviours (Liutikas, 2017). The action – travelling – is related to the demonstration, expression, development or confirmation of values and identity. The main difference distinguishing a recreational and a values-rich journey is the motivation of the traveller and his/her inner disposition. Spiritual and religious motives are primary in a religious pilgrimage; cultural, national, community-based and personal motives are primary in a secular pilgrimage. The idea of a values-rich journey is reaching the objective (destination) and at the same time expressing personal or social identity and values, defining the features which identify the self (Liutikas, 2012). Internal motivators (such as attitudes, perceptions, beliefs, preferences and expectations) influence the wish and motivation for travelling and external motivators (such as cultural traditions and destination image) influence the place where to travel (Fig. 1.3).

Through the metaphor of the paths to a mountain peak, we can understand the essence of travelling. Pilgrimage is like the climb to a mount of values. A traveller can choose the path among different values. At the bottom of the mount, there are dislocated values related to pleasure and life vitality, higher are cultural and spiritual values, and on the top is the holy value identified by Scheler (1973). Some people prefer the higher values and others the lower ones. The climbing height is decided by the act of preferring (Scheler, 1973, p. 87). The satisfaction of manifesting higher values is deeper.

The manifestation of social identity during the pilgrimage depends on specific contexts. Social bonds and responsibilities help to create the meaning of the journey to themselves and to send a message to others. During the journey, pilgrims perceive themselves in action, they develop and construct relationships with the religious environment and other travellers.

Bond and Falk (2013) present that many researchers 'continue to struggle with how best to operationalize the term "identity" within their research'. They presented the Identity Related Tourism Motivation model, which has three components: identity development, identity maintenance, and identity moderation and reconstruction (Bond and Falk, 2013). To explore a new or developing self-aspect, the individual travels to a specific destination or undertakes a specific activity. When a person wishes to maintain a particular role or particular aspect of identity, he seeks out other individuals with a similar identity and interests. In communities of practice, a person could find a more experienced individual and learn from him. If the person seeks out a new or alternative experience, it could lead to identity moderation and reconstruction. This model could also be applied to pilgrimage.

The research on the impact of personal values to tourist motivation and journey motives are based largely on the work of Rokeach (1968, 1973). The Rokeach Value Survey (RVS) consists of 18 instrumental values (ideal modes of behaviour) and 18 terminal values (ideal end-states of existence). The first set contains such attributes as ambitiousness, broadmindedness, politeness and so on. The second set defines the individual's desired end-state of existence and includes such conditions as a sense of accomplishment, equality, happiness and so on (Mehmetoglu *et al.*, 2010). Kahle (1983) proposes a List of Values (LOV), which consist of nine terminal values (warm relationships with others, self-respect, sense of accomplishment, fun and enjoyment in life, self-fulfillment, being well respected, security, a sense of belonging and excitement). Mitchell (1983) developed the Values and Life-styles (VALS) instrument. This model is also useful for segmentation purposes. Nine lifestyle groups were identified based on demographic and attitudinal questions: survivors, sustainers, belongers, emulators, achievers, I-am-me, experiential, societally conscious and integrated. Mitchell (1983) grouped lifestyles into three categories: need-driven (consumer behaviour is driven more by needs than values and attitudes as survivors and sustainers), outer-directed (consumer behaviour is influenced by social factors – belongers, emulators, achievers) and inner-directed (consumer behaviour is predicated on personal needs – I-am-me, experientials, societally conscious individuals). Integrateds are characterized by psychological maturity and self-actualizing philosophy.

The strength of motivation correlates to the amount of effort and depends on the value of a reward of a possible journey. The strength of motivation is the main factor for overcoming various obstacles and constraints to go on a pilgrimage journey. The satisfaction of the journey is greatly influenced by an individual's knowledge, preparation and his or her perception of what the required goal is. The understanding of behaviour during the journey differs, e.g. performance of rituals, other activities related to a goal.

Motivation to travel to a pilgrimage site is usually integrated into a comprehensive and steady circle (Fig. 1.4). The values and identity that are formed require certain goals and actions. Life objectives must be clear, achievable and verifiable. Objectives should motivate for actions. For example, if a pilgrim has Catholic values and identity and has an established objective to strengthen them, his actions may include increased participation in the life of the religious community, reading religious and spiritual books, and pilgrimage.

Preparation for a pilgrimage is different from preparation for a cognitive or recreational

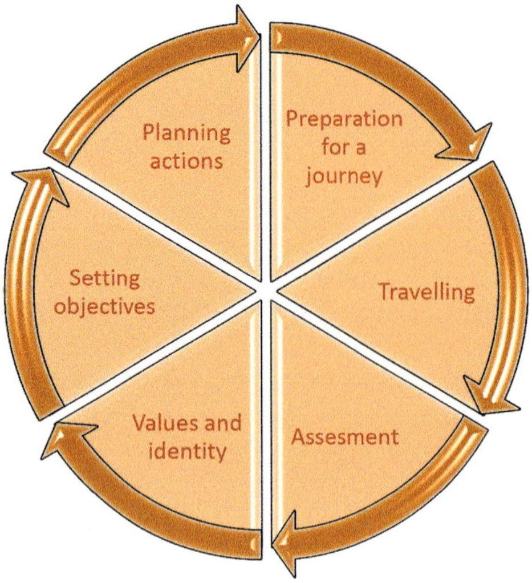

Fig. 1.4. Values and identities in the decision-making circle for pilgrimage.

trip. The true pilgrim contemplates the purpose and intentions of the journey. Pilgrims travelling in an organized group participate in a common prayer and holy mass before the journey. Pilgrims also confess and repent of past sins. Such preparation leads to thoughts of a preparation for the last trip that each person will perform after his or her death. So serious reflection on one's life helps one to comprehend pilgrimage as the moment from which the direction of life can change significantly. A blessing from the spiritual leader is very important for a pilgrim before going on a journey. He also thinks about what steps to take to make the trip meet his expectations.

The journey itself is full of rituals (participation in a holy mass, receiving an indulgence, leaving a vote, touching a relic, acquiring devotionals, meditation in a holy place or participation in a religious procession), prayer and spiritual discoveries. An organized trip provides a sense of community or *communitas* (Turner and Turner, 1978). Travellers on a pilgrimage become related to each other through a unifying experience and the sharing of common values. Communication with other participants in the journey in an organized group allows travellers to consolidate social and personal identity and to increase self-respect and self-consciousness.

Pilgrims from organized groups discuss the trip in their groups and share impressions and experiences. Such sharing seems to perpetuate the travel experience and opens new paths for the development of values and identity. During the assessment of the journey, pilgrims often see the contradiction between the emptiness of the secular everyday world and the spiritual pilgrimage opposed to it. Pilgrims say that returning from a pilgrim journey changes their attitude, understanding of problems and religious life itself (Liutikas, 2003). The thoughts and attitudes formed in the sacrum space remain in a pilgrim's mind for a while after returning from the trip (Fig. 1.5).

However, a pilgrim's motivation is much more complex than these particular theories. Motivations can change over time due to internal and external circumstances. Frey (1998) shows that some pilgrims form their motives or change them after their journey has started.

Book Structure

In terms of organizational structure, this book is divided into four parts. Part I consists of four chapters, which focus on the more general aspects of the pilgrim's identity and values.

In Chapter 2, Darius Liutikas provides a theoretical framework of pilgrimage based on values and identities. An overview of the concept of values and identity reveals their interrelationships and importance in a person's choices. Religiosity could be measured by different dimensions and indicators. However, a person's behaviour and performed rituals could help to express and affirm their identity. Typologies of religious- and cultural-motivated travellers are comprised of a variety of categories. Religious travellers are divided into 12 groups ranging from religious to secular motivation and from a low level of religious identity and involvement to a high level of religious identity and involvement. A secular pilgrimage is very heterogeneous. Different reasons for travelling create different behaviour and implemented activities. However, based on the motivation of the travellers and their approach to the trip, cultural travellers are divided into three groups: cultural pilgrims, cultural tourists and peripherical culture consumers. So, values fostered by pilgrims could be totally different from our stereotypical

Fig. 1.5. Hill of Crosses, Lithuania. (Photo: the author)

imagination. We must be careful about considering pilgrims as a homogeneous segment. Quite often their motivation and behaviour manifest in a tourist identity.

In Chapter 3, Kip Reddick analyses the connection between liminal places and hospitality. Pilgrims traverse liminal places as they embark on journeys to sacred destinations. Their identity requires an openness to unconditional hospitality. Tourists, in contrast to pilgrims, receive conditional hospitality within the boundaries of the recreation industry. This chapter reveals the values of tourists versus pilgrims in the interplay of hospitality. Tourists feel as consumers among locals who sell both adventure and the exotic; pilgrims want to receive hospitality freely rather than from the tourism industry. The author stresses that pilgrim values and identity rise out of their liminality. The difference between pilgrim and tourist thinking becomes apparent in the engagement of rituals (pilgrims want transformation) and recreation (tourists want a good vacation).

In Chapter 4, Slawoj Tanas focuses on thanatourism, which involves travelling to death spaces. The author notices that a significant number of religious pilgrimages are made to sanctuaries, which could be churches, graves, chapels or cemeteries. The majority of the destinations in the medieval period were to tombs (St James in Santiago de Compostela, Thomas Becket in Canterbury or St Peter and St Paul in Rome). Pilgrimage to graves is noticeable in Judaism and Islam as well. The author also describes journeys to see body relics, war memorials and war cemeteries, public death penalties, judicial duels, mummified bodies and ossuaries. Art presenting death for visitors arouses the fear of death and the respect of life. The author discusses motivations, values and identities of thanatourism. The motivation to visit death spaces could be related to religion, personal or social identity, empathy and remembrance. However, the author also identifies the secular sets of motivations, including education and cognition, curiosity, entertainment, and to experience the emotions of fear. The whole chapter tries to answer the question: What is the meaning of death in the process of shaping tourist spaces? Slawoj Tanas points out that the way a death space is organized comes from the adopted cultural norms. This space has religious, educational, socializing, emotional and entertaining value. The development of pop culture has led to the growth of post-modernist interest in death spaces, as a result of which death-related tourist products have been created.

In Chapter 5, Larry Russell presents his personal experiences on pilgrimage to Chimayó. He uses his personal experience and narrative as a way of developing his arguments. He points out that the practice of pilgrimage nurtures discrete identities. His phenomenological approach forces him to track what is happening on the way and make his response to the understanding of the pilgrim journey. Phenomenology presumes a sense of wonder as a research framework, and the theory evolves from experience. He doesn't presume to know what a pilgrim is or where a pilgrimage leads. He examines the etymology of 'pilgrim' as a stranger and identifies the conditions structuring his experience. He discusses such states of mind experienced on walking a pilgrimage as strangeness, solitude, momentum of transformation, turbulence, burdens, yielding, reaching frontier, and risk. The author confesses that the only way to understand pilgrimage is to participate rather than observe. Each pilgrim discovers her or his goal along the way before the destination is reached. The author tries to answer such questions as 'Why do we call certain journeys a pilgrimage?'. Larry Russell brackets the definition of pilgrimage while he carefully examines the conditions that make the pilgrimage possible. The author claims that the narrative told in this chapter is a value-rich experience that has shaped his identity. During his pilgrimage, he sometimes felt like a tourist but sometimes like a pilgrim. However, he understands that he can reach his identity through brokenness. The final understanding is that it is risky to be certain about one's own identity.

Part II provides an overview of motivations, values, identities and behaviour of the pilgrims travelling to Santiago de Compostela. In Chapter 6, Luis Alfonso Escudero Gómez, Lucrezia Lopez, María Ángeles Piñeiro Antelo and Rubén Camilo Lois González examine the motivations and impact of the pilgrimage on the personal lives of pilgrims. The authors develop an argument that pilgrimage to Santiago de Compostela is inspired by a complex series of different motivations. However, the pilgrimage continues after coming back home. Based on semi-structured interviews, the authors discover that motivating factors of the pilgrimage could be sport, personal promise, friendship, culture, religion, spirituality, adventure and landscape. However, pilgrims quite often have more than one reason for pilgrimage. The pilgrimage to Santiago changes the daily lives of the pilgrims and forms a new outlook to their identities and the surrounding environment. The authors conclude that the Way to Santiago could transform pilgrims' lives and priorities because the pilgrimage experience continues and revives after the physical journey.

In Chapter 7, Xosé M. Santos also offers an analysis about the motivations, values and identities of pilgrims walking to Santiago de Compostela. The author analyses the agents that manage the different routes to Santiago, such as international agencies, public administration, the Church and the Associations of the Friends of the Way. The author presents the responses to an article in the newspaper which challenges the traditional understanding of the Way. Commentators on the newspaper article acknowledge the performance of pseudo- or self-styled pilgrims (Liutikas, 2015). However, religion and spirituality play a special role in the complexity of the Way motivations. Those comments were the result of and reflection on the views of the institutions involved in the governance of the Way and presented their perspectives. The author stresses the impact of post-secularism to the current discourse of the role of the Way.

In Chapter 8, Augusta X. Thomson discusses the phenomenon of commensality and its relationship to pilgrims' *communitas* (Turner, 1969), their motivations and identities. Commensality was less explored in Christian pilgrimage, except for Eucharist in the Catholic tradition. Communal meals shared by pilgrims form feelings of connection among pilgrims and between pilgrims and their route. Some key theories related to commensality and exchange are presented in the chapter. Mealtime is a possibility to make community with co-travellers and share daily impressions and reflections. The author argues that cultural and socio-economic distinctions between pilgrims contribute to their collective experience of *communitas*. An example of such collective experiences of commensality or *communitas* could be taking a common photograph. However, contestation in *communitas* informs different pilgrims' values, motivations and identities.

In Chapter 9, Pedro Azevedo and Xerardo Pereiro present their research on the identity and values of pilgrims travelling on the Portuguese Inner Way to Santiago de Compostela. In the research, they use the term 'tourist-pilgrim', which signalizes the mixed motivation of travellers. The authors compare the identities and behaviour of the Portuguese and foreign tourist-pilgrims. They notice that there are differences between the motivations of tourist-pilgrims according to their nationality, gender, age, education, social capital and other variables such as shorter time in travelling, and the way pilgrims undertake their journey. The Portuguese Inner Way to Santiago de Compostela is a new project finished in 2012. The motivation to go this Way is also influenced by different discourses of various stakeholders. The pilgrimage route becomes multi-motivational, linking culture, nature, sport, adventure and religion.

Part III provides key studies on pilgrims' values and identities in different regions. In Chapter 10, Ricardo Nicolas Progano and Kumi Kato examine the shift of values and motivation at a pilgrimage in Japan. First, the authors present the concept of pilgrimage in Japan, which differs from western pilgrimage. Later, the characteristics of the development of pilgrimage in Japan are discussed. The authors claim that con temporary spirituality, including visiting 'powerspots' and other places for physical and mental restoration and self-discovery, reinterprets pilgrimage sites. Secularization has added such factors as the motivation to visit pilgrim sites as heritage, wellness, sightseeing and cuisine. Also, tourism includes the element of hiking to the sacred peaks; however, the motivation of these journeys is not necessarily related to religion. International visitors also travel to Japanese pilgrimage sites. According to the authors, western visitors are motivated by a variety of reasons including leisure, intercultural learning and individualized spirituality.

In Chapter 11, Shin Yasuda focuses on *goshuin* culture and religious mobility in contemporary Japan. *Goshuin* (shrine or temple seal) collection has become a massive social phenomenon in Japanese society and is related to religious commodification. The author discusses the characteristics of religious mobility in *goshuin* practice and social significance of individual investments in this movement. Collecting *goshuin* could help to create new religious and cultural routes. However, this practice moved from the tradition to approve the authenticity of religious experiences to the consumption and commodification of religion or culture. Various tourism stakeholders are involved in the promotion of the *goshuin*. Some religious shrines and temples also actively advertise *goshuin* to attract visitors, e.g. they introduced particular supplications, such as for business success or wishes for good fortune. There is a variety of discussions about what constitutes a real practice of *goshuin*. According to the author, this practice could be cost-effective in religious investment and religious capital enhancement.

In Chapter 12, Rana P.B. Singh and Pravin S. Rana provide an in-depth examination of the contemporary perspectives of Hindu pilgrimage in India. Pilgrimage has a long historical tradition in India. It starts from the inner space (home) to the outer territory and vice versa. Motivation for Hindu pilgrims could be related to spirituality, purification and bliss. Pilgrims want to acquire spiritual merits, to be blessed by a holy man, to fulfil a vow or to perform ancestral ritual. The authors acknowledge that pilgrimage links different inherent and perceived meanings. It is an important notion of being at the sacred place. Kumbha Melā (a riverside festival) – the largest religious gathering in the world – attracts tourists even from different religions. Pilgrimage becomes a network of different relationships of cultural actors, traditionally and culturally constructed systems of norms and transformative channels.

In Chapter 13, Dane Munro offers a discourse on pilgrimage to Malta's ancient temples. The neolithic Maltese temples are used by a variety of travellers to Malta. Some of them, and especially the adherents of the New Religious Movements, use them for direct purposes. They find these temples inspiring, and they provide them with a full sensory experience. The usage of temples varies from contemplation, meditation, or pagan weddings, to baby-naming ceremonies or asking for blessings from the Mother Goddess. The benefit of valuistic journeys is related to a person's fulfilment. Pilgrims seek authenticity or fulfilling experiences. The sacred could be felt from different objects of veneration, or from the spirit of place. In such temples could be the overlap between the religious and touristic approach. Even a specialized policy exists for using neolithic sites as places of worship. The author states that New Religious Movements searching for authenticity and spiritual fulfilment have found their way into faith-based tourism.

In Chapter 14, Luciana Thais Villa Gonzalez explores the phenomenon of the World Youth Days. The particular object of the chapter is World Youth Days in Rio (2013). After reviewing the scientific literature examining the differences between pilgrims and tourists, the author analyses the World Youth Days as the concept of shifting pilgrimage. These trips are not fixed in space and time, and they include tourist attractions and symbols of tourism. The tourism and pilgrimage dimensions overlap with each other in such journeys. The conception of tourism is used to attract young pilgrims, and the conception of pilgrimage is used to persuade the local population to welcome them.

In Chapter 15, Elyor Karimov and Darius Liutikas outline a pilgrimage tradition to sacred sites of modern central Asia. The authors present a review of the main pilgrimage sites in central Asia. The majority of them are associated with burials and tombs. Visitors to the shrines request special favours from God and make donations. Some visits involve commemorative rituals and feasts. A difficult situation regarding visiting sacred places occurred during the Soviet era. The number of pilgrims decreased due to the closure of mosques and other shrines; various other bans were implemented. Some traditions have survived. This chapter describes the ritual of 'Mushkel Kushod' of central Asian women. This ritual, intended for women only, illustrates the intertwining of Islamic and ancient local traditions. This ritual carries the possibility to express women's religious identity and values.

In Chapter 16, Rami K. Isaac presents the 'come and see' phenomenon in Palestine. This idea is related to the targeting of the pilgrimage market for Christian pilgrims. The author examines the motives and different experiences of these pilgrimages that visit Palestine and Israel. The chapter presents the Palestinians' view of the situation in the Holy Land. 'Come and see' is a politically motivated initiative seeking to attract 'justice tourists'. The author uses netnographical methods to try to understand the impressions of the travellers to Palestine. The analysis of semantic networks indicates three themes: appreciation, destination and experience – the most frequently used words of Tripadvisor reviewers. According to the author, the tours to Palestine may involve personal transformations in the travellers and teach them about the historical and political situation in the Holy Land.

In Chapter 17, Derek Dalton offers an analysis of the dichotomy of visits to the Auschwitz-Birkenau extermination camp. The camp is visited by pilgrims who are, on the one hand, well-behaved, educated, and possess a sound spiritual motivation, and on the other by tourists who are consumers and motivated by voyeurism. Drawing on tourism literature and field research, the author seeks to show how these two supposedly different types of visitors experience the camp complex. Auschwitz-Birkenau, as a space, accommodates lots of different purposes and a variety of different visitors. The author admits the antithesis between a spiritual and contemplative pilgrim and the tourist who understands his visit as a museum visit. Derek Dalton seeks to touch upon tourist practices and experiences and reflects on the controversies that attend the practice of tourist photography at Auschwitz. This tourism is a complex phenomenon. Some tourists are facile and superficial, and others have much more in common with pilgrims. Accordingly, the author compares the sorts of things that tourists and pilgrims do, and how we might better appreciate their motivations and what they get from their visit. However, it is unhelpful to construct the 'tourist' as the antithesis of the 'pilgrim'. According to the author, we need to adopt a less antagonistic model for appreciating how these two types of visitors coexist at this profoundly important site of remembrance. Every respectful visitor plays an important role in witnessing the Holocaust.

In Chapter 18, Josephine Pryce presents insights from her sabbatical journey throughout the USA. This chapter brings together the sabbatical journey and train travel to explore her transformative experiences on spiritual enlightenment. Train travel experience into liberated time and liminal space challenges values and identity, and projects powerful and constant changes. The author explains that such a journey was a useful time for personal and professional development in holistic ways. It was a time to slow down, and it helped her to gather fragmented parts of her personality. Using autoethnography, the author explains why the sabbatical journey was a pilgrimage. Transformation, self-discovery, healing, learning and extraordinary experiences are the main elements which can lead to understanding a journey as a pilgrimage.

Concluding insights and discussions are included in Part IV. Daniel H. Olsen and Dallen J. Timothy discuss contemporary perspectives of pilgrimage in Chapter 19. The authors analyse the role pilgrimage plays in a post-modern and post-secularized world, focusing on such challenges as commodification and overtourism, sacred site management, and undertourism. In presenting various statistics on visits to sacred places, the authors state that religion has survived the challenges of secularization and retained its importance in the global space. The authors notice that many cultural objects on UNESCO's World Heritage List are associated with a religion or faith. This title increases visits and consumption of the places. A growing number of visitors to religious sites could lead to overtourism and other negative ecological and social outcomes. Overcrowded religious destinations could reduce the quality of spiritual experience for pilgrims and create management challenges. However, the 2020 situation with the COVID-19 pandemic illustrates that undertourism can also be a problem for religious destinations.

In the Conclusions chapter, Darius Liutikas discusses the adaptability and resilience of the phenomenon of pilgrimage. This was particularly evident during the pandemic crisis. Alternative forms of pilgrimages, such as spiritual pilgrimage, virtual pilgrimage and hyper-real reality, replacements of travellers, and creation of sacred place in the home environment, are introduced. The author claims that pilgrimages will be affected not only by the consequences of the global health situation and management of COVID-19, but also by various technological, social, environmental and legal factors in the near future.

Acknowledgement

This book is part of CABI's Religious Tourism and Pilgrimage Series (http://www.cabi.org/). This book would not have been possible if it were not for the support from the series' general editors, Dr Razaq Raj (Leeds Beckett University, UK) and Dr Kevin Griffin (Dublin Institute of Technology, Ireland). I am also grateful for all the authors whose research revealed the multi-faceted and vivid world of pilgrims.

References

Badone, E. (2004) Crossing boundaries: exploring the borderlands of ethnography, tourism, and pilgrimage. In: Badone, E. and Roseman, S.R. (eds) *Intersecting Journeys. The Anthropology of Pilgrimage and Tourism*. University of Illinois Press, Chicago, Illinois, pp. 1–23.

Badone, E. and Roseman, S.R. (eds) (2004) *Intersecting Journeys: The Anthropology of Pilgrimage and Tourism*. University of Illinois Press, Chicago, Illinois.

Beaver, A. (2012) *Oxford Dictionary of Travel and Tourism*. Oxford University Press, Oxford.

Bhandari, K. (2014) *Tourism and National Identity: Heritage and Nationhood in Scotland*. Channel View Publications, Clevedon, UK.

Blackwell, R. (2007) Motivation for religious tourism, pilgrimage, festivals and events. In: Raj, R. and Morpeth, N.D. (eds) *Religious Tourism and Pilgrimage Festivals Management: An International Perspective*. CAB International, Wallingford, UK, pp. 35–47.
Bond, N. and Falk, J. (2013) Tourism and identity-related motivations: why am I here (and not there)? *International Journal of Tourism Research* 15(5), 430–442. DOI: 10.1002/jtr.1886.
Brandon, S.G.F. (ed.) (1970) *A Dictionary of Comparative Religion*. Weidenfeld and Nicolson, London.
Burns, M.P. and Novelli, M. (2006) Tourism and social identities: introduction. In: Burns, M.P. and Novelli, M. (eds) *Tourism and Social Identities: Global Frameworks and Local Realities*. Elsevier Ltd, Oxford, UK, pp. 1–12.
Clift, J.D. and Clift, B.W. (1996) *The Archetype of Pilgrimage*. New York.
Cohen, E. (1972) Toward a sociology of international tourism. *Social Research* 39, 164–172.
Cohen, E. (1979) A phenomenology of tourist experiences. *Sociology* 13(2), 179–201. DOI: 10.1177/003803857901300203.
Cohen, E. (1992) Pilgrimage centers, concentric and excentric. *Annals of Tourism Research* 19(1), 33–50.
Cohen, E. (2001) Who is a tourist? A conceptual clarification. In: Cohen, E. (ed.) *Contemporary Tourism: Diversity and Change*. Elsevier Science, pp. 17–36.
Coleman, S. and Elsner, J. (eds) (1995) *Pilgrimage: Past and Present in the World Religions*. Harvard University Press, Cambridge, Massachusetts.
Crompton, J.L. (1979) Motivations for pleasure vacation. *Annals of Tourism Research* 6(4), 408–424. DOI: 10.1016/0160-7383(79)90004-5.
Dann, G.M.S. (1981) Tourist motivation an appraisal. *Annals of Tourism Research* 8(2), 187–219. DOI: 10.1016/0160-7383(81)90082-7.
Desforges, L. (2000) Travelling the world: identity and travel biography. *Annals of Tourism Research* 27(4), 926–945.
Digance, J. (2003) Pilgrimage at contested sites. *Annals of Tourism Research* 30(1), 143–159. DOI: 10.1016/S0160-7383(02)00028-2.
Dransart, P. (2004) Pilgrimage. In: Salamone, A.F. (ed.) *Encyclopedia of Religious Rites, Rituals, and Festivals*. Routledge.
Eade, J. and Sallnow, J.M. (2000) *Contesting the Sacred: The Anthropology of Christian Pilgrimage*. Urbana, Chicago, Illinois.
Espin, O. (2007) Pilgrimage. In: Espin, O. and Nickoloff, B.J. (eds) *An Introductory Dictionary of Theology and Religious Studies*. Liturgical Press, Collegeville, Minnesota.
Frew, E. and White, L. (eds) (2011) *Tourism and National Identities: An International Perspective*. Routledge.
Frey, N.L. (1998) *Pilgrim Stories: On and Off the Road to Santiago*. University of California Press.
Frochot, I. and Morrison, A.M. (2000) Benefit segmentation: a review of its applications to travel and tourism research. *Journal of Travel & Tourism Marketing* 9(4), 21–45. DOI: 10.1300/J073v09n04_02.
Graburn, H.H.N. (1989) Tourism: the sacred journey. In: Smith, L.V. (ed.) *Hosts and Guests: The Anthropology of Tourism*. University of Pennsylvania Press, pp. 21–36.
Haley, R. I. (1968) Benefit segmentation: a decision orientated research tool. *Journal of Marketing* 32, 30–35.
Iso-Ahola, S.E. (1982) Toward a social psychological theory of tourism motivation: a rejoinder. *Annals of Tourism Research* 9(2), 256–262. DOI: 10.1016/0160-7383(82)90049-4.
Kahle, R. L. (ed.) (1983) *Social Values and Social Change*. Praeger, New York.
Koontz, H. and Weihrich, H. (2008) *Essentials of Management: An International Perspective*. Tata McGraw-Hill Publishing Company Limited, New Delhi.
Lepp, A. and Harris, J. (2008) Tourism and national identity in Uganda. *International Journal of Tourism Research* 10(6), 525–536. DOI: 10.1002/jtr.693.
Liutikas, D. (2003) Katalikiškosios piligrimystės sociologiniai aspektai. *SOTER* 9, 117–132.
Liutikas, D. (2012) Experiences of valuistic journeys: motivation and behaviour. In: Sharpley, R. and Stone, Ph. (eds) *Contemporary Tourist Experience: Concepts and Consequences*. Routledge, New York, pp. 38–56.
Liutikas, D. (2013) Experiences of pilgrimage in Lithuania: expressions of values and identity at new destinations. *Journal of Tourism Consumption and Practice*5(2), 43–60.
Liutikas, D. (2014) Lithuanian valuistic journeys: traditional and secular pilgrimage. *Journal of Heritage Tourism* 9(4), 299–316. DOI: 10.1080/1743873X.2014.890205.

Liutikas, D. (2015) Indulgence feasts: manifestation of religious and communal identity. In: Jepson, A. and Clarke, A. (eds) *Managing and Developing Communities, Festivals and Events*. Palgrave Macmillan, pp. 148–164.

Liutikas, D. (2017) The manifestation of values and identity in travelling: the social engagement of pilgrimage. *Tourism Management Perspectives* 24, 217–224. DOI: 10.1016/j.tmp.2017.07.014.

Louw, E.M. (2007) *Everyday Islam in Post-Soviet Central Asia*. Routledge.

MacCannell, D. (1973) Staged authenticity: arrangements of social space in tourist settings. *American Journal of Sociology* 79(3), 589–603. DOI: 10.1086/225585.

MacCannell, D. (1999 [1976]) *The Tourist: A New Theory of the Leisure Class*. University of California Press.

Mansfeld, Y. (1992) From motivation to actual travel. *Annals of Tourism Research* 19(3), 399–419. DOI: 10.1016/0160-7383(92)90127-B.

Margry, J.P. (ed.) (2008) *Shrines and Pilgrimage in the Modern World: New Itineraries into the Sacred*. Amsterdam University, Amsterdam.

Maslow, A. (1954) *Motivation and Personality*. Harper and Row, New York.

Mehmetoglu, M., Hines, K., Graumann, C. and Greibrokk, J. (2010) The relationship between personal values and tourism behaviour: a segmentation approach. *Journal of Vacation Marketing* 16(1), 17–27. DOI: 10.1177/1356766709356210.

Mitchell, A. (1983) *The Nine American Life Styles*. Warner, New York.

Morinis, A. (1984) *Pilgrimage in the Hindu Tradition: A Case Study of West Bengal*. Oxford University Press, Delhi.

Morinis, A. (1992) Introduction: the territory of the anthropology of pilgrimage. In: Morinis, A. (ed.) *Sacred Journeys: The Anthropology of Pilgrimage*. Greenwood Press, Westport, Connecticut and London, pp. 1–28.

Norman, A. (2011) *Spiritual Tourism: Travel and Religious Practice in Western Society*. Continuum, London.

Norman, A. (2020) Pilgrimage and tourism. In: Possamai, A. and Blasi, J.A. (eds) *The Sage Encyclopedia of the Sociology of Religion*. Sage Publications.

Norman, A. (ed.) (2013) *Journeys and Destinations: Studies in Travel, Identity and Meaning*. Cambridge Scholars Publishing.

Olsen, D.H. (2010) Pilgrims, tourists and Max Weber's "ideal types". *Annals of Tourism Research* 37(3), 848–851. DOI: 10.1016/j.annals.2010.02.002.

Olsen, D.H. (2012) Negotiating identity at religious sites: a management perspective. *Journal of Heritage Tourism* 7(4), 359–366. DOI: 10.1080/1743873X.2012.722642.

Page, J.S. (2019) *Tourism Management*. Routledge.

Pearce, P.L. and Lee, U.-I. (2005) Developing the travel career approach to tourist motivation. *Journal of Travel Research* 43(3), 226–237. DOI: 10.1177/0047287504272020.

Preston, J.J. (1992) Spiritual magnetism: an organising principle for the study of pilgrimage. In: Morinis, A. (ed.) *Sacred Journeys: The Anthropology of Pilgrimage*. Greenwood Press, Westport, CT and London, pp. 31–46.

Pearce, P.L. (1988) *The Ulysses Factor: Evaluating Visitors in Tourist Settings*. Soringer, New York.

Pearce, P.L. (1993) Fundamentals of tourist motivation. In: Pearce, D.G. and Butler, R.W. (eds) *Tourism Research: Critiques and Challenges*. Routledge, London, pp. 113–134.

Pearce, P.L. (2005) *Tourist Behaviour: Themes and Conceptual Schemes*. Channel View Publications, Clevedon, UK.

Reader, I. and Walter, T. (eds) (1993) *Pilgrimage in Popular Culture*. Macmillan, Basingstoke.

Rokeach, M. (1968) *Beliefs, Attitudes, and Values*. Jossey-Bass, San Francisco, California.

Rokeach, M. (1973) *The Nature of Human Values*. Free Press, New York.

Reader, I. (2015) *Pilgrimage: A Very Short Introduction*. Oxford University Press.

Scheler, M. (1973) *Formalism in Ethics and Non-formal Ethics of Values: A New Attempt toward the Foundation of an Ethical Personalism. Translated by Manfred S. Frings and Roger L. Funk*. Northwestern University Press, Evanston.

Smed, M.K. (2012) Identity in tourist motivation and the dynamics of meaning. In: Sharpley, R. and Stone, P.R. (ed.) *Contemporary Tourist Experience: Concepts and Consequences*. Routledge, New York, pp. 130–146.

Smith, L.V. (1992) Introduction: the quest in guest. *Annals of Tourism Research* 19(1), 1–17.

Smith, M. (2003) Holistic holidays: tourism and the reconciliation of body, mind and spirit. *Tourism Recreation Research* 28(1), 103–108. DOI: 10.1080/02508281.2003.11081392.

Smith, M.K. and Kelly, C. (2006) Journeys of the self: the rise of holistic tourism. *Journal of Tourism Recreation Research* 31(1), 15–24.

Timothy, D.J. (1997) Tourism and the personal heritage experience. *Annals of Tourism Research* 24(3), 751–754. DOI: 10.1016/S0160-7383(97)00006-6.

Timothy, J.D. and Olsen, H.D. (eds) (2006) *Tourism, Religion and Spiritual Journeys*. Routledge.

Turner, V. (1969) *Ritual Process: Structure and Anti-Structure*. Routledge, New York.

Turner, V. and Turner, E. (1978) *Image and Pilgrimage in Christian Culture*. New York.

Voye, L. (2002) Popular religion and pilgrimages in Western Europe. In: Swatos, W.H. and Tomasi, L. (eds) *From Medieval Pilgrimage to Religious Tourism*. Praeger, Westport, Connecticut, pp. 115–136.

Vroom, V.H. (1964) *Work and Motivation*. Wiley, New York.

Wearing, S., Stevenson, D. and Young, T. (2010) *Tourist Cultures: Identity, Place and the Traveller*. Sage.

Winkleman, M. and Dubisch, J. (2005) *Pilgrimage and healing*. The University of Arizona Press, Tucson, AZ.

Yoon, Y. and Uysal, M. (2005) An examination of the effects of motivation and satisfaction on destination loyalty: a structural model. *Tourism Management* 26(1), 45–56. DOI: 10.1016/j.tourman.2003.08.016.

2 The Expression of Identities in Pilgrim Journeys

Darius Liutikas*
Lithuanian Social Research Centre, Vilnius, Lithuania

Introduction

The purpose of this chapter is to contribute to the literature by discussing the theoretical framework of pilgrimage based on values and identities. As Higham and Hinch (2009) point out, tourism is an instrument that manifests or constructs one's identity through contact with others. Tourism influences the values and identity of both hosts and guests. Destinations serve as reference points in the activities related to the manifestation of identities and values. Social identities always remain an open subject for negotiation; identities are symbolic, open, political and dynamic (Ruiz Ballesteros and Hernández Ramírez, 2007). Identities are catalysts of social actions; they stimulate discourses and are linked to particular places in the landscape (Tuan, 1977) or imagined spaces (Anderson, 2006).

So tourism and pilgrimage are two of the constructing activities of identity, shaping the ways in which one relates to and understands self and the other (Palmer, 2005). The phenomenon of pilgrimage is very heterogeneous. Researchers try to find the right motivators of a desire to travel (see Ch. 1). However, a valuistic approach could be such a common framework which is suitable for different types of pilgrimages. Valuistic journeys (Liutikas, 2012) related to the expression of an individual's values and identity offer the opportunity to consider the meaning of life and personal goals.

Travelling to pilgrimage places helps people discern between the relative values of different beliefs. It helps manifest and/or reaffirm existing personal or social identity. Values fostered by contemporary pilgrims can be related to religion and a search for a personal spiritual route, as well embodying national, cultural or other collective ideals; or they may of unique value to an individual.

This chapter analyses the phenomenon of pilgrimage based on valuistic motivation. Typologies of religious and culturally motivated travellers comprise of a variety of categories showing different approaches. Values fostered by pilgrims could be totally different from our stereotypical imagination. Quite usually, pilgrims' motivation and behaviour manifest in a tourist identity. In the concept developed in this book, the relevant contexts for the expression of identities are behaviour and ritualistic meaning-making acts. The identity is guided by meaning and meaning has to be relevant for the realization of the self (Fuhrer, 2004). Pilgrims behave to express their real goals, intentions, motives, values and identities.

Values and Identities

Mitchell (1983, p. vii) understands values as 'the entire constellation of a person's attitudes,

*darius.liutikas@gmail.com

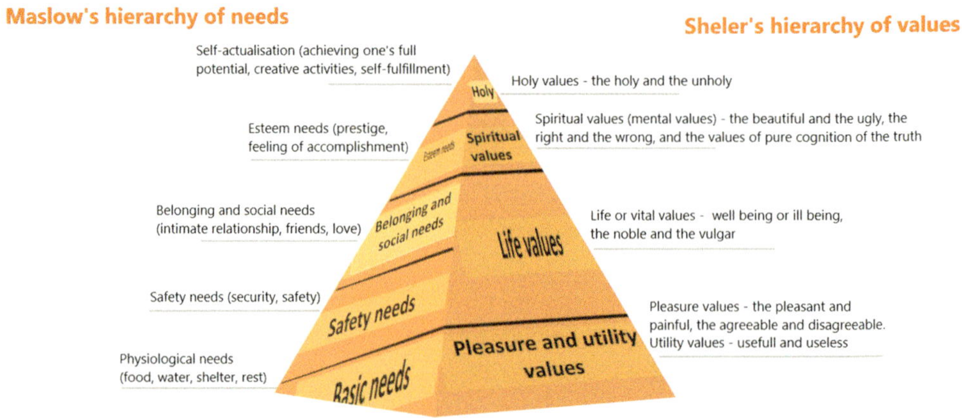

Fig. 2.1. Maslow's hierarchy of needs and Sheler's hierarchy of values.

beliefs, opinions, hopes, fears, prejudices, needs, desires, and aspirations that, taken together, govern how one behaves'. Reynolds (2017) defines values as 'the conceptions of social action that denote certain actions as desirable or good and certain actions as undesirable or bad'. Values could be grouped by personal importance and the hierarchy of values could be formed (Roccas and Elster, 2014). The individual chooses for himself which values are the most important. The meaning of values and the hierarchy of values depend on the subjective attitudes of the individual. Values directly affect an individual's behaviour because people consider more attractive actions that are consistent with their values (Feather, 1995; Roccas and Elster, 2014).

The main source of values formation is social and cultural environment. This arena includes family, friends, peers, professional colleagues, community, work organizations and the whole society.

Max Scheler (1874–1928) divides values into different hierarchical levels. Scheler's value theory is based on the notion of personality, how one acts and, in acting, demonstrates his or her values (Fig. 2.1). According to Scheler, 'The person is only in and through his acts. The psychic act is not self-contained. It is an event "in" time which, in principle, we can observe from the centre of spirit and which we can objectify through introspection' (Scheler, 1995). At the lowest rank of hierarchy is sensible values (pleasure and utility values). These values fall within the range of agreeable and disagreeable, pleasure and pain, enjoyment and suffering. In other considerations, utility values form a separate value rank. They range from useful to useless and include the practical, the efficient and the economic sphere. Life or vital values are related to well-being or ill-being, the noble and the vulgar, health and disease, and strength and weakness. Spiritual values (mental values) include aesthetic values (beauty to ugly), values of justice (right and wrong) and values of pure cognition of the truth. The highest values constitute the sphere of the holy and the unholy. The holy values are essential values of the person. Gottlöber (2013) states, that values for Scheler are not rationally known. Values appear 'in between' the loving/hating act of the person and the object. However, values have some objective quality. Each person chooses to act according to his or her value preferences (Gottlöber, 2013). An ideal group type for the value of the holy is a community of saints and believers. The holy values range from blissfulness to despair, and from faith to lack of faith. Blissfulness fills out the entire person, but holiness appears only in absolute objects and is independent of anything that has been considered holy by any people any time. The act of love ultimately leads to the holy.

Some other values models also include metaphysical components. Hodgkinson (1978, 1991) identifies three different levels of values:

- the sub-rational (values are based on personal preferences and feelings and are rooted in emotions. They could be asocial and amoral, and refer to what is enjoyable);

- the rational (values are grounded within a social context of norms, customs, standards and expectations. They depend on social justification and consensus); and
- the trans-rational (values are based on beliefs, faith, ethical codes, commitments and moral insights. They are conceived as metaphysical).

In tourism and pilgrimage, values appear in definitions of travelling types, e.g. religious tourism and religious pilgrimage, ecotourism, sustainable tourism, business tourism, music tourism and heritage tourism. Since many types of values-rich journeys exist, there are as many groups of people who have different values.

Following Yamin (2008), social and personal identity is defined as consciousness of oneself and others' perception of one's individuality. In this perspective it is important to know that identity is related to the representation of one's values, personal experience, memory and intelligence.

Tajfel (1981) states that the identification with a group and the emotional value of membership in a group are integral to the individual's sense of self. Some affiliations are especially important to self-concept and could provide the senses of self-esteem, stability, security or power. Individuals appreciate the experience of belonging and acceptance, solidarity in sharing common values and life goals. Socialization is the process of the acquisition of a social identity. People learn specific social and cultural norms and religious beliefs. Various rituals, feasts and festivals, and lectures and sermons, help in the socialization and re-socialization process and provide a strong guideline for values. In this context, pilgrimage could serve as an instrument for socialization and formation of social identity.

The more developed a person's social ties, the more diverse his or her personal and social identity becomes. The more these relationships are developed with people who take a different position in the structure of society, the more different attitudes, beliefs and forms of behaviour a person can form. Such a difference in relationships also determines the versatility of personal values, as well as the number of symbolic places that mark these values in space.

Dominant physical landmarks such as churches, temples or mosques help to define a religious group and guide people to a cognition of social structure. Tuan (1977) notices that identity of place is achieved by dramatizing the aspirations, needs and functional rhythms of personal and group life.

Personal identity distinguishes the person as a unique individual. The readiness to act and use a specific identity within and across situations is defined as identity salience (Stets and Serpe, 2013). Individuals may seek out opportunities to perform those identities which have higher salience. So identity salience is related to the behaviour of individuals and is based on commitments. According to Stets and Serpe (2013), commitment brings in social structure because it refers to the degree to which people are tied to particular social networks. For example, a religious pilgrim has a high salience manifesting in religious identity. Commitment to a religious community could be measured along the number of other social contacts sharing the same identity, the amount of time spent employing this identity, and the depth of relationships based on this identity.

Behaviour of Religious Pilgrims – Identity Manifestation in Actions

Different forms of a traveller's behaviour and performed rituals could be indicated in contemporary pilgrimages. Religion is a highly complex phenomenon (Beit-Hallahmi and Argyle, 1997) for which it is difficult to apply a single theory on religious behaviour. Pilgrims' behaviour could be a mix of social learning (behaviour models are regularly transmitted from generation to generation), expression of social identity (Tajfel and Turner, 1979) or collective consciousness (Durkheim, 2008 [1915]), rite of passage (Turner, 1973; Turner and Turner, 1978) or ritual work for social change (Goffman, 1971 [1959]).

Research on religiosity could help to answer the question of how important religion is in one's life. The level of religiosity impacts an individual's attitudes, values and actions. Religiosity could be defined as 'the degree to which beliefs in specific religious values and ideals are held and practised by an individual' (Delener, 1990). However, religiosity could be the source of other values and one aspect of a whole system of values (Reynolds, 2017). Personal values are likely to affect the level of religiosity because people are motivated to act in accordance with their values (Roccas and Elster, 2014).

Different dimensions and indicators have been attributed to measure religiosity (Lenski, 1961; Stark and Glock, 1968; Verbit, 1970; Stark and Glock, 1968; Cornwall *et al.*, 1986; Pearce *et al.*, 2017). In all dimensions, the three main components are recognized as knowing (associated with cognition, ideology, doctrinal orthodoxy, dogmatism and meaning system), feeling (associated with religious beliefs, devotion and feelings towards religious objects) and doing (associated with behaviour, rituals, religious community and religious practices). Stark and Glock (1968) identify five dimensions of religious commitment: the belief dimension, religious practice, the experience dimension, the knowledge dimension, and the consequential dimension. Pilgrimage is related to religious practice, which includes two forms, ritual and devotion. In religions, rituals are 'specific formalized religious actions that the adherents of a religion are expected to perform' (Furseth and Repstad, 2006). The main forms of manifestation of religiosity and religious identity include participation in various religious services and events, social and communal life, keeping traditions and performing religious rituals, and financial contributions. Personal prayers, study of religious literature and demonstration of religious symbols are also important elements of the demonstration of religious identity.

The Fetzer Institute and the National Institute on Ageing Working Group (John, 1999) developed the Brief Multidimensional Measure of Religiousness/Spirituality (BMMRS). This tool was used to research health outcomes and involved such factors as spiritual experiences, religious practices and congregationally based social support (Johnstone *et al.*, 2009). Spiritual experiences were divided into emotional experience of communication to a higher power, a sense of meaning in life, religious values and beliefs, the degree of forgiveness of self and others (and belief in the forgiveness of a higher power), and the religious/spiritual beliefs related to coping with life's problems. Religious practices were divided into private (which measures religious behaviours) and organizational (which measures involvement in a formal public religious institution). Congregational support was understood as the degree to which a local congregation provides help, support and comfort (John, 1999).

Religiosity is connected with religious identity, which refers to people's ways of relating to religion, how strongly they feel about their faith, and how they choose to demonstrate their beliefs during their daily lives and journeys (Liutikas, 2016). Religious identity, as other identities, is constructed through the internalization of certain values and beliefs and/or the aspiration to accept certain ideas and values.

Religious salience (Pearce *et al.*, 2017) represents religion's position among other identities. Religious salience describes 'the potential level of influence that religion might have on other realms of life' (Pearce *et al.*, 2017). Religious involvement (Edgell, 2006) and religious commitment (Cornwall *et al.*, 1986) are also related to the representation of one's values and identity.

Some dimensions of religious life along with personality traits can influence the motivation for pilgrimage. Religious beliefs (e.g. the belief in non-recognized apparitions of the Virgin Mary) and the wish to perform a particular ritual (to walk the Way of Crosses) could direct the pilgrims' destinations. Active participation in one's religious community leads to communal travels and the manifestation of communal identity.

Identity nurturing is related to repeated visits. Some pilgrims have the tradition of travelling to the same sacred site at a specific time (e.g. every month, every year, every second year). Catholic Church feasts are usually celebrated during a liturgical year. The annual tradition of travelling to the indulgence feasts is an important social tradition. Pilgrimages are an opportunity to meet friends and relatives and to spend time communicating with them (Liutikas, 2016). In general, social interaction becomes more and more important for developing contemporary identities (Wearing *et al.*, 2010)

Religious institutions encourage behavioural patterns and rituals symbolizing communal identity and values. Clift and Clift (1996) notice that rituals are a way of bringing symbolic meanings into our everyday reality. Through the ritual actions, the pilgrim makes a connection with that which was the goal. Rituals allow pilgrims to express and affirm their identity and make sense of their beliefs. Three key roles of rituals can be identified:

1. **Ritual as the main motive of values-rich journeys**. Ritual becomes the main motive for the travelling because it helps to create, preserve, develop and modify religious identity. The meaning of the ritual is constantly reviewed. Even religious beliefs could be transformed and changed through the creation and practice of rituals (Bell, 1997).
2. **Ritual as part of the travel experience**. In one case, the pilgrim knows exactly the traditional, sometimes archaic, elements of the ritual. He would like to follow a strict, often repetitive, pattern of ritual. In other cases, the pilgrim is open to experience and to perform ritual actions which are new for him. The performance of rituals allows travellers to experience transformation.
3. **Ritual as part of socialization**. In this case we discuss the social functions of rituals and their potential for social integration (Turner, 1969; Graburn, 1983; Smith, 1987; MacCannell, 1999 [1976]; Durkheim, 2008 [1915]).

The Catholic pilgrimage tradition includes a broad range of devotional practices and rituals. Some of them can be applied both during the journey itself and in the holy place, while other rituals are performed only in the holy place. Such Catholic practices like praying the rosary, singing hymns, participation in a holy mass, penitence or confession of sins could be performed both on the trip and in the sacred place. Others, such as placing flowers and lighting candles before a holy image; touching or kissing statues, images or relics; participation in special processions; crawling on the knees in special places (Fig. 2.2.); and having special family or community gatherings, could be performed only at the sacred sites.

The research shows typical behavioural features of Catholic pilgrims at the destination site. As an example, half of the Catholic pilgrims in Lithuania receive the holy sacrament (Communion); one third donate to the Church, community or beggars; and a quarter partake in the sacrament of Confession (Liutikas, 2012).

Religious journeys constitute different models of the specific pilgrims' behaviour. The most common Catholic religious practices are: (i) participation at the holy mass and receiving indulgence (including confession (Fig. 2.3), Eucharistic communion and prayers for the intentions of the Holy Father); (ii) praying (including rosary, singing hymns and private prayers in silence); (iii) processions and the Way of the Cross (including processions carrying a cross or a holy image, candle processions, and walking on the knees); (iv) usage of water (including baptism, drinking, taking baths, collecting holy water to bring home, wading, washing faces, and blessing); (v) leaving votive amulets (including leaving written or internal intentions, prayers, leaving inscriptions of gratitude on tiles or walls, lighting a candle, making offerings and donating money); and (vi) obtaining souvenirs and devotionalias (including holy images, badges, and taking photographs) (Liutikas, 2018). Some other models of behaviour could be noticed as donation to beggars, praying for the dead and communal eating after the religious service.

Liutikas (2014) identified six types of rites, rituals and characteristic behaviour of pilgrims. All these types could be applied to religious pilgrims, sports fans, music fans, pilgrims of national identity and pilgrims of the search of personal identity. They are:

1. internal (praying, meditation, contemplation, reading, internal promises, etc.);
2. external oral (praying, singing, confession, chants, slogans, communication with other travellers, etc.);
3. external physical actions (walking by foot, barefoot or on the knees, touching and kissing the relics, making the sign of the cross, positioning the body in a certain way for prayer, touching hands of sportsmen and musicians, clapping and trampling, etc.);
4. external using attributes (playing musical instruments, burning incense, wearing special clothes and symbols, carrying crosses, flags, banners, body painting, etc.);
5. complex actions (participation at the holy mass, processions, sport event, concert, commemorative event); and
6. actions to memorize a journey (photography, shooting, buying souvenirs, devotionals, leaving votive amulets, collecting signs of sportsmen, musicians or politicians, etc.).

Fig. 2.2. Climbing the Holy Stairs (*Scala Sancta*) on the knees in Rome. (Photo: the author)

Sacrifice was one of the most common rituals in earlier times, and today pilgrims believe that their personal sacrifice can help to appeal to supernatural beings to their advantage. Some pilgrims seek to leave special offerings (votive amulets at the miraculous image of the Virgin Mary, a cross at the Hill of Crosses, a donation to the Church) in the holy place. Some religious pilgrims seek physical challenges on their journey. Hardship or various restrictions symbolize the process of purification, sacrifice and transformation. It is believed that ascetism, austerities and danger lead to better rewards and merits in daily life. In some cases, pilgrimage is understood as a voluntary act of penance and religious devotion. In the Middle Ages, pilgrimage

Fig. 2.3. Confession. National Congress of Mercy, Vilnius. 2016 (Photo: the author)

as an act of penance was also imposed by ecclesiastical authorities (Webb, 2001). The most contemporary common hardships are walking long distances on foot, crawling on the knees, fasting, and walking. Long travel by bus can also cause other inconveniences, such as spending the night in the bus, travelling in hot buses with no air conditioning or just a long and tiring trip. Pilgrims also apply other personal restrictions, such as not listening to favourite music and not using the phone or the internet.

Some ritual forms are related to piety and the honour of the shrine (circumambulation, walking on the knees the last metres), self-renewal (walking in silence, Eucharistic adoration) and strengthening collective identity (common prayer and singing, holding each other's hands). Pilgrims have different identity markers, which help to foster communal identity. These external markers include pilgrims' clothes, headscarves, neck scarves, wristbands, necklaces, hats, medallions, crosses, small flags, badges and others. Organized pilgrims travelling by bus often affix the image of the destination site to the windshield of the bus. Different pilgrim badges were very popular in the Middle Ages. They mostly were made from metal and depicted symbols of the pilgrimage place.

Identification with a social group affects the person's preparedness to define themselves using the social values of that group (e.g. pilgrim group). When the person's identification is strong with a particular social group, he or she will be more prepared to interpret the world, and their own place within it, in a manner consistent with that organization's values, ideology and culture (Haslam and Ellemers, 2005).

Segmentation of Religious Travellers

Many scholars noticed that not all pilgrims have the same motivation to travel (Morinis, 1992; Smith, 1992; Cohen, 1992; Fleischer, 2000; Amos, 2009; Collins-Kreiner, 2010; Liutikas, 2009, Liutikas, 2012). This is most noticeable in the organized pilgrims' groups. Recognition of the official and real motivation of the journey form the basis for the concept of pseudo or self-styled pilgrimage (Liutikas, 2012, Liutikas, 2016). Self-styled persons travel with a group of pilgrims on motives different from the expression of values and identity.

Travellers could be segmented according to socio-demographic characteristics (age, gender), social status and financial abilities, lifestyle characteristics, change acceptance, depth of experience, needs, and others. The majority of travellers' segmentation models are related to their motivation and behaviour (activities) (McKercher *et al.*, 2002). Travellers are grouped into segments which have, more or less, homogeneous motivation, behaviour patterns or reactions to the travel activities. McKercher *et al.* (2002) notice that such factors as time availability, travel partners, tour group participation, level of awareness, level of interest, education, ethnic background and a variety of other factors may affect the individual's ability to engage with a place.

The Values and Lifestyles (VALS) model can also be an effective tool of pilgrimage research and the segmentation of values-rich travellers. Segments are based on two dimensions: the consumer's resources and the consumer's self-motivation (Strategic Business Insights, 2020). For values-rich journeys, such resources could be education, social capital, income, health, mobility level, and pilgrims' values and identities. Three consumer's self-motivations could be identified as ideals (traditions), achievement and self-expression (Strategic Business Insights, 2020). In values-rich journeys, travellers motivated by ideals could make choices based on their values, identity and principles. They resist impulses, try to get more information, and prefer traditions and customs. Travellers motivated by achievement are goal-orientated, seek approval from a valued social group, prefer symbols, sometimes lack information, and sometimes are not highly committed to their chosen identity and values. Travellers motivated by self-expression are spontaneous; they seek adventure and novelty and try to get new personal experiences.

First we could define three different types of the religious travel conceptions: (i) based on searching or the manifestation of religious values and identity; (ii) based on partly religious or non-religious motivation; and (iii) based on instrumental motivation. These conceptions provide a good framework for understanding the specific motives of travellers (Fig. 2.4). Second, the values-based framework led to distinguishing seven groups of travel motives: (i) pure religious and spiritual motives; (ii) ritual-orientated motives; (iii) a family's or community's tradition; (iv) the possibility for inner change and new social relations; (v) the possibility to understand religion; (vi) a new form of holiday travelling (good leisure time, sightseeing); and (vii) an instrument for seeking other objectives.

Travellers' identities and values could help to classify them. The majority of authors (Stoddard, 1997; Santos, 2003) indicate three types of travellers: religious pilgrims on one side, secular tourists on the other side and religious tourists in the middle. Smith (1992) also identified two more types: religious tourists who are more pilgrims and less tourists, and also religious

Fig. 2.4. Pilgrims and tourists at the Temple of the Sacred Tooth Relic, Kandy, Sri Lanka. (Photo: the author)

tourists who are more tourists than pilgrims but use pilgrimage as an opportunity for travelling. Baeva (2014) divides different types of visitors to the Russian Church in Sofia into three categories – parishioner, pilgrim and tourist.

Based on the travellers' purpose, Navruz-zoda and Navruz-zoda (2018) segmented Sufi-pilgrimage in Uzbekistan into five groups: (i) Muslim-Sufis (their goal is the realization of Sufi rituals); (ii) Muslim non-Sufis (their goal is visiting Sufi shrines as a part of religious ceremonies); (iii) representatives of other religions (they also visit Sufi shrines as a part of religious ceremonies); (iv) scientists, intellectuals, researchers and educators (they visit Sufi shrines in order to study the philosophy of Sufism); and (v) other categories of tourists (they visit these places as a part of another tourism package).

The religious involvement (Edgell, 2006) approach can also be applied in the pilgrims' segmentation context. In general, a construct of involvement is widely accepted to examine the motivation and participation in other leisure activities (Funk and Bruun, 2007). Religious involvement could be public (e.g. frequency of church attendance; frequency of participation in public religious rituals, events and pilgrimages) and private (e.g. frequency of private prayer, frequency of reading of religious literature, watching or listening religious programmes). Individuals with higher levels of running involvement tend to demonstrate an increased frequency of participation in pilgrimages.

Various socio-demographic differences affect the extent of religious involvement (e.g. age, education, gender, race, marital status, health, socio-economic status, region). For example, women are more religiously involved than men (Stark, 2002). As compared to whites, black respondents demonstrate higher levels of both public and private religious behaviours (Taylor *et al.*, 2004), and the young are less religious than the elderly (Deaton, 2011). Subjective religiosity involves various perceptions, attitudes

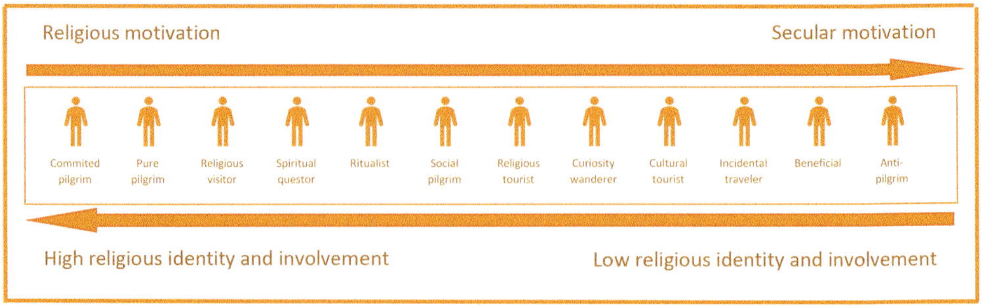

Fig. 2.5. Types of religious-related travellers.

and experiences related to religion. A pilgrimage can be determined by how important people consider religion to be in their lives, and what the role of religious beliefs in the daily life of a pilgrim is.

Here, the model of pilgrim types based on their values and identity can be presented. Twelve types of religious-related travellers can be identified ranging from religious to secular motivation and from a low level of religious identity and involvement to a high level of religious identity and involvement (Fig. 2.5). Starting from the first, every type reflects a decreasing level of personal involvement and a weaker ratio with religious values and identity.

1. **Committed pilgrims**. Some individuals are intensely involved with religion, e.g. clergy (priests and members of religious orders), converts (especially those who had sudden and dramatic conversions), fundamentalists (people with a strong prejudice), people who had intense religious experiences (mystical experiences, ecstasy) and psychological disorders or having psychological difficulties (Beit-Hallahmi and Argyle, 1997). Pilgrimage for them is one element of their intense religious life.
2. **Pure pilgrims**. The journey of these pilgrims is based on religious and spiritual motive. They devote their journey to prayer or meditation, communication on religious topics and analysis of issues of the sense of life. Usually, they are involved in religious communities, attend religious services and participate in church-related social events. They manifest religious values and identity during the pilgrimage.
3. **Religious visitors**. These pilgrims take the opportunity to visit a pilgrimage place while visiting a city or area for other reasons. Their primary goal of the trip could be business, participation at a conference, seminar, or cultural event. However, they see such trips as a chance to visit a nearby sacred site. The pilgrimage place is related to their religious identity and values. Some religious visitors are likely to be repeat visitors from the neighbourhood area. They can either have moderate or high level of religious identity and involvement.
4. **Spiritual questors (seekers)**. Sometimes a pilgrim journey could serve as an opportunity to renew religious identity or to strengthen it. The specific religious practices are performed at a pilgrimage site. The place itself leads to the development of an identity or to cause a real religious conversion from seeker to believer. Spiritual seekers may have a formal involvement in the religious community. Either they have religious creeds but seek to refresh them, or they want to change their spiritual life through spiritual guidance and meanings obtained at the pilgrimage site.
5. **Ritualists**. Schnell and Pali (2013) emphasize that contemporary pilgrims to Santiago de Compostela can be understood to perform a personal ritual. The meaning-making potential of the ritual unfolds transformative experience of pilgrimage. Life-transforming experiences are the core elements of both traditional and more contemporary forms of pilgrimage (Winkelman and Dubisch, 2005).

Pilgrimage is considered as the ritual of transformation during which new insights are given and deeper understanding is attained. A pilgrim journey could mark a transition (Turner and Turner, 1978) in the life of a traveller. It could help them to recover from serious illness or physical or psychological trauma (betrayals or losses). Pilgrims also travel to participate or to observe rites of passage like baptism, confirmation (in the Catholic Church), marriage and funerals.

6. **Social pilgrims.** For these pilgrims, the most important elements are communitas and socialization with other pilgrims. Usually, they are traditional (cultural) believers. The journey is a possibility to spend leisure time with family members, friends or community members who share the same values and identities. These pilgrims foster security and conformity values which express the need for a certain and stable social environment. These pilgrims also choose the journey for extrinsic religious motivation (e.g. social prestige, affirmation of one's lifestyle). Sometimes main community events are related to the annual pilgrimages to local shrines.
7. **Purposeful religious tourists.** Cohen (2006) notices that travel, religion and education share a possible goal – a quest for understanding. Religious education or acquaintance with a religious tangible and intangible heritage is the primary motive for visiting a pilgrimage site. Travellers learn about religious traditions, architecture and art and observe religious feasts and rituals. They are happy to get new knowledge and new experiences. Such travel could be part of a larger religious education programme. Sometimes they participate in some religious rituals and activities. They can either have formal relationships to religious communities in their daily lives, or not.
8. **Curiosity wanderers.** Their main motive is to experience what's happening in the pilgrimage site and how it looks. Travellers are motivated by curiosity, interest and the possibility to observe. They seek new impressions at the new environment and to take a break from the routine of everyday life. The pilgrim way or site could stimulate their imagination and the search for self-development. Some of them 'collect' visited places and enjoy the authenticity of place or event. The decision to go to a pilgrimage place could be made spontaneously. This is a way to diversify leisure time. The level of religious identity of these travellers is quite low.
9. **Sightseeing cultural tourists.** The religious objects and pilgrimage places are only a part of the route. They are interested in collecting different experiences. Religious involvement and religious identity are weak. Sightseeing, relaxation and recreational motives are primary. They also engage in other secular activities during the journey. These tourists don't travel for pilgrims' or purposeful religious tourists' reasons. However, they can end their journey having had a deep religious or spiritual experience.
10. **Incidental travellers.** They understand that they come to the pilgrimage site under certain circumstances. The decision to visit this place is made by somebody else with whom the casual traveller is temporarily or permanently connected. These travellers usually accompany other travellers – friends, business partners, family members with serious illnesses and disabilities. The site could be just a stop on the way to a further destination. It also can be a part of an excursion organized by business partners. Religious identity and involvement of such travellers is very low. However, they can participate in some activities at the destination site.
11. **Beneficials.** One of the motives could be related to providing various services for pilgrims (especially during religious feasts). The travel to the site is related to economic benefit (transportation, guiding, catering, security service, or devotionalias trade). They also can be politicians and officials who seek out visibility and notoriety. Advertisers, promoters of various ideologies, can also travel to places that attract crowds of pilgrims. We can also include media representatives and photographers in this group.
12. **Anti-pilgrims** are a special type of traveller. The reasons for travelling to the sacred site for these travellers is totally to

oppose religious values. Their main motives relate to criminal activities: terrorism, robbery, stealing, cheating, prostitution and drug distribution. Various robbers attacked pilgrims in the Middle Ages as well.

This model is partly based on pilgrims' empirical research in Lithuania (Liutikas, 2012). Pilgrim types were identified according to these principles: they must be easily identified and measured, each group must have a sufficient number of representatives and the motivation and behaviour of the representatives of each type have empirical evidence in various other studies. Other important questions like shifting from one type to another during the same journey should be answered after further studies.

Secular Pilgrims and their Identity

Secular pilgrimage is very heterogeneous. Most forms of secular pilgrimage relate to social identity affiliation (nationality, culture, sport) and fan-affinity groups, as well as the search for personal identity. Clift and Clift (1996) notice that pilgrimage journeys could be to sites that are held in high esteem and represent values served, 'values that people have even been willing to die for'. Different areas of identity are not equally important to a person at a particular time. Individuals can choose secular pilgrimage when different cultural identities become dominant.

Journeys of national identity (patriotic pilgrimage) include visiting the graves of heroes, an ethnic homeland, battlefields and war memorials, places important to the history of a nation, or monuments of national pride. Such trips can also be journeys to meet political leaders. Some pilgrims identify themselves with strong political leaders and their political programmes. By taking on personal responsibilities, charismatic leaders motivate followers to raise collective identities and values and associate such behaviour with them. Gecas (2000) noticed that charismatic leaders are attracted by the personal strength and perception that they symbolize or personify in the values and goals of the social movement. Such leaders, based on their moral authority, are able to transform the needs and interests of followers from selfish to collective interests, values and goals, and their trust to symbolic and inspiring activities.

Journeys of national identity are very important in the face of a threat to the nation. The need to defend themselves from a common enemy, to liberate a country, or to commemorate the historic events of the nation could attract millions of people. On 23 August 1989, political pilgrims from Lithuania, Latvia and Estonia flocked to the Baltic Way. Pilgrims joined their hands to form a human chain spanning 675.5 km in the three Baltic countries. Approximately 2 million people travelled to this event which became a symbol of the aspiration for the independence of the Baltic states. The trip to this political action expressed clear values of the people regarding the fate of their nations. The euphoric atmosphere of community when the names of the countries were chanted or their anthems were sung was the culmination of this pilgrimage.

Journeys of personal identity include the search for genealogical roots or visiting the graves of relatives. Places important to the history of a nation or family and possible travel routes are often known from the stories of relatives. Such trips bring emigrants closer to their relatives and friends. Cohen (2001) note that the tourist nature of such trips becomes more pronounced when the country of the parents or ancestors is visited by second or later generations of emigrants.

A secular pilgrimage is also a trip to meet people who share the same values or have the same worldview (scout jamborees, nature observers' meetings), or who have the same life experience (war veterans' or Chernobyl firefighters' meetings). It is important to emphasize here that a meeting or conference does not have to be a work or day-to-day business trip. The journey must be associated with holidays, which are a kind of sacred time (Graburn, 1989, pp. 21–36).

Stone (2020) notice that secular pilgrimages related to visits to sites associated with the dead, an atrocity or a disaster have emerged from the act of a traditional pilgrimage. The places where the family members died become the personal places of pilgrimage. At such places, pilgrims can express their grief or remember the loved one. Examples of such places are places of tragic car accidents, places of genocide and places of executions. Pilgrims at such places lay flowers or light candles. In Christian countries such places often are marked by the symbol of the Cross.

The graves are associated with the perception of the afterlife and respect for loved ones and ancestors. The cemeteries and other burial sites are undoubtedly associated with transcendence. In such places people raise questions about the meaning of life and death. Meditation on death and understanding is a common theme of dark tourism and pilgrimage (Korstanje and Olsen, 2020). The thoughts and the way of life of dead loved ones influence the destinies and choices of the living. Therefore, often visiting relatives in cemeteries, which are often located in another city or remote rural area, becomes an important act of pilgrimage.

Sometimes, for Holocaust survivors and their loved ones, small monuments marking places of torture are the only physical objects that link tragic memories to the place. Places of various catastrophes (wars, genocide, natural disasters etc.) once again attract those who have survived these events. Survivors and witnesses of the tragic events seek to visit the places that determined their later lives before death.

Sport fans could be classified according the level of identification. Such team identification refers to the extent to which a fan feels psychologically connected to a team. This extent appears to be relatively stable (Wann *et al.*, 2001). Fans with a high level of team identification are more likely to choose the journey to support the team outside of their place of residence. Such a role of a team follower could be the central component of a fan's identity. Their motivation could be group affiliation (opportunity to spend time with others), family (opportunity to spend time with family members), aesthetic (artistic beauty and grace of sport movement), self-esteem (opportunity to feel better about himself or herself), economic (sport gambling), eustress (excitement and arousal felt while watching sporting events), escape (diversion from the rest of his or her life) and entertainment (enjoyable pastime) (Wann *et al.*, 2001).

One of the most common factors of the construction of social identity is music. People, especially the young, want to identify themselves with pop or rock stars and their music. When people express information about their music preferences, it places them into a particular social category or group (Rentfrow *et al.*, 2009). Musical preferences can reveal a person's values, attitudes and beliefs (Fig. 2.6). Identification with a social group might provide information about a person's lifestyle, opinions and preferences. Researchers (Rentfrow *et al.*, 2009) also point out that individuals prefer those styles of music that reinforce and reflect various aspects of their identities and personalities (e.g. energetic music is associated with extroverts, classical music with intelligence, pop music with physical attractiveness).

Certain actions become ritualized in a secular pilgrimage. In this case, simple actions like spoken words, bending, touching or placing flowers assume greater meaning and significance. Pilgrims give importance to symbolism. They generate special symbols to achieve travel goals. A complex of symbolic actions develop a traveller's identity. However, secular symbols and rituals are more amorphous and more flexible. Schnell and Pali (2013) point out that personal rituals cannot be identified from an outside perspective; only the actor's subjective interpretation of meaning enables that formalized pattern of action to become a personal ritual.

Sport and music fans, and other secular pilgrims, also have different motivations. Different reasons of travelling create different behaviour and implemented activities. We can try to divide travellers into different segments according their primary motive; however, the boundaries between such segmented groups could be vague. In addition, secular pilgrims must be distinguished from tourists or entertainment seekers. Liutikas (2009) notes the intertwined motivation of cultural pilgrims and cultural tourists. The majority of cultural travellers who travelled to concerts desired to have a good leisure time and experience entertainment. Such a motive for travel as the primary motive was mentioned by two thirds of cultural travellers. However, about one third of travellers believed that such a trip is of value to them (Liutikas, 2009). So, based on the motivation of the travellers and their approach to the trip, cultural travellers could be divided into three groups: cultural pilgrims, cultural tourists and peripherical culture consumers (Fig. 2.7).

The first group is dominated by motives related to the expression of values and identities. Cultural pilgrims could be subdivided into additional groups:

Fig. 2.6. Participants and spectators of the National Song Festival: the dance night, Vilnius, Lithuania, 2014. (Photo: the author)

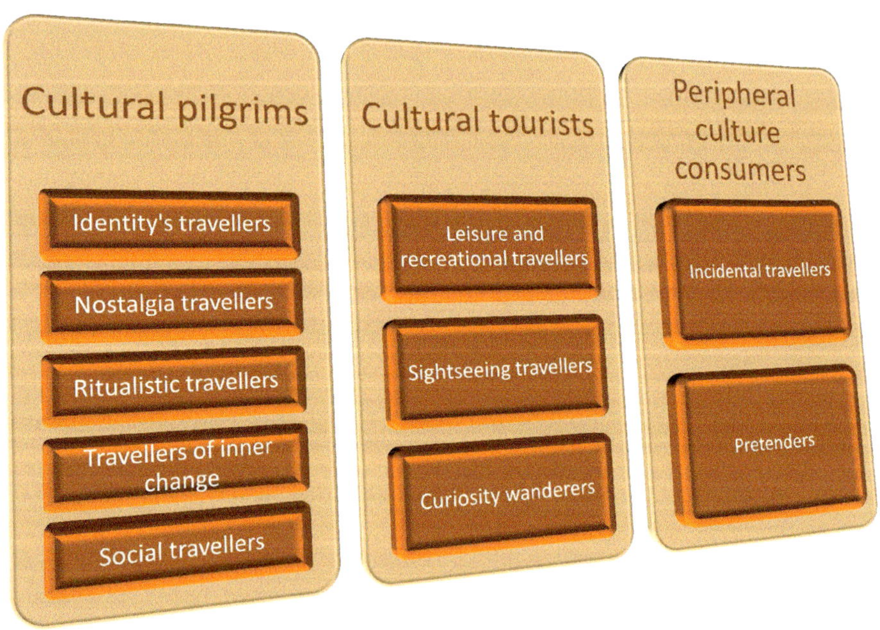

Fig. 2.7. Types of cultural travellers.

- **identity' travellers** (motivation related to the expression of values, attitudes and identities, various hobbies);
- **nostalgia travellers** (motivation related to nostalgia, previous identities, memories of youth);
- **ritualistic travellers** (motivation related to the tradition or wish to implement a special ritual);
- **travellers of inner change** (motivation related to the opportunity for inner change, emotional and spiritual renewal, and psychological solace);
- **social travellers** (motivation related to the identity of a social group, community and family traditions).

The second group shares tourist motives and the aim to have a good time. Cultural tourists are subdivided to such groups as:

- **leisure and recreational travellers** (motivation related to a way of spending leisure time, seeking new impressions and the possibility to relax, getting a rest from the routine);
- **sightseeing travellers** (motivation related to learning about cultural heritage and traditions; visiting an event or location is just one component of a travel programme);
- **curiosity wanderers** (motivation related to the satisfaction of curiosity; interest and the possibility to observe; the main wish is to experience the site or atmosphere of an event).

The motives of the third group are focused on peripheral culture consumption. This group is subdivided into:

- **incidental travellers** (visiting a cultural venue or event is totally incidental, for example by receiving a flyer or at the suggestion of a hotel administrator or a fellow traveller);
- **pretenders** (culture is just a secondary non-essential motive; the primary motive could be accompanying a disabled family member, to pick up a girl, or testing a newly obtained camera).

Like in the segmentation of religious-related travellers, the travellers' groups **beneficials** and **anti-cultural tourists** could be also identified. Cultural tourists' segmentation could be based on other criteria. Du Cros and McKercher (2015) divide cultural tourists into five segments: serendipitous cultural tourist, purposeful cultural tourist, incidental cultural tourist, casual cultural tourist and sightseeing cultural tourist. These segments were based on the role that culture plays as a trip motivator and the depth of experience (level of engagement).

Perhaps the main difference between cultural tourism and cultural pilgrimage is that cultural places (architectural heritage, museum) and events (art exhibition, concert, opera) are visited by tourists simply to meet cultural, cognitive and leisure needs, while for pilgrims such a trip and a visit to a specific place or event is an expression of their values and social or personal identity. For secular pilgrims the date is more often important for the journey (personal commemorations, national and cultural anniversaries, meetings days, sport and music events, festivals), the length of stay (pilgrims' length of stay at the destination is greater than that of the general travelling public) and self-reflection.

Conclusions

This chapter has presented different expressions of religious and secular identities in travelling. Pilgrimage belongs to the main forms of the manifestation of religiosity and religious identity. Various dimensions and indicators of how to measure the level of religiosity can be applied. Rituals allow pilgrims to express and affirm their identity and make sense of their beliefs. A broad range of rituals and devotional practices are applied in the Catholic tradition. They show the ways to the internal transformation of pilgrims. Even modern religious practices and the use of symbols (identity markers) of Catholic pilgrims have deep historic traditions.

Religious identity and religious involvement become a key component in the identification and distinguishing of different types of religious-related travellers. This chapter presented 12 types of religious-related travellers ranging from religious to secular motivation and from low level of religious identity and involvement to high level of religious identity and involvement. This classification shows a wide set of motives that drive people to visit religious places. The understanding of the variety of religious-related traveller and pilgrim types could be useful for other research and pilgrimage marketing.

Secular pilgrimage is much more heterogeneous. The rituals of secular pilgrims are looser, with more individuality, most frequently without uniform structure and form. Every realm of secular pilgrimage could be subdivided into additional groups. All these travellers' groups, having different needs and goals, can be simultaneously found at the pilgrimage site or event. It becomes evident that the expression of values and identity is only one of the motivational sets. The people travelling with a group of pilgrims could manifest tourist identity as well. In those cases, the main goals of travellers are spending leisure time, and the possibility to seek out new impressions and new social contacts. Typologies of religious and cultural travellers show that sometimes the motives of travellers and their behaviour may be completely opposite to the cultural or religious values. A value-rich journey could be an instrument for seeking other objectives. We have to be careful about considering pilgrims as a homogeneous group.

References

Amos, R.S. (2009) Towards a typological model of contemporary Christian travel. *Journal of Heritage Tourism* 4(4), 287–297.

Anderson, B. (2006) *Imagined communities: Reflections on the origin and spread of nationalism*. Verso.

Baeva, V. (2014) Parishioners, pilgrims, tourists: the visitors of an orthodox Christian shrine in Sofia. In: Katič, M., Klarin, T. and McDonald, M. (eds) *Pilgrimage and Sacred Places in Southeast Europe: History, Religious Tourism and Contemporary Trends*. LIT, pp. 79–96.

Beit-Hallahmi, B. and Argyle, M. (1997) *The Psychology of Religious Behaviour, Belief and Experience*. Routledge.

Bell, C. (1997) *Ritual: Perspectives and Dimensions*. Oxford University Press, Oxford.

Clift, J.D. and Clift, B.W. (1996) *The Archetype of Pilgrimage*. New York.

Cohen, E. (1992) Pilgrimage centers, concentric and excentric. *Annals of Tourism Research* 19(1), 33–50.

Cohen, E. (2001) Who is a tourist? A conceptual clarification. In: Cohen, E. (ed.) *Contemporary Tourism: Diversity and Change*. Elsevier Science, pp. 17–36.

Cohen, E. (2006) Religious tourism as an educational experience. In: Timothy, J.D. and Olsen, D.H. (eds) *Tourism, Religion and Spiritual Journeys*. Routledge, pp. 78–93.

Collins-Kreiner, N. (2010) Current Jewish pilgrimage tourism: modes and models of development. *Tourism* 58(3), 259–270.

Cornwall, M., Albrecht, S.L., Cunningham, P.H. and Pitcher, B.L. (1986) The dimensions of religiosity: a conceptual model with an empirical test. *Review of Religious Research* 27(3), 226–244. DOI: 10.2307/3511418.

Deaton, A. (2011) Aging, religion and health. In: Wise, A.D. (ed.) *Explorations in the Economics of Aging*. University of Chicago Press, pp. 237–268.

Delener, N. (1990) The effects of religious factors on perceived risk in durable goods purchase decisions. *Journal of Consumer Marketing* 7(3), 27–38. DOI: 10.1108/EUM0000000002580.

Du Cros, H. and McKercher, B. (2015) *Cultural Tourism*. Routledge, London and New York.

Durkheim, E. (2008 [1915]) *The Elementary Forms of the Religious Life*. Dover Publications, Inc.

Edgell, P. (2006) *Religion and Family in a Changing Society*. Princeton University Press, New Jersey.

Feather, N.T. (1995) Values, valences, and choice: the influences of values on the perceived attractiveness and choice of alternatives. *Journal of Personality and Social Psychology* 68(6), 1135–1151. DOI: 10.1037/0022-3514.68.6.1135.

Fleischer, A. (2000) The tourist behind the pilgrim in the Holy Land. *International Journal of Hospitality Management* 19(3), 311–326. DOI: 10.1016/S0278-4319(00)00026-8.

Fuhrer, U. (2004) *Cultivating MindsM Identity as Meaning-making Practice*. Routledge.

Funk, D.C. and Bruun, T.J. (2007) The role of socio-psychological and culture-education motives in marketing international sport tourism: a cross-cultural perspective. *Tourism Management* 28(3), 806–819. DOI: 10.1016/j.tourman.2006.05.011.

Furseth, I. and Repstad, P. (2006) *An Introduction to the Sociology of Religion: Classical and Contemporary Perspectives*. Ashgate, Farnham, UK.
Gecas, V. (2000) Value identities, self-motives and social movements. In: Stryker, S.E., Owens, J.T. and White, W.R. (eds) *Self, Identity, and Social Movements*. University of Minnesota Press, Minneapolis and London, pp. 93–109.
Goffman, E. (1971 [1959]) *The Presentation of Self in Everyday Life*. Penguin Books, Harmondsworth, UK.
Gottlöber, S. (2013) The role of aesthetic values, art, and the artistic genius in Scheler's ethical personalism. In: Ward, A. (ed.) *Socrates and Dionysus: Philosophy and Art in Dialogue*. Cambridge Scholars Publishing, pp. 159–174.
Graburn, N.H.H. (1983) The anthropology of tourism. *Annals of Tourism Research* 10(1), 9–33. DOI: 10.1016/0160-7383(83)90113-5.
Graburn, H.H.N. (1989) Tourism: the sacred journey. In: Smith, L.V. (ed.) *Hosts and Guests: The Anthropology of Tourism*. University of Pennsylvania Press, pp. 21–36.
Haslam, A.S. and Ellemers, N. (2005) Social identity in industrial and organizational psychology: concepts, controversies, and contributions. In: Hodgkinson, P.G. and Ford, K.J. (eds) *International Review of Industrial and Organizational Psychology*. Vol. 20. Wiley, pp. 39–118.
Higham, J. and Hinch, T. (2009) *Sport and Tourism: Globaliztion, Mobility and Identity*. Elsevier.
Hodgkinson, C. (1978) *Towards a Philosophy of Administration*. Basil Blackwell, Oxford.
Hodgkinson, C. (1991) *Educational Leadership: The Moral Art*. State University of New York Press.
John, E. (1999) *Multidimensional Measurement of Religiousness/Spirituality for Use in Health Research*. John E. Fetzer Institute / National Institute on Aging Working Group, Kalamazoo, Michigan.
Johnstone, B., Yoon, D.P., Franklin, K.L., Schopp, L. and Hinkebein, J. (2009) Re-conceptualizing the factor structure of the brief multidimensional measure of religiousness/spirituality. *Journal of Religion and Health* 48(2), 146–163. DOI: 10.1007/s10943-008-9179-9.
Korstanje, E.M. and Olsen, D.H. (2020) Negotiating the intersections between dark tourism and pilgrimage. In: Olsen, D.H. and Korstanje, E.M. (eds) *Dark Tourism and Pilgrimage*. CAB International, pp. 1–15.
Lenski, G.E. (1961) *The Religious Factor: A Sociological Study of Religion's Impact on Politics, Economics, and Family Life*. Greenwood Press, Westport, Connecticut.
Liutikas, D. (2009) *Piligrimystė. Vertybių ir tapatumo išraiškos kelionėse Vilnius*. Lietuvos piligrimų bendrija.
Liutikas, D. (2012) Experiences of valuistic journeys: motivation and behaviour. In: Sharpley, R. and Stone, P. (eds) *Contemporary Tourist Experience: Concepts and Consequences*. Routledge, New York, pp. 38–56.
Liutikas, D. (2014) Lithuanian valuistic journeys: traditional and secular pilgrimage. *Journal of Heritage Tourism* 9(4), 299–316. DOI: 10.1080/1743873X.2014.890205.
Liutikas, D. (2016) Indulgence feasts: manifestation of religious and communal identity. In: Jepson, A. and Clarke, A. (eds) *Managing and Developing Communities, Festivals and Events*. Palgrave Macmillan, pp. 148–164.
Liutikas, D. (2018) Catholic pilgrimage in Europe: contemporary traditions and challenges. In: El-Gohary, H., Edwards, D.J. and Eid, R. (eds) *Global Perspectives on Religious Tourism and Pilgrimage*. IGI Global, pp. 84–103.
MacCannell, D. (1999 [1976]) *The Tourist: A New Theory of the Leisure Class*. University of California Press, Berkeley, California.
McKercher, B., Ho, P.S.Y., Cros, H.D. and So-Ming, B.C. (2002) Activities-based segmentation of the cultural tourism market. *Journal of Travel & Tourism Marketing* 12(1), 23–46. DOI: 10.1300/J073v12n01_02.
Mitchell, A. (1983) *The Nine American Life Styles*. Warner, New York.
Morinis, A. (1992) Introduction: the territory of the anthropology of pilgrimage. In: Morinis, A. (ed.) *Sacred Journeys. The Anthropology of Pilgrimage*. Greenwood Press, Westport, Connecticut and London, pp. 1–28.
Navruz-zoda, B. and Navruz-zoda, Z. (2018) The destination marketing development of religious tourism in Uzbekistan: a case study. In: Yasuda, S., Raj, R. and Griffin, K. (eds) *Religious Tourism in Asia: Tradition and Change through Case Studies and Narratives*. CAB International, Wallingford, UK, pp. 149–157.
Palmer, C. (2005) An ethnography of Englishness: tourism and national identity'. *Annals of Tourism Research* 32(1), 7–27.
Pearce, L.D., Hayward, G.M. and Pearlman, J.A. (2017) Measuring five dimensions of religiosity across adolescence. *Review of Religious Research* 59(3), 367–393. DOI: 10.1007/s13644-017-0291-8.

Rentfrow, P.J., McDonald, J.A. and Oldmeadow, J.A. (2009) You are what you listen to: young people's stereotypes about music fans. *Group Processes & Intergroup Relations* 12(3), 329–344. DOI: 10.1177/1368430209102845.
Reynolds, C. (2017) Religion and values in the ESS: individual and societal effects. In: Breen, J.M. (ed.) *Values and Identities in Europe: Evidence from the European Social Survey*. Routledge.
Roccas, S. and Elster, A. (2014) Values and religiosity. In: Saroglou, V. (ed.) *Religion, Personality, and Social Behavior*. Taylor & Francis, pp. 193–212.
Ruiz Ballesteros, E. and Hernández Ramírez, M. (2007) Identity and community—Reflections on the development of mining heritage tourism in southern Spain. *Tourism Management* 28(3), 677–687. DOI: 10.1016/j.tourman.2006.03.001.
Santos, M. (2003) Religious tourism: contributions towards a clarification of concepts. In: Fernandes, C., Mcgettigan, F. and Edwards, J. (eds) *Religious Tourism and Pilgrimage*. Atlas: Special Interest Group, Tourism Board of Leiria/Fatima, Portugal, pp. pp. 27–42.
Scheler, M. (1995) Die Stellung des Menschen im Kosmos. In: Frings, M.S. (ed.) *Späte Schriften*. Bouvier Verlag, Bonn, pp. 9–71.
Schnell, T. and Pali, S. (2013) Pilgrimage today: the meaning-making potential of ritual. *Mental Health, Religion & Culture* 16(9), 887–902. DOI: 10.1080/13674676.2013.766449.
Smith, R. (1987) Leisure of tourist with a disability: barriers to travel. *Annals of Tourism Research* 14, 376–389.
Smith, L. V. (1992) Introduction: the quest in guest. *Annals of Tourism Research* 19(1), 1–17.
Stark, R. (2002) Physiology and faith: addressing the 'universal' gender difference in religious commitment. *Journal for the Scientific Study of Religion* 41(3), 495–507. DOI: 10.1111/1468-5906.00133.
Stark, R. and Glock, Y.C. (1968) *American Piety*. University of California Press, Berkeley, California.
Stets, E.J. and Serpe, T.R. (2013) Identity theory. In: DeLamater, J. and Ward, A. (eds) *Handbook of Social Psychology*. Springer, pp. 31–60.
Stoddard, R.H. (1997) Defining and classifying pilgrimages. In: Stoddard, R.H. and Morinis, A. (eds) *Sacred Places, Sacred Spaces: The Geography of Pilgrimages*. Geoscience Publications, Louisiana State University, Baton Rouge, Louisiana, pp. 41–60.
Stone, R.P. (2020) Preface. In: Olsen, H.D. and Korstanje, E.M. (eds) *Dark Tourism and Pilgrimage*. CAB International, Wallingford, UK, pp. i–0.
Strategic Business Insights (2020) US Framework and VALSTM. Available at: http://www.strategicbusinessinsights.com/vals/ustypes.shtml
Tajfel, H. (1981) *Human Groups and Social Categories: Studies in Social Psychology*. Cambridge University Press, Cambridge, UK.
Tajfel, H. and Turner, J. (1979) An integrative theory of intergroup conflict. In: Austin, W. and Worchel, S. (eds) *The Social Psychology of Intergroup Relations*. Brooks/Cole, Pacific Grove, California, pp. 33–47.
Taylor, J.R., Chatters, M.L. and Levin, J. (2004) *Religion in the Lives of African Americans. Social, Psychological, and Health Perspectives*. Sage Publications.
Tuan, Y.-F. (1977) *Space and Place. The Perspective of Experience*. University of Minnesota Press, Minneapolis and London.
Turner, V. (1969) *The Ritual Process*. Aldine, Chicago, Illinois.
Turner, V. (1973) The center out there: pilgrim's goal. *History of Religions* 12(3), 191–230. DOI: 10.1086/462677.
Turner, V. and Turner, E. (1978) *Image and Pilgrimage in Christian Culture*. New York.
Verbit, M.F. (1970) The components and dimensions of religious behavior: toward a reconceptualization of religiosity. In: Hammond, P.E. and Johnson, B. (eds) *American Mosaic*. Random House, New York, pp. 24–39.
Wann, D.L., Melnick, M., Russell, G. and Pease, D. (2001) *Sport Fans: The Psychology and Social Impact of Spectators*. Routledge, London.
Wearing, S., Stevenson, D. and Young, T. (2010) *Tourist Cultures: Identity, Place and the Traveller*. Sage.
Webb, D. (2001) *Pilgrims and Pilgrimage in the Medieval West*. I.B. Tauris Publishers.
Winkelman, M. and Dubisch, J. (2005) Introduction. The anthropology of pilgrimage. In: Dubisch, J. and Winkelman, M. (eds) *Pilgrimage and Healing*. The University of Arizona Press, pp. ix–0.
Yamin, S. (2008) Understanding religious identity and the causes of religious violence. *Peace Prints: South Asian Journal of Peacebuilding* 1(1), 1–21.

3 The Connection Between Liminal Places and Hospitality in Manifesting Pilgrim Values and Identity

Kip Redick*
Christopher Newport University, Virginia, USA

Introduction

An archetype of Christian pilgrimage can be traced back to Judaism and the various visits to Jerusalem during the festivals of Pesach (Passover or Feast of Unleavened Bread), Shavuot (Weeks or Pentecost) and Sukkot (Tabernacles or Booths). These feasts had their origins in Israel's mythic desert sojourn. The Hebrew exodus from Egypt involved a *liminal* journey in that they were undergoing transformation between their former dwelling in Egypt and the future promised land. In addition, their trek passed through the *liminal* space of wilderness, wherein they encountered hospitality.

However, the story unfolding in the exodus account shows the ephemerality of liminality. When Moses stays too long on the mountain, meeting with God, the people make a god for themselves, a golden calf (Exodus 32: 1–6). In so doing, they exit liminality and attempt to create a world wherein they retain control. In constructing the god and altar, in declaring the golden calf a god who rescued the people from Egypt, they have erected a shrine that is the centre of power, what Gerardus Van der Leeuw refers to as 'a world in itself' (Van der Leeuw, 1963, p. 397). The Hebrews are no longer between worlds; they enter their own world.

In addition to showing the ephemeral nature of liminality, the account also indicates the dynamic of hospitality. Earlier in the story Moses had received hospitality from Jethro but ends up marrying one of Jethro's daughters (Exodus 2: 16–22). While the Hebrews journey through the wilderness, they receive hospitality from God, who gives them some strange substance, manna, for their sustenance (Exodus 16). They grumble about the strange food and express desires for the foods they had eaten in Egypt, an affront to the hospitality they encountered. Having received a gift, they were complaining about it (Numbers 11: 4–9). In wilderness, visitors are all strangers who receive the hospitality of the inhabitants who dwell there. In wresting control, regaining mastery, the visitor negates hospitality and liminality, transforming the boundary places into a centre, into a world where the visitor becomes the centre of power.

While this account indicates an archetype of pilgrimage and its dynamics of liminality and hospitality, further elaborations occur as Christian pilgrimage evolves from medieval to modern expressions. With modernity comes a rising middle class and tourism begins to manifest more and more along established pilgrimage

*kredick@cnu.edu

routes. In some instances tourists remain clearly distinct from pilgrims. However, as the recreation industry, communication technology and mass transit expand, the blending of tourist and pilgrim becomes inevitable. Is it possible to draw distinctions between pilgrim and tourist when a particular traveller might engage in both practices while on the same journey, even in the same day?

While I have explored some distinctions in previous publications,[1] this essay follows a phenomenological exploration of pilgrim values and identity as they are related to the ephemerality of *liminal* places in the interplay of conditional and unconditional hospitality, both in the context of wilderness trails and the pilgrimage to Santiago de Compostela. Pilgrims and long-distance hikers traverse liminal places and become *liminal personae* as they embark on journeys through wilderness and to sacred destinations. Their identity as strangers versus tourists requires openness to what Derrida *et al.* (2000) refers to as unconditional hospitality. Tourists, in contrast to pilgrims, receive conditional hospitality within the boundaries established by institutions of the recreation industry. Just as Derrida shows the aporia of hospitality, contemporary journeys of pilgrims, tourists and long-distance hikers reveal the ephemerality of *liminal* places, identity and the values manifest along the way. A rapid rise in the popularity of the Appalachian Trail and Camino de Santiago, along with the ubiquity of smartphones, the increase of private hostels and albergues and the increasing availability of convenient transportation, interacts with pilgrim values and identity as places shift from existential boundary to comfort and familiarity.

Whereas those who identify as pilgrims negotiate strangeness in boundary regions, walking as strangers among those who dwell, those who identify as tourists seek adventure and the exotic, which are mediated by a market industry, evolving and rising in the same places where pilgrims tread. The values of tourist versus pilgrim are revealed in the interplay of hospitality, freely given or within the conditional range of the marketplace. Tourists travel as customers among dwellers who sell both adventure and the exotic. Value is negotiated as in any market place. Pilgrims discover value in receiving hospitality freely rather than from a 'hospitality industry'. Value transcends negotiation and is manifested as gift.

In many instances the distinction between pilgrim and tourist is as ephemeral as the liminal places through which they sojourn. The marketplace has colonized the Camino de Santiago and is pervasive throughout the many routes. Pilgrim identity shifts in the interplay of conditional and unconditional hospitality. A similar existential tension happens on wilderness trails where smartphones and other technologies mediate long-distance hikers' experience of places beyond culture, domesticating those wild environments. Domesticating the wild involves the practice of colonizing and transforming it into a site of adventure or aesthetic tourism.

I will begin this exploration with a discussion of liminality in the context of ritual and play. I will then examine liminality as it unfolds on contemporary wilderness trails. Wilderness as it manifests in mythology will then be outlined. From there, an examination of varying sojourns will show divergent ways of approaching wilderness hiking, both as pilgrimage and tourism in relation to hospitality. Finally, pilgrim identity and value will be explored in the context of the Camino de Santiago and the Appalachian Trail (hereafter AT).

Liminality, Ritual and Play

Liminal places are intertwined with symbolic, ritual action, and a comparison of pilgrimage and play, both employing symbolic action, sheds further light on the ephemeral and existential nature of such places. Johan Huizinga writes, 'Ritual grew up in sacred play' (Huizinga, 1950, p. 173). Jane Ellen Harrison draws a picture of the intertwining of ritual and play. Dancing a representation of the hunt, the tribe enacts a playful participation and relives the celebrated event (Harrison, 1969, pp. 44–45). Harrison notes that Greek dramas such as those of Aeschylus, Sophocles and Euripides 'were not played on the stage, and not in the *theatre*' but 'in the *orchestra*', the 'circular *dancing-place*', the focal point: 'The chorus danced and sang that Dithyramb we know so well' (1969, p. 123). In ancient times, long before the Greeks performed tragedy, 'men and boys, tillers of the earth, who danced when

Fig. 3.1. Delphi Theater. (Photo: the author)

they rested from sowing and ploughing' engaged in play within a circular space (Fig. 3.1).

That circular space, the *orchestra*, changes as the ritual shifts towards art and performance. The *theatre*, writes Harrison, 'to the Greeks was "simply the place of seeing, the place where the spectators sat"' (1969, p. 124). She notes that if 'we want to realize the primitive Greek orchestra or dancing-place, we must think these stone seats away. . . There is no division at first between actors and spectators; all are actors, all are doing the thing done, dancing the dance danced' (1969, p. 126). As the place of drama shifts, it takes on the new name of *theatre*. The play of dancing becomes the aloof and sedentary act of seeing. The players become professionals and eventually the audience's connection to symbolic action in ritual is severed and changed to entertainment. Harrison writes, 'whereas in old days all or nearly all were worshippers acting', the shift to theatre changes the identity of the participants so that they become 'spectators, watching, feeling' (1969, p. 126). The liminality of the dancing-place has been co-opted as one more institution of the marketplace.

Those who danced in the orchestra shared in both play and ritual and in the symbolic action found communion, *communitas*. In the same way, pilgrims step into an imaginative world of potentiality, of the unknown; their dance unfolds along the pathway toward some telos that transcends the marketplace. This distinguishes them from tourists, who compare to the audience of the theatre. Both spectators and tourists also travel to some exotic destination on both internal and external journeys, but the destinations couple with the marketplace. William Cavanaugh notes that in the medieval period 'Pilgrimage was not a for-profit industry, and was available to all members of society, including the poor' (2011, p. 80). The coupling of tourism and pilgrimage complicates contemporary practices, the liminal and the marketplace ambiguate one another. Tourism is available to those who have enough surplus wealth to afford a distraction from their everyday life. But the distraction does not plunge the tourist into liminality; the tourist remains in control, a master of the situation versus the pilgrim who is homeless and in need of true hospitality.

Children at play are closer to the situation of the pilgrim than the tourist. As the child plays, she steps beyond the world of her parents. Paul Shepard writes that play is 'an alternative to everyday utilitarian activities. To "go out to play" is to go out of the banal envelope of the daily routine of the adults, to leave the system of profane rules' (Shepard, 1967, p. 197). Tourism is not a release from the system of profane rules. Huizinga writes, 'play is not "ordinary" or "real" life. It is rather a stepping out of "real" life into a temporary sphere of activity with a disposition all of its own' (1950, p. 8). 'Not being "ordinary" life it stands outside the immediate satisfaction of wants and appetites; indeed it interrupts the appetitive process' (1950, p. 9). Tourists are thoroughly immersed in the appetitive process.

Huizinga adds that in addition to this limitation of time, there is also a bounded space for play: 'All play moves and has its being within a playground marked off . . . Just as there is no formal difference between play and ritual, so the "consecrated spot" cannot be formally distinguished from the play-ground' (1950, p. 10). These bounded spaces are liminal and ephemeral: 'All are temporary worlds within the ordinary world, dedicated to the performance of an act apart' (1950, p. 10). The liminal nature of the space rests in the idea that 'one of the most important characteristics of play was its spatial separation from ordinary life' (1950, p. 19). Pilgrims as *liminal personae* engage in ritual. Huizinga describes this, writing, 'Nearly all rites of consecration and initiation entail a certain artificial seclusion for the performers and those to be initiated' (1950, p. 20). Just as the theatre encroached on the orchestra and divided the identity of spectator from actor, the marketplace has entered the 'playground', the pilgrimage, and divides the identity of participants into tourists and *liminal personae*.

The marketplace has begun to alter the liminality of pilgrimage sites. The Camino de Santiago offers an assortment of accommodations from private guesthouses to albergues run by religious orders or municipalities (Fig. 3.2). It is becoming more and more common for pilgrims/tourists to book their journey. Wilderness trails are subject to the same marketplace phenomena: hostels, private and public parks; and such apps as Booking.com have made it possible to walk the trail by day and sleep inside at night.

Fig. 3.2. Queue to Albergue. (Photo: the author)

Fig. 3.3. Camping in the Meseta. (Photo: the author).

Cellphones make it possible to communicate with fellow travellers and select shelter locations based on preferred trail friends. The spontaneity and anti-structure of liminal action is co-opted by marketplace structure.

Two more examples illustrate the distinction between liminal and utilitarian placement and the ephemerality of such places. It is illegal to camp on private land in Spain. But pilgrims do engage in a practice known as stealth camping. As the sun drops low on the horizon, those pilgrims engaged find sites beyond cities or villages to bed down for the night. Much of the land between the villages is agricultural. A farm road between wheat fields makes a perfect place to lay out under the canopy of stars (Fig. 3.3).

In Galicia, western Spain, groves of oaks and chestnut trees as well as agricultural fields abound. These fields and groves are clearly utilitarian spaces, and in light of laws concerning camping in such places, off-limits to camping. Pilgrims who do find rest beyond both the law and utilitarian function enter a liminal space that only remains temporarily.

A rustic café in between Santabinez de Valdeiglesia and Astorga provides another example. A Spaniard named David occupied an old barn in the wheat fields about five kilometres east of Astorga in 2010. The Camino climbs a hill after Santabinez and enters a patchwork of oak groves and wheat fields, rising and falling in the ridges and valleys. The last rise brings pilgrims onto a mesa and the barn appears small in the distance, about 500 metres to the west. The mountains of Leon loom far beyond the barn. David set up a small café stand just outside the barn. He decorated the outside walls of the barn with slogans inviting pilgrims to feast on organic food and drink (Fig. 3.4).

I have returned to encounter David's hospitality in subsequent years – 2013, 2015 and 2017. His café is itself liminal. He has no running water but draws buckets from a well that is located about 500 metres from the barn. He asks for no compensation, offering everything gratis. He does have a box for donations but will not call attention to it. If one asks David for a suggestion on the donation, he indicates that he does not want to know how much you give. Everything about this café exists beyond the boundaries of the marketplace. He finds fellowship with pilgrims in his own liminality. The fields just there

Fig. 3.4. David's café food cart. (Photo: the author)

surrounding the barn are utilitarian but the pilgrims passing by exist in another economy.

Contemporary Wilderness Trails and Liminality

Henry David Thoreau symbolized journeying in the wild as a restorative endeavour. The journey toward salvation takes the pilgrim westward, away from civilization's cradle, away from the alienating influences of the old world. He writes, 'We go eastward to realize history and study the works of art and literature, retracing the steps of the race; we go westward as into the future, with a spirit of enterprise and adventure' (Thoreau, 1957, p. 604). Thoreau's version of the promised land, 'the West', is, he writes, 'another name for the Wild' (1957, p. 609). So, his often-quoted axiom, 'in Wildness is the preservation of the world', comes to us in the context of an extended metaphor that calls for a journey to the west, which is also a journey into the wild, the land of salvation. He continues in this vein, writing that no culture is able to provide this restoration, and that we must look to myth rather than literature to find inspiration. The appeal goes out to follow a new mythology, an 'American mythology', that will inspire the poets who will in turn inspire the rest of us to journey in this direction. Of course, this American mythology rises out of the journey to the west and therefore to the wild. Thoreau concludes his essay with an apocalyptic passage, a culminating vision of the salvation to which he has been pointing: 'So we saunter toward the Holy Land, till one day the sun shall shine more brightly than ever he has done, shall perchance shine into our minds and hearts, and light up our whole lives with a great awakening light, as warm and serene and golden as on a bankside in autumn' (1957, p. 630).

Thoreau's reference to a plurality sauntering toward the celestial anti-city, the wilderness, raises questions about an aporetic relation between the individual and the social, the wild and culture. When humans start marching into the wild searching for salvation, how long will the wild itself be preserved? Humans tend to domesticate the places they traverse and inhabit. But if the journey through the wild is liminal, a pilgrimage of preservation, maybe the aporia works itself out? However, when cultures establish a journey as something regular, a route often travelled even if it is liminal, the places through which pilgrims pass may become domesticated. We are back to the aporia. The journey on a 'wilderness trail' happens only in name. Wilderness evades us, moving further to the west, just ahead of those who would go into it.

In this light the AT, a proclaimed pathway through the wilderness, a way 'for those who seek fellowship with wilderness',[2] becomes aporetic. The AT appeared in the east after America's advance west came to the vast wilderness of the Pacific Ocean. The writings of Emmanuel Levinas, especially his focus on the face as the manifestation of alterity (2013, p. 50), show the possibility of a fellowship with wilderness that does not at the same time tame and thereby destroy that which those who saunter seek. Just as a phenomenological approach brackets conceptual practices in order to return to the things themselves, a liminal wilderness journey can also involve the practice of bracketing cultural tendencies of colonization in order to return to the constituent's act of hospitality as they give themselves. Just as entering the liminal space of wilderness facilitates a turning from cultural enframing, phenomenology opens us to reducing conceptual framing. Maurice Merleau-Ponty's phenomenological description applied to this study reveals the possibility of turning from the conceptual landscapes of human habitation and opening ourselves to the wild that is beyond our projection: 'To turn back to the things themselves is to return to that world prior to knowledge of which knowledge speaks' (Merleau-Ponty, 1956, p. 60). A phenomenological approach in the context of a liminal wilderness journey opens those who would seek wildness (or pilgrimage to a sacred place) to discover it as it gives itself from beyond sedimented and objectified experience.

Jean-Luc Marion has proposed a 'fourth and last formulation of a possible first principle of phenomenology: "As much reduction, as much givenness"' (2002, p. 17). Reduction is a distancing of the natural attitude; that is, a setting aside of our ordinary way of lived experience. One setting of this reduction as distancing from the ordinary happens in the liminal space of a fully embodied encounter. Liminal space in its distancing us from the natural attitude creates a gap in which the phenomenon given both gives itself and is received, a kind of interaction of hospitality. Marion points to the creation of such a gap 'between the (appearing, transcendent) thing and (immanent) lived experience (in which the thing would appear)' (2002, p. 55). In bracketing the natural attitude, liminality opens the sojourner to an alternative attitude, such as the attitude of the body climbing the steep slope and finding communion with the boulders along the path. One particular example of an alternative attitude happens in flow. When flow happens, the pilgrim's full attention is focused on the present moment and their full awareness encompasses all of the particular constituents of the surrounding milieu. Mihaly Csikszentmihalyi writes, 'one is very aware of one's actions, but not of the awareness itself' (Csikszentmihalyi, 1975, p. 45). Shifting the awareness beyond self-consciousness opens the pilgrim to an embodied dialogue wherein the phenomenological given gives fully in the gap between itself as transcendent and the pilgrim's (immanent) lived experience.[3]

Liminality opens us to an alternative meaning, a distancing from the usual or natural attitude. The natural attitude gives the wilderness as being composed of things ready to hand. This could unfold as an objectification of the constituents of a survivalist's march or an aesthetic hiker's scene, which combines with established conceptualizations and reinforces them. As the survivalist or the aesthetic hiker surveys the scene with eyes as the principal means of gathering, an individuation of senses imposes a self-derived meaning upon the given. This hiker stands as master of all that is experienced as a scene. In this case the hiker's gaze puts her in the centre of the prospect and gives rise to a conceptual frame that lacks the alterity of the face that gives itself in the way Levinas (2013, p. 50) discovers. The scene comes into its meaning as a thing, ready to hand rather than as an 'I know not what' that I must 'seek', 'await', 'touch', with my whole being (Marion, 2002, p. 46).

In pointing to wilderness, Thoreau references something beyond the influences of civilization. Levinas shows us that approaching an alterity such as wildness requires care: 'To think the infinite, the transcendent, the Stranger, is hence not to think an object' (2013, p. 49). The infinite, transcendent, stranger and wilderness appear from beyond cultural adaptations and all human conceptualizations. Levinas shows us a way to approach transcendence. He writes: 'But to think what does not have the lineaments of an object is in reality to do more or better than think' (2013, p. 49). The other as transcendent presents herself, the face appearing before me, and the face exceeds 'the idea of the other in me' (2013, p. 50). 'The face of the Other at each moment destroys and overflows the plastic image it leaves me, the idea existing to my own measure' (2013, pp. 50–51). Thoreau's wilderness mythology opens us to a 'world prior to knowledge of which knowledge speaks' (1957, p. 60).

Wilderness in Mythology

Paul Ricoeur writes, 'In losing its explanatory pretensions the myth reveals its exploratory significance and its contribution to understanding, . . . its power of discovering and revealing the bond between man and what he considers sacred' (Ricoeur, 1967, p. 5). In this way myths are like dreams that manifest a preconceptual and meaning-making experience. Myths, understood this way, measure us, act upon us, giving us a world of meaning. Myths open us to collaboration with the other, both the human and extra-human other, in discovering a mutual habitation. Philosophy and science through theory reverse the arrangement so that it is the human who acts on the world, creating meaning from conceptual practices that colonize the other.

Wilderness as a liminal space appears in various myths and becomes especially meaningful in both the Jewish and Christian traditions. Wilderness is the English term Bible translators used most often to express the Hebrew *midbar* and the Greek *eremos*, as well as other Hebrew and Greek words meaning desert, barren and solitary places (Kittel, 1976, p. 657). Max Oelschlaeger calls attention to the meaning of *midbar* that nomadic peoples ascribed to their dwelling beyond the cultivated fields of more fertile lands.

> The shepherd-farmer mythology 'bespeaks a deeper psychic conflict' than any simple antipathy between herders and farmers. It represents a persistent opposition to civilization. In this mythology the wilderness assumes a deep symbolic meaning, representing both (in Genesis) the shepherd's departure from the detested city (itself symbolic of the high cultures) and the exodus from slavery into the desert to face Yahweh's challenge' (Oelschlaeger, 1991, pp. 48–49)

The particular mythology depicting Moses leading the people into the solitary openness of *midbar* distinguishes their temporary dwelling, a sojourn, from that of the shepherd nomads who had sheltered Moses earlier in his own exile from Egypt. These shepherd nomads may not have referenced their countryside as *midbar*. The story is told from the point of view of people who are leaving high culture and entering a perceived wasteland. The story implies this when the people complain to Moses that they have been led to the *midbar* to perish and that they long for the fish, melons, leeks, garlic and onions of Goshen, their previous home. In the mythic account, God fed them via something called manna. This food substance appeared each day and was gathered up to be eaten on the day it was collected from the ground. The people complained about the manna, longing for the rich food they had enjoyed back in Egypt. The combination of receiving manna and not having the food that both shepherd nomads and agriculturalists regularly eat illustrates that these sojourners where not living from the land, domesticating the land and thereby changing it. Rather, the wilderness was changing the people, measuring them. This *midbar* is liminal for the Hebrews, between their enslavement by the Egyptians and their future permanent dwelling in the promised land. In this liminal country the Hebrews were transformed into the Chosen People, who became ethically opened to a new relationship, rising out of liminality and the hospitality received during their journey.

This new ethical relationship expresses itself in the form of a command, a *mitzvah*. Exodus 22: 21 (*The Holy Bible*, 1971) reads: 'You shall not wrong a stranger or oppress him, for you were strangers in the land of Egypt.' Deuteronomy 10: 19 (*The Holy Bible*, 1989) reads: 'You shall also love the stranger, for you were strangers in Egypt.' Leviticus 19: 34 (*The Holy Bible*, 2016) reads: 'You shall treat the stranger who sojourns with you as the native among you, and you shall love him as yourself, for you were strangers in the land of Egypt: I am the LORD your God.' From these passages we learn that during Israel's wilderness sojourn, in the experience of liminality and receiving hospitality, they identify with the stranger and find a capacity to give hospitality, especially to the stranger.

Levinas comments on this relationship. Reflecting on Psalm 119: 19, 'I am a stranger on the earth; do not hide your commandments from me', Levinas disagrees with exegetes who place this text in the Hellenistic period 'when oriental spirituality may have been seduced by the Platonic myth of the soul exiled in the body' (Levinas, 2002, p. 66). Rather, Leviticus Chapter 25: 23 is pointed to as predating both Socrates and Plato: 'No land will be alienated irrevocably, because the land is mine, because you are but strangers, housed in my land' (2002, p. 66). Levinas points out that the reference to the commandments in Psalm 119 emphasizes obligations one has to the other, writing, 'the condition – or incondition – of strangers and slaves in the land of Egypt brings man closer to his fellow man. Men seek one another in their incondition of strangers. No one is at home' (2002, p. 66). The wilderness not only provides the space/place of receiving God's commandment but also the liminality through which the people of God find their responsibility one to another. In the wilderness we are all strangers who receive the hospitality of the extra-human inhabitants who dwell there.

Varying Sojourns in the Wilderness

Given the mythic interpretation versus a conceptual understanding of wilderness, how do contemporary hikers experience liminality on a wilderness trail? Wilderness survivalists go out into the wild, but do they encounter the face of alterity? The well-trained survivalist travels within the conceptual provinces of technique. They study edible flora and fauna of particular environments, the shelter-making possibilities, and other ways of domesticating, even temporarily, the land. Rather than receiving the hospitality of the constituents of a particular environmental milieu, the survivalist imposes technique and masters the other who is already there. There is no looking into the face of the other, no establishment of relationship where the other is the host. Levinas writes, 'The Other becomes my neighbor precisely through the way the face summons me, calls for me, begs for me, and in so doing recalls my responsibility, and calls me into question' (Levinas, 1989, p. 83). The survivalist never finds a face that summons, calls for help. Rather, everything in the wilderness is ready to hand.

The survivalist may counter, pointing to the very motivation for such an endeavour as desiring to be tested, and so to be measured by the wilderness. But, such a test points back to the techniques that the survivalist uses in approaching the wild. The measure fathoms the application of technique, the practice of successfully implementing a conceptual frame.

The same can be said of the recreational or aesthetic hiker. José Ortega y Gasset points to the way such a hiker approaches wilderness, writing: 'For the tourist, the countryside, as landscape, is no less human than the others: it is a "painting" and its existence depends on the lyric conditions that man wishes and is able to mobilize. . . Poets and painters are the ones who have formed it' (1972, p. 140)

Landscape painters introduced wild scenery. They opened culture to a new conceptualization of wilderness: 'Only by the beginning of the nineteenth century was there enough force behind the human impulse which leads man to convert a piece of ground into the ideality of a landscape' (Ortega y Gasset, 1972, p. 141). Such landscape differs from the *midbar* of Jewish and Christian mythology. Wilderness as landscape results from enframing and is revealed as cultural rather than an alterity. We measure it rather than being measured by it.

This objectification of landscapes isolates scenery. That is, scenery does not manifest ecologically. The Claude Glass, carried by aesthetic tourists and used to frame scenery during walks, illustrates the practice of isolating prospects and creating a landscape. The countryside, as Ortega y Gasset would say, hides beneath conception, buried under the sediment of objectification. Wilderness is experienced as a framed painting or through glass. The camera is yet another imposition of technology to transform the wilderness into landscape. Artists teach us an aesthetic valuation and train us to peer through frames.

Aesthetic hikers experience scenery through windows that separate them from the constituents of wilderness, from the ecology of the life world. The gaze that falls upon the beautiful, the picturesque or the sublime removes the aesthetic hiker from the place itself. The orienting frame composing the scene distances them. Landscape is an object of aesthetic value and mediates a subjective experience.

Some who trek long distances do experience liminality and receive hospitality. The following account illustrates the reception of hospitality. While hiking through Maine on the AT, Truckin, Soul Slasher and I were offered a ride and detour to enjoy some real food for lunch.[4] As we sat eating hamburgers and enjoying beers at the Kennebec River Brewery, a woman at the next table interrupted us and asked, 'Are you thru-hiking?' When she discovered that we indeed had walked from Georgia all the way to Caratunk, Maine, she invited us to her home for cupcakes. She lived near Moxie Pond, about 12 miles from our lunch place further north on the AT. It was about 2 pm, and we planned to get back on the trail and walk till evening. She also invited us to pitch tents in her yard and enjoy those cupcakes if we made the 12 miles that day. We headed out and climbed over Pleasant Pond Mountain and then descended to Moxie Pond, finding her cabin before dusk. She cooked us a spaghetti dinner and supplied us with beer. She then offered us a room in the cabin for the night.

The three of us were recipients of what hikers on the AT call 'trail magic'. Both the giving and receiving of this kind of hospitality and the title 'trail magic' illustrate the aporia of pilgrims as *liminal personae* versus tourists. Pilgrim identity as strangers versus tourists engaged in travel requires openness to what Derrida refers to as unconditional hospitality. Tourists receive conditional hospitality within the boundaries established by institutions of the recreation industry. 'Trail magic' ambiguates hospitality because magic involves manipulation. Martin Buber writes that those who practise magic want 'to be effective without entering into any relationship' (Buber, 1970, p. 131). Those who receive such 'magic' are then implicated in bringing it about and so enter into an economy of exchange, thus negating the gift of hospitality. Marion writes, 'The gift begins and in fact ends as soon as the giver envisions that he owes something to someone, when he admits that he could be a debtor, and thus a recipient' (Marion, 2008, p. 91). In magic, the giver is bound to giving because the giver has become a debtor through the manipulation of the one who knows the formula. Marion shows that the gift is annulled in such a situation, 'it does not belong to the economy of exchange' (2008, p. 96). Giving and receiving magic is shown to be very similar to any other economic exchange where a token of power brings about the desired result. The marketplace extends its economic influence onto the AT and hikers become tourists. Hikers begin to value the exchange of goods rather than the gift of hospitality.

An interpretation consistent with pilgrim values and identity understands the gift that my hiking companions and I received as grace. Neither we nor the woman who invited us into her home participated in any economy of exchange. Her gift was freely given without condition, and we freely received without feeling any need to reciprocate. In that our identity as long-distance hikers cast us as *liminal personae* rather than tourists, we were asked to give nothing in return for her hospitality. On the other hand, we could step out of our liminality, just as already had happened during lunch at the Kennebec River Brewery, and become tourists engaging in an economy of exchange. Derrida writes:

> Absolute hospitality requires that I open up my home and that I give not only to the foreigner . . . but to the absolute, unknown, anonymous other, and that I *give place* to them, that I let them come, that I let them arrive, and take place in the place I offer them without asking of them either reciprocity (entering into a pact) or even their names. (2000, p. 25)

This woman opened her home without condition. She merely asked us if we were thru-hikers before offering her gift. Our identity as thru-hikers placed us beyond any status: pilgrims and long-distance hikers exist beyond such classifications. Victor Turner writes, 'these persons elude or slip through the network of classifications that normally locate states and positions in cultural space' (Turner, 1969, p. 95). He goes on to describe these persons has having 'no status, property, insignia, secular clothing indicating rank or role, position in a kinship system' (1969, p. 95). We were anonymous in that we only indicated our 'trail names', which located us as characters who have a kind of dual existence: we travel beyond the boundaries of the marketplace reality and at the same time seem to inhabit that same marketplace, sometimes entering into an economy of exchange.

Tourists, on the other hand, exist within the boundaries of the marketplace and don't have a dual existence. Comparing tourism with contemporary journeys of pilgrims and long-distance hikers on wilderness trails reveals the ephemerality of liminal places and pilgrim identity. Though both pilgrims and tourists leave home, travelling beyond the boundaries of their places of dwelling, they enter different spaces, and boundaries are experienced uniquely. Even though they may occupy the same physical space, their orientation to that space differs and thus the space itself becomes existentially different.

Existential Difference, Identity and Value

William Cavanaugh illuminates further differences between pilgrim and tourist describing 'the tourist gaze' as cosmopolitan, whereas travelling by foot does not give 'a commanding view of the globe from above' (2011, p. 82). The identity of the pilgrim is that of a stranger, and so liminal, in the very places through which they walk. The tourist gaze prevents entry into liminal space for it co-opts environments, making them into a centre of desired adventure. Cavanaugh writes:

> Unhindered by borders, the tourist scans the globe and imagines herself entering into the experience of otherness in any part of the globe. At the same time, however, borders do not simply disappear, for the maintenance of borders is crucial to the maintenance of the otherness that the tourist seeks. The maintenance of the center and periphery remain important for the tourist gaze. (2011, p. 75)

Tourists cross borders but never remain in the margins. Pilgrims who cross borders remain marginal, strangers away from home and processing toward a destination that is also beyond borders, a sacred place. The ambiguity of liminality counters the maintenance of boundaries. Contrasted with tourist values, 'The pilgrim seeks to transgress all artificial borders that impeded the quest for communion with God and with other people' (Cavanaugh, 2011, p. 82). Those engaged in tourism remain in control of both time and place, their border crossings become integral to the experience. They have an established itinerary and most likely have booked accommodations along the way. They are not engaged in ritual action, though some of their activities may appear ritualistic. Tourist values rise out of a marketplace interaction rather than from liminality.

Pilgrims, on the other hand, being liminal are, as Turner describes, participating in the in-between world of religious ritual, the mode of being where potentiality dominates, where change is necessary. Tourist values work to eliminate potentiality – one would want their booking to be reliable rather than potential. Liminality involves distancing the ritual participant from their former life situation in both time and place. The life situation of the tourist is contemporaneous and emphasized as it makes the venture possible: tourist identity empowers them to have sought-after experiences. Pilgrims values rise out of leaving their own particular places of habitation and cultural entanglements behind and walking together through the in-between world. Their previous identities have no part in the encounters along the way, evidenced by the tradition of trail names on the AT. Even though they walk through towns and villages, pilgrims are separated into their own ritual practices. They are, as Turner and Turner describe, 'either permanently and by ascription set outside the structural arrangements of a given social system', or 'voluntarily' set 'apart from the behavior of status-occupying, role-playing members of that system' (Turner and Turner, 1978, p. 223).

Pilgrim values rise out of their liminality. Turner and Turner write that 'pilgrimage is exteriorized mysticism' (1978, p. 7). In this light, Cavanaugh notes that pilgrimage, unlike tourism, is 'a kenotic movement, a stripping away of the external sources of stability in one's life .... It is 'the way of the cross' (2011, p. 80). As a shared mystical experience pilgrimage is also a communal journey and a social event. The values of shared ordeal continue to manifest along the journey. Cavanaugh writes, 'For this reason, pilgrims welcome other pilgrims, but tourists regard other tourists with disdain' (2011, p. 81). In welcoming this fellowship, this communion, 'The pilgrim thus rejoices when others join with her on pilgrimage, because communion with God is also communion with other people' (2011, p. 83). As tourism and marketplace values invade the Camino, fellowship is replaced by the values of competition. Communion and communitas rise out of the mystical encounters along the way. Mysticism is negated in the journey of the tourist.

In addition to a distinction in journeys, the places through which tourists and pilgrims move are existentially different. Cavanaugh writes that 'The vacation vacates particular locations, so the tourist must constantly be on the move, seeking out the unspoiled, only to spoil it with his presence' (2011, p. 81). The constant movement of the pilgrim does not arise from seeking unique experiences. As noted, pilgrimage is kenotic and therefore antithetical to the acquisition of experience. Cavanaugh continues:

> The cosmopolitan gaze of the tourist seeks to connect with others, but ends up vacating their otherness, and thus destroying the connection. The pilgrim, on the other hand, sees all as potential brothers and sisters on a common journey to God. The pilgrim preserves otherness precisely by not seeking otherness for its own sake, but moving toward a common center. (2011, p. 83)

A centre outside one's self.

Conclusion

This phenomenological exploration has shown that pilgrim values and identity rise out of liminality and hospitality. Pilgrims as *liminal personae* differ from tourists, who engage in recreation rather than ritual. However, the ephemeral nature of liminality and the manifestation of new technologies implicate a shifting identity, pilgrims become tourists sometimes: pilgrims may exit liminality and act in ways consistent with tourists. Most recent advances in communication technology – smartphones, cellular coverage, and wifi – add an extra layer of mediation, interfering with the experience of religious walking or hiking in the wilderness. In the same way, new private albergues and hostels along pilgrimage routes facilitate exiting liminality and interacting in the marketplace.

The distinction between tourist and pilgrim identity has been shown in that tourists remain their own centre of power through their command within a marketplace economy, whereas pilgrims move towards a centre of power beyond themselves and receive gifts of unconditional hospitality along their journey. Tourist interaction with hospitality is conditional and based on a seeming oxymoronic 'hospitality industry'.

A distinction in tourist versus pilgrim values has also been shown. Tourists negotiate value, interacting in a utilitarian way and projecting extrinsic value, which in some cases manifests when purchasing goods along the journey. Pilgrims may purchase goods as well but discover intrinsic value, especially when benefitting from unconditional hospitality wherein goods become gifts received. When tourists receive conditional hospitality, it is interpreted as an economic exchange coming from within the context of the marketplace. This is evidenced when the tourist thinks of the gift as a 'good deal'. Gifts that transcend the marketplace reveal the valuation of personal encounter rather than an exchange of goods. The purchased good reveals an economy of exchange, which remains an exchange of conditional hospitality.

Value and identity rise out of engagement in ritual versus recreation. The tourist returns home after recreation invigorated, but the pilgrim has been transformed. That transformation involves both identity and values. *Liminal personae* share in *communitas* and learn the value of caring for the stranger. The camaraderie of the tourist rises out of a common experience with fellows, rather than communion with strangers. A key value given and received on pilgrimage is the ethic of giving hospitality to the stranger, of finding responsibility one for another. This then leads to a more inclusive community, including human and extra-human constituents.

Notes

[1] See 'Interpreting contemporary pilgrimage as spiritual journey or aesthetic tourism along the Appalachian Trail' in *International Journal of Religious Tourism and Pilgrimage* 6, 78–88, for an exploration of the tourist as scenic adventurer versus the long-distance hiker as similar to pilgrim (Redick, 2018). See 'Profane experience and sacred encounter: journeys to Disney and the Camino de Santiago' in *Environment, Space, Place* 5(1), 46–72, for comparing travel to Disney World and walking the Camino de Santiago (Redick, 2013).

[2] Quote from a plaque on Springer Mountain, Georgia, at the southern terminus of the Appalachian Trail.

[3] I have more fully examined flow in the context of pilgrimage in Redick (2016) 'Spiritual rambling: long distance wilderness sojourning as meaning-making' in *Journal of Ritual Studies* 30(2), 41–51.

[4] Trukin and Soul Slasher are trail names. There is a tradition on the AT wherein hikers receive trail names.

References

Buber, M. (1970) *I and Thou. Walter Kaufmann (trans.)*. Simon and Schuster, New York.

Cavanaugh, W.T. (2011) *Migrations of the Holy: God, State, and the Political Meaning of the Church*. Eeermans Publishing Co., Grand Rapids, Michigan.

Csikszentmihalyi, M. (1975) Play and intrinsic rewards. *Journal of Humanistic Psychology* 15(3), 41–63.

Derrida, J., Dufourmantelle, A. and Bowlby, R. (2000) *Of Hospitality*. Stadford University Press, Stanford, Connecticut.

Harrison, J.E. (1969) *Ancient Art and Ritual*. Greenwood Press, New York.

Huizinga, J. (1950) *Homo Ludens: a Study of the Play-Element in Culture*. Beacon Press, Boston, Massachusetts.

Kittel, G. (1976) *Theological Dictionary of the New Testament.*Eerdmans, Grand Rapids, Michigan.

Levinas, E. (1989) *The Levinas Reader (ed. Hand, S.)*. Blackwell Publishing, Malden, Massachusetts.

Levinas, E. (2002) *Humanism of the Other*. University of Illinois Press, Urbana, Illinois.

Levinas, E. (2013) *Totality and Infinity: An Essay on Exteriority*. Duquesne University Press, Pittsburg.

Marion, J.-L. (2002) *In Excess: Studies of Saturated Phenomena*, trans. Horner, R. and Berraud, V. Fordham University Press, New York.

Marion, J.-L. (2008) *The Visible and the Revealed*. Fordham University Press, New York.

Merleau-Ponty, M. (1956) What is phenomenology? *Cross Currents* 6(2), 59–70.

Oelschlaeger, M. (1991) *The Idea of Wilderness*. Yale University Press, New Haven, Conneeticut.

Ortega y Gasset, J. (1972) *Meditations on Hunting (trans. Wescott, H.B.)*. Charles Scribner & Sons, New York.

Redick, K2013) Profane experience and sacred encounter: journeys to Disney and the Camino de Santiago. *Environment, Space, Place* 5(1) , 46–72 DOI: 10.7761/ESP.5.1.46.

Redick, K. (2016) Spiritual rambling: long distance wilderness sojourning as meaning-making. *Journal of Ritual Studies* 30(2), 41–51.

Redick, K. (2018) Interpreting contemporary pilgrimage as spiritual journey or aesthetic tourism along the Appalachian Trail. *International Journal of Religious Tourism and Pilgrimage* 6(ii), 78–88.

Ricoeur, P. (1967) *The Symbolism of Evil*. Harper & Row, New York.

Shepard, P. (1967) *Man in the Landscape: A Historic View of the Esthetics of Nature*. A&M University Press, College Station, Texas.

The Holy Bible (1971) New American Standard Bible. Lockman Foundation.

The Holy Bible (1989) *New Revised Standard Version*. Harper Bibles, San Francisco, California.

The Holy Bible (2016) *English Standard Version*. Crossway Bibles, Wheaton.

Thoreau, H.D. (1957) Walking. In: Bode, C. (ed.) *The Portable Thoreau*. Penguin Books, New York, pp. 597–598.

Turner, V. (1969) *The Ritual Process: Structure and Anti-Structure*. Aldine De Gruyter, New York.

Turner, V. and Turner, E.L.B. (1978) *Image and Pilgrimage in Christian Culture*. Columbia University Press, New York.

Van der Leeuw, G. (1963) *Religion in Essence and Manifestation*. Princeton University Press, Princeton, New Jersey.

4 Changes in Motivations for Journeys to Death Spaces: The Case of Thanatourism

Slawoj Tanas*
Faculty of Geographical Sciences, University of Lodz, Poland

Introduction

The intensive growth of global tourism determines the organization of tourism spaces which develop as a result of travellers' changing needs. A common element of cultural tourism, where the traveller visits places representing tangible and intangible historical and artistic heritage, is the symbolism of death.

Attitudes towards human death have developed throughout the history of mankind and are expressed by cultural heritage represented by customs, standards of conduct, symbolism and artistic creation. Death spaces have always been a part of pilgrimage journeys, both religious and historical, and related to a specific cult of the dead. Medieval Europe was a space for pilgrimages to the graves of those who played a significant role in the history of Christianity at places such as Jerusalem (Jesus' sepulchre), Rome (St Peter's grave) or Santiago de Compostela (St James' grave).

In travelling (a pilgrimage, a tourist journey), death is perceived through the stories that have been told about a person, his/her suffering and death, or through its institutionalization based on specific laws, beliefs and rituals regulating human life. The journey is often an impulse to think about one's own mortality and death. The experience of someone's death makes us reflect on the sense of life, suffering or mortality.

As modern developed societies are departing from the conventional model of conduct as regards the dead, they are often curious about traditional rituals. A traveller's attitude to death is often only emotional, without deeper thought being given to the source of the ritual, and this leads to a superficial perception. Motivations for travel, as a result of which a traveller moves from the taboo of death to a discourse about it, are determined by individual factors such as needs, interests and the traveller's personality; as well as external ones, socio-cultural (e.g. norms, morality) or geographical (distance, physiographic obstacles) (Stone and Sharpley, 2008; Stone, 2009).

Death space (Tanaś, 2006) includes tangible and intangible elements collected throughout history, related to the commemoration of death in geographical space, as well as actions resulting in death; it will be referred to here as the culture of death. These can be divided into the following groups:

tangible/material elements – symbolism, art, sculpture, painting, literature, film, architecture, sites which document death (museums), places which commemorate death (monuments, plaques, individual

*slawoj.tanas@geo.uni.lodz.pl

sites), souvenirs, publications, entertainment facilities;
intangible/non-material elements – traditions, practices, rituals, music;
events – acts of violence, murders, wars, acts of terrorism, disasters, accidents;
other (culinary products, clothes).

Cognitive journeys and pilgrimages to death spaces are brought together in thanatourism, a type of cultural tourism which involves travelling to death spaces for cognitive, educational, religious and entertainment purposes where the object of the traveller's interest is the culture of death (Seaton, 2002; Tanaś, 2006, 2013a, b).

Just like pilgrimages to sacred relics, tourist journeys to death spaces may cause a change in the traveller's attitude towards the culture of death, its interpretation, understanding and, subsequently, in an approach to another human being, the end of life, faith or a way of contemplating death (*thanatopsis*) and in everyday life behaviour (Seaton, 1996).

Pilgrimages (Spiritual Journeys)

A pilgrimage is a journey taken for religious reasons to a place considered to be sacred due to a particular agency of God or a deity, in order to perform specific religious acts there (Jackowski, 2003). It is a desire to experience the presence of the *sacrum*. A pilgrimage is also a 'journey to a place of cult', which may concern historical events, people (usually deceased), animals, plants, rocks, objects, etc. Looking from this perspective, pilgrimages to places related to martyrs' deaths (e.g. those who died for their motherland), tragic death (e.g. in an accident, act of violence), the death of a family and community member, or one which has evoked the need for commemoration, should also be considered. Thus, pilgrimage will be treated as a journey to the *sacrum* – a sacred place, which can be an element of the natural environment (e.g. sacred rocks, rivers, grottos) or a man-made object (e.g. a temple, mausoleum, sepulchre, or another object of a cult). The aim of a pilgrimage may, in addition, be a place connected with the life, activity and the death of those who have played a significant role in building a religion, often celebrated as saints, or those who will be remembered by future generations for their merit in shaping statehood, civic attitudes, or because they died a martyr's death, giving their lives for a nation, ideas or beliefs.

A significant number of pilgrimages are made to sanctuaries, which could be churches, graves, chapels or cemeteries. Some pilgrims travel to sacred places in order to die there; the intention is to rest the body in a grave, cremate it (Mecca, Medina, Varanasi), or perform another ritual for the deceased person. These are the 'dead people's pilgrimages' to places which are recognizable in the landscape as vast burial grounds or flaming funeral pyres. They originate in the faith that burying or cremating the body at a holy site will guarantee the dead person's salvation and entrance to paradise. This phenomenon is observed primarily in Islam (Mecca, Medina) and Hinduism (holy cities and rivers – Varanasi). In Christianity, the tradition of burying the dead near sacred sites was known from the very beginning, but did not develop until the Middle Ages when the faithful were buried in or near temples. The moment a saint or a blessed person is buried there, such places became pilgrimage destinations. Newly created sanctuaries remained under the care of religious orders (e.g. in the city of Ampurias). In the literature, terms like *death tourism* or *death journey* are found, which define a journey to the place of one's own death, not necessarily by natural causes, sometimes resulting from euthanasia (Srinivas, 2009; Martin, 2010; Safyan, 2011).

The history of pilgrimages includes journeys made to participate in specific rites, such as ritual sacrifices, throughout the ages; pilgrimages were a part of religious life and a social phenomenon of immense cultural significance. The temples to which pilgrims travelled were usually tombs, ranging from Jesus' holy sepulchre in Jerusalem to the relics in St Andrew, Scotland. The grave of St James the Greater in Santiago de Compostela 'competed' with the grave of Thomas Becket in the English city of Canterbury.

From the early days of Christianity, the most numerous pilgrimages were those related to the cult of martyrs' and saints' graves. The most important centres associated with this cult included Ephesus, Antioch, Edessa, Seleucia, Chalcedon, Mira, Constantinople, Abu Mina and Patras. Many others could be found in Europe:

Santiago de Compostela (the grave of St James the Greater), Rome (the graves of St Peter and St Paul), Assisi (the grave of St Francis), Padua (the grave of St Anthony), Canterbury (the grave of Thomas Becket), San Giovanni Rotondo (the grave of Padre Pio) and Kraków (the grave of St Faustina).

Journeys motivated by religion play an important role in many countries and are a regular religious practice. We should turn our attention to the significance of the special character of journeys to places related to death as regards shaping future tourist needs. The early confirmation that places of death were treated as a spectacle or cognitive asset evoking the need to travel, were martyrs' graves called *martyrium* or *memoriae*. They were destinations of pilgrimage or final journeys made in order to rest a dead body near a martyr's grave. Sanctuaries devoted to martyrs were built at the sites of worshipped graves situated at cemeteries outside towns (*extra muros*). Later, there were temples, gradually enlarged to accommodate pilgrims attracted by a given saint's fame. With time, the presence of holy relics not only drew pilgrims but also became a reason to bury the faithful near them. Examples of such sites are the sepulchral basilicas in Hippona, Carthage, Ampurias, Saint-Sernin (near Toulouse) and Saint-Victor (near Marseilles) and Tablea (Aries, 1985).

In the history of Christianity, one of the first directions of pilgrimage connected with the cult of saints was Jerusalem with Golgotha and the Holy Sepulchre. The activity of Helena, Constantine's mother, started the custom for pilgrimages to the Holy Land, with the aim of visiting the place where Christ had died and was buried, above which the Holy Sepulchre Church was erected in the fourth century. The Holy Sepulchre Sanctuary became an inspiration to build many churches as in some of these the Jerusalem Holy Week liturgy was celebrated, and the cult of relics of the Passion flourished. In the seventh century, after the conquest of Palestine by Arabs, pilgrimages to the Holy Sepulchre were rare. As journeys to the Holy Land were exhausting and dangerous, in the tenth century, churches with replicas of the Holy Sepulchre chapel started to appear in Europe (24 such replicas had been built by the end of the 15th century). Replicas of the Sepulchre itself were constructed (over 30) (Jackowski, 2003), some of them erected and guarded by the knights of the Order of the Holy Sepulchre in Jerusalem.

One of the oldest and most famous pilgrimage centres is Santiago de Compostela with St James' grave. In the 12th century, it was regarded as one of the three most important Christian centres (next to Jerusalem and Rome). Since then, it has been visited by pilgrims from all over Europe. Many trails led to this sanctuary and were mainly used by pilgrims from France, Germany, Italy, The Netherlands and England. Along the trails and in Santiago de Compostela itself there were hospices with cemeteries nearby used for burying those who died on their way to the grave of St James. The phenomenon of burying the body *ad sanctos*, i.e. as close to the saint's remains as possible, was widespread in the Middle Ages. In a way, for a pilgrimage centre to develop, it had to have a saint's grave or his/her remains. The cult of personality, and pilgrimages related to it, formed a European sepulchral space, which is currently an object of tourists' interest.

Because of the Jewish Diaspora, pilgrimage centres based on the graves of famous rabbis and Hasidic leaders play an important role. The Necropolis, which is most frequently visited by pilgrims, is the cemetery on the Mount of Olives in Jerusalem. Other destinations include the Patriarchs' Mausoleum in Hebron with the grave of Abraham, as well as the graves of rabbis in Meron, Safed and Katshima. In Poland, the world's cradle of Hasidism, the main pilgrimage destinations are the graves of Hasidic leaders, especially those at the Remuh cemetery in Kraków, as well as in Leżajsk and Dynów.

Muslim holy men's graves were surrounded with settlements, which were gradually transformed into cities performing a religious function. Muhammed's grave in Medina became a pilgrimage destination; though Muhammed condemned the cult of the dead, it is currently a significant element of Islam. One of the most famous and largest Islamic cemeteries (the Valley of Peace – Wadi-us-Salaam) is situated in Najaf, Iraq, at the back of the mosque where Ali – Muhammed's cousin and son-in-law – was buried. Many Islamic cemeteries grew due to the cult of saints, because its followers believe that burying a person near a martyr or saint gives them a pass to paradise. Therefore, cemeteries were set up very close to saints' graves.

Some pilgrimages to the graves of Muslim saints (*ziara* or *mussem*) are believed to equal the pilgrimage to Mecca. The most renowned centres related to the cult of the dead, with the most important Islamic necropolises, include Medina (Saudi Arabia), Mashhad (Iran), Kumm (Iran), Karbala, Baghdad and Najaf (Iraq), Kairuan (Tunesia) and Mazar-i Sharif (Afghanistan).

The element regarded by Hindus as important from a religious point of view is the place of cremation, usually situated on rivers, as well as pilgrimages related to the journey of the ashes after cremation. The most popular pilgrimage centres, called *Shraaddha*, are Varanasi, Hardwar, Allahabad and Gaia. In Buddhism, the destination of a pilgrimage is the place of Kushingara Buddha's death. The most important pilgrimage centre for the followers of Confucianism is Qufu, the place where Confucius died and was buried.

The faithful also take part in pilgrimages to places where they can find relics consisting of parts of bodies or mummified bodies, exhibited in reliquaries in the shape of glass coffins and sarcophagi. Since the 19th century, people have been travelling to such places for religious and also cognitive and emotional reasons. Tourists want to see human corpses or their fragments shrouded in mystique. Mummies appeared on church altars in the Middle Ages, when the bodies of saints were dressed in cloaks, locked in ornamented glass coffins and exhibited in temples. If a body had decomposed, the faithful were shown decorated skeletons, or wooden or wax figures. In churches in many parts of the world, over 200 bodies saved from decomposition are presented to the public in glass coffins (reliquaries).

When it was impossible to preserve the human body, its fragments were turned into relics, usually the heart, hand or head. In many temples around the world, reliquaries can be seen, which are particularly popular in the Eastern Orthodox Church. Collections of relics can also be found in museums, e.g. the Victoria and Albert in London.

The cult of saints' bodies was perceived by the Church as a manifestation of religiosity. Nowadays, relics still attract pilgrims, though to a much lesser extent than in the past. Relics were the focus of religious life and many destinations developed due to an increased volume of pilgrims. Christians were encouraged to go on pilgrimages by belief in the healing power of sacred remains. In the 17th century, a particular kind of industry developed which involved manufacturing wax figures containing fragments of human bodies; they were dressed in fine clothes and placed in glass coffins (Fig. 4.1).

The pilgrims also subconsciously wished to discover the world they were travelling through and this curiosity inspired travellers to record their pilgrimages. Their descriptions slowly became a kind of guide for future pilgrims and travellers, and many Christians dreamed of travelling to places connected with Jesus and his painful death. Unfortunately, for various reasons, it was not always possible; thus, in the late Middle Ages, there appeared Calvaries and Ways of the Cross. Setting up a Calvary was followed by organizing paschal mysteries which attracted many pilgrims who wanted to take part in the staging of Christ's last journey and death during Holy Week. The best-known mysteries, which are currently tourist attractions, include those held in Seville, Malaga and Murcia (Spain); Oberammergau (Germany); Popayan (Columbia); on Marinduque (the Philippines); and in Poland in Kalwaria Zebrzydowska.

The American Civil War had an influence on how soldiers killed in battle were commemorated. Their bodies were buried in cemeteries set up near battlefields, such as the National Cemetery in Gettysburg, or in graves at the place of death, e.g. in the *USS Arizona* wreck in Pearl Harbor. Unfortunately, it was not always possible to visit the soldiers' graves, so monuments or memorials with the names of the victims inscribed on them, erected away from the battlefield, were an alternative. In this way, a fashion for cenotaphs, symbolic graves and tombs of unknown soldiers spread widely.

World Wars I and II decimated communities of Europe, Asia and North America. As a consequence, whole societies were in mourning. Suffering from a collective trauma, many countries created systems of support for the families of the deceased by commemorating them. This was mostly expressed by erecting monuments, funding memorial plaques and organizing graves and cemeteries.

Travelling to the graves of those who died in wars, acts of terrorism, during mass ethnic purges or as a result of genocide usually takes the form of educational, didactic, patriotic and

Fig. 4.1. The incorrupt body of Saint Catherine Labouré – Paris. (Photo: the author)

commemorative journeys. Such journeys can be treated as pilgrimages.

Cultural (Secular) Journeys

The spectacularity and the macabre nature of death became a motivation to embark on journeys whose aim was usually to take part (passively, as a spectator) in a public execution, though occasionally the spectator participated in this death actively. Therefore, we may talk about cultural journeys motivated by real contact with death. The reason depends on the object of cognition and cognitive motives. In the case of journeys motivated by the need to learn about death culture, the primary object of interest will be the form of death and its artefacts.

In the course of human history, death has changed its face many times. Sometimes, the way a person died was a spectacle. Public death, which took different forms through the centuries, for a long time was considered an entertaining performance.

From the earliest times, administering punishment in public, including the death penalty, played a number of roles. First of all, it prevented people from committing or repeating an offence, but another aim could be revenge or vengeance. Until the end of the 19th century, an additional quality of public punishment by death was that it was treated as a show providing entertainment for people who normally did not have any opportunity to participate in cultural events that were reserved for the 'upper classes'.

Western civilization experienced the spectacularity of punishment as early as prehistoric times during ritual murders and burnt offerings (e.g. Celts burnt their victims locked in wicker cages). In the Roman Empire, gladiator fights or victims eaten alive by animals were a significant element of early entertainment. Punishment by crucifixion administered by Romans to criminals attracted crowds. The spectacle usually took place outside the city walls, where the bodies of the crucified stayed in sight for weeks to deter others from committing the same crime.

In the Middle Ages, a popular event was a judicial duel where opponents fought using the weapon they had agreed on, and the loser, who was often wounded or killed, was found guilty. Burning alive was in those times a punishment administered to arsonists, heretics and those suspected of witchcraft. It was commonly used in Europe where the estimated number of people burnt over the period from the 11th to the 18th century was 200,000 (Farrington, 2000). In Europe, burning was a popular way of punishing witches and those suspected of lycanthropy. The place most famous for executing witches in the USA is Salem, where in 1692 witchcraft trials were held in which even children were sentenced to death.

Punishment in the Middle Ages was usually extremely cruel, because of its spectacular character. The most popular form of punishment in Europe was hanging, reserved for common criminals, as well as beheading, considered to be a better way to die and reserved for the nobility. The dead bodies of those who had been hanged, or severed heads on poles, were exposed at the city gates, performing a preventive function, warning against the consequences of committing crime. Public punishments and executions were accompanied by festivals, festivities or fairs included in the execution scenario which started with leading the convict to the scaffold and ended with the act of taking their life. Those spectacles gathered not only local inhabitants but also people who had travelled from afar (tourists?). In England, the spectacles at the infamous Tyburn, where convicts were hanged between 1220 and 1783, gained such popularity that stands had to be built around the gallows.

In 1789, the guillotine was introduced in France. Although similar devices had already been used by Persians in the 10th century and later, and it found its way to other countries where it was used in public executions. In France it was commonly used until the 1920s. Gadgets in the shape of the guillotine made popular souvenirs from executions which were a popular spectacle, often ticketed. Small, guillotine-shaped trinkets were also sold as toys for children. Due to this particular invention, Madam Tussaud's Wax Figures Museum became famous. Maria Grosholtz, born in Switzerland, moved to Paris ten years before the French Revolution with the intention of organizing a wax figure exhibition. In 1783, Dr Philip Curtis opened *Caverne des Grands Voleurs* and exhibited portraits of well-known criminals of that time, and working as his assistant, Maria Grosholtz made wax masks of execution victims. After Curtis died, Maria moved to London and travelled across England, Scotland and Ireland for many years, showing her exhibition, and at the same time making masks of infamous British criminals. Such were the beginnings of the first, pop-cultural museum of terror in England.

In the 20th century, public executions became forbidden in western countries. At present, quasi-public executions are still performed in the USA, limited to very small groups of witnesses. The public death penalty is maintained in countries such as Pakistan, Libya, Saudi Arabia, Burundi, Guinea and China.

Based on the history of the death penalty and torture, in the 19th century facilities which documented this, as well as stored related artefacts, started to appear. With time, they turned into tourist attractions and products.

Common death artefacts are human remains (bones and flesh) shown to the public. They also include paintings, sculptures, pieces of literature and music, performances and art events, conveying a message about the vanity of human existence. With respect to this, the development of art, whose products can be classified as death culture creations that interest tourists, can be analysed.

In the 17th century, the phenomenon of death returned to public discourse in the form of a dead body (corpse) which became the main image of death. Autopsy and the embalmment of dead bodies made a corpse something to admire. In this way, apart from a dead body, skeleton or decomposing flesh, art opened to anatomy. In the 14th century, in Padua (*c.*1340), probably the first public autopsy was performed. The Renaissance popularized autopsies during which doctors learnt about the human body, students expanded their medical knowledge and the audience delighted in the scenes. The fashion for public presentations of anatomy 'live' started to develop in the 16th century, culminating in the 17th and 18th centuries. Public autopsy was in those times a kind of theatre and entertainment for the spectators gathered around. Operating theatres were built, interiors were turned into auditoriums, and tickets and souvenirs were sold. It was a type of entertainment which exists in modern times as well, but takes a slightly different form. A good example is a 'Body Worlds' exhibition of specially prepared bodies.

The belief that a corpse should not simply disappear, that it should be preserved, displayed and viewed by people, changed the approach to the dead in the 17th century. Monks, Capuchins and Franciscans, in particular, dealt with the mummification of bodies, which, after being specially prepared and dried, were exhibited in 'mummy mortuaries' visited by thrill-hungry people. This method of attracting the public's attention lasted until the 19th century.

Since earliest times, dead bodies have been subject to conservation procedures, mostly for practical or religious reasons. A body could be accidentally prevented from decay, intentionally conserved, or sometimes simply did not undergo decomposition. In this last case, people resorted to religious beliefs, usually attributing the lack of decay to the interference of supernatural powers.

There are many places around the world where conserved bodies are found. Naturally mummified bodies, for instance, from the peat bogs of north-western England, western Ireland, The Netherlands, Germany, Denmark, in the dry areas of north Africa (Egypt), western and central Asia or the cold regions of northern Asia and northern Europe. Egypt is famous for pharaohs' mummies in the graves in the Valley of the Kings. Here, visited by thousands of tourists, researchers found very well preserved and embalmed bodies of pharaohs, priests and members of the aristocracy from that time. Egyptian mummies can be seen in many places in the world.

Mummies are also discovered in many places in North and South America, usually in mountains and caves, or on dry plains. The area is famous for the mummies of Indians and tribes once inhabiting the Andes, Cordilleras and American plains including Arizona, Alaska, Columbia, Ecuador, Peru, Chile, Argentina and Bolivia. In the dry air of the Andes, as in Egypt, the process of mummification occurred naturally.

Those found in China, in the Taklamakan Desert, have been conserved by the dry desert climate of the Tarim Basin. Mummies have also been discovered in Japan, the Canary Islands, the Pacific Islands, Greenland and Europe, where mummification became popular in the early Middle Ages.

Mummified bodies were shown at fairs and sought by collectors. A human body which has not decomposed has always caused a sensation. However, in contemporary times there is usually a negative association with death and a dead body. There have been many cases worldwide when the dead bodies of embalmed politicians were presented to the public (e.g. the bodies of Vladimir Lenin and Josef Stalin in Moscow, Ho Chi Minh, Mao Zedong or Kim Il Sung). Many bodies can be found in temples, mausoleums and museums; there are also mummies resting in monasteries (Palermo, Pechersk Lavra in Kiev, Reformed Franciscan Monastery in Kraków). Nowadays, many human bodies are valuable museum exhibits; they are often the main attraction for visitors and bring significant profits.

Death-related rites allow the living to experience direct contact with the phenomenon of death. A dead body is a constant reminder of the deceased person and at the same time the subject of numerous expressions of death. A dead body is also central to a variety of rituals whose character depends on their cultural space. Popular since the Middle Ages, the cult of saints, and the resulting tradition of creating reliquaries, encouraged mummification. In Italy alone, there are at least 25 mummified bodies of people who are considered to be saints by Christians.

In Europe, a traveller will most typically find mummified bodies in church crypts and mausoleums, e.g. the mummies from Sicily kept in the Capuchins' church in Palermo (*c.*800), or in the churches in Savoca, Venzone, Comiso and Naples.

The functions performed by ossuaries have changed. At first, they were storehouses, where the bones dug out at cemeteries or from graves were stored. In the 14th century, due to the growing fascination with the macabre, they started to be used for the religious message of *memento mori*. Ossuaries, such as Santa Maria Della Concezione in Rome, became a kind of decoration in which human bones became a part of a Baroque and Rococo artistic installation and where human remains were an element of religious performance (Fig. 4.2).

There is a long tradition of discovering death through art. Initially, in 11th century European art, a dead person personified death.

Fig. 4.2. All Saints' Cemetery Church, Kutna Hora, Czechia. (Photo: the author)

In the Middle Ages, when the sight of a dead body was commonplace, the sense of sight played an important role in the perception of death. It was not until the 15th and 16th centuries that images such as a skeleton or a decomposing body (*vanitas*) became common. The medieval *danse macabre* was inspired by the everyday morbidity of imagination: people dying in the streets, victims of war and disease, the commonness of exhumations conducted at cemeteries, funeral customs, public torment and execution. All these made death a common thing, experienced through different senses. At the same time, dance, as a folk expression of the dark side of human sensuality, was present everywhere in sacred and non-sacred places alike. The motif of *danse macabre*, the image of death dancing, spread through Europe along with the appearance of church frescoes at the La Chaise Dieu Abbey in Auvergne (1390–1410), or the paintings in the cloisters of the Holy Innocents' Cemetery in Paris (1424). The motifs of *ars moriendi*, the triumph of death and the dance of death, decorated the walls of monastery cloisters, ossuaries and churches. The medieval *ars moriendi* was the beginning of a discourse about images of personified death, an artistic vision transferred to the sphere of an image. Starting in the 12th century, iconography showed death themes for the next four centuries including scenes of the Last Judgment, Apocalypse, overcrowded hell and guardian angels. In the 14th to the 16th centuries, art was full of repulsive images of decay; pictures of naked dead bodies covered in worms (*transi*) 'decorated' tombs in those times. Paschal performances, calvaries and holy sepulchres were found everywhere in medieval and early modern Europe. Nowadays, they are religious-artistic artefacts affecting the senses of the faithful.

The Renaissance, despite its apotheosis of life, introduced death into art, especially when commemorating the dead in a posture that expressed their lives most appropriately. Baroque art was full of eschatological symbolism, imperatives and teachings. Attributes of death – the skull, skeleton, demons, devils, angels and saints – were set in paintings as a particular way of teaching the rules of proper conduct. Baroque *vanitas* art was also reflected in painting, sculpture and sacred applied art: vestments, banners, coffins or catafalques. In the 17th century, another leading iconographic motif was resurrection. The Enlightenment introduced into painting the theme of the medicalization of human anatomy, a demonstration involving a dead body (e.g. Rembrandt's paintings: Dr Tulpe's and Dr Deijman's anatomy lessons), while Romanticism, with its fascination with the macabre, introduced the sensational character of death and phantasms. Gruesome decorations evoked fear of death, and portraits of the deceased created an illusion of death. This leads to the conclusion that already in the Middle Ages, as well as later, art presenting death evoked emotions similar to those evoked nowadays by images of cruelty and a gory death. The aim has always been the same – to arouse the fear of death and respect for life.

An expression of cultural and social attitudes to death in the 19th century was sepulchral art, presented in the form of tombstone sculpture from simple graves to decorative chapels and allegoric epitaphs. In Poland, a particular role was played by the Sarmatian coffin portrait. The 18th and 19th centuries were a time when artists presented heroic deaths, for instance dying for the motherland or battle scenes. However, towards the end of the 19th century, a taboo on death gradually started to appear in art, replacing its expressive picture with allegory.

World War I showed the senselessness of death which remained in memory both through individual graves and mass war cemeteries. Similarly, World War II became the reason why the theme of death was rejected in European art, with the exception of commemoration space. Death reappeared in art, particularly in film and literature, in the second half of the 20th century. However, contrary to the historical picture, it presented cruel, spectacular, macabre death, an admiration for the atrocities of the 1960s, and later a fascination with blood, crimes, road accidents and disasters.

The theme of death is still popular in contemporary art. The artistic vision of death is often controversial, transcending accepted boundaries of morality, and shocking. Gunther von Hagen's 'Body Worlds' exhibitions, presenting 'plastinated' human and animal bodies, evoke repulsion and admiration at the same time. The experience is reinforced by the opportunity to buy postcards, albums and T-shirts.

The artistic presentation of 'fictional' death, historical but not realistic, includes spectacular reconstructions showing death: killing, fighting and suffering.

Works of art that are a particularly valuable tourist asset often determine the attractiveness of a tourist visit. They are viewed for different purposes and motivations: commemorative, educational, relational, aesthetic, emotional and entertainment. These examples of the vision of death confirm its spectacular, often morbid character, as well as the links between death and art, discovered during cultural journeys and pilgrimages.

Motivations, Values and Identity in Thanatourism

Travel is human spatial mobility which results in changing the geographical, social and cultural environment. It can be inspired by someone's interest in the culture of death, and its material and spiritual heritages; it may arise from social and cultural needs (cognitive, educational, religious, patriotic or related to entertainment).

A traveller may come across an object or place which had not functioned in their awareness before but, considering its significance, originality or by curiosity, it becomes the aim of a journey. It is not only information and experience gained that have an influence on the motivation to travel, but also the traveller's cultural and personal features as well. Death space and the products of death culture are not frequent aims of travel and they often evoke negative feelings, repulsion and objection. By nature, people tend to visit positive places when they travel.

Motivations to visit death space have been analysed by the author before (Tanaś, 2008). He distinguished two groups: *sacrum* (the sacred) and profane (Fig. 4.3).

The motivations referring to the *sacrum* of death space include the following:

- religious motivation – resulting from religious needs, strictly connected with the cult of the dead and relics, expressed by pilgrimages to death sites, graves and places where relics are kept;
- identity motivation – seeking one's ancestors, discovering history, shaping the

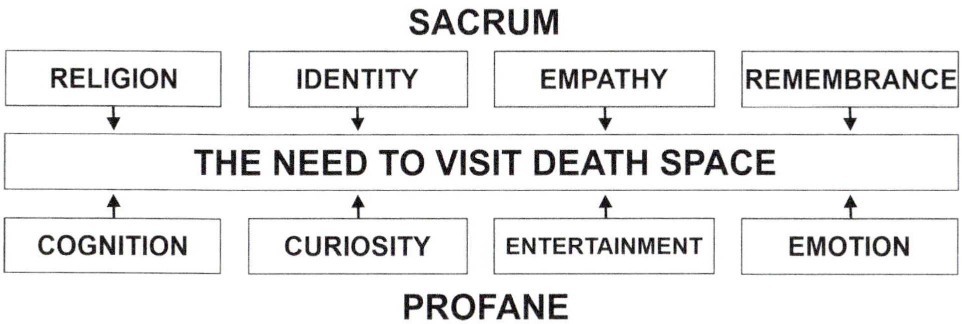

Fig. 4.3. Motivations for visiting death spaces. (From Tanaś, 2008)

family, local or national identity; it is connected with sentimental tourism, and alternatively referred to as 'roots tourism';

- empathy motivation – resulting from the need to share grief with those who were close to the deceased; it involves journeys to the graves of the victims of wars, genocide, disasters as well as to places documenting tragic events;
- remembrance motivation – an effect of cultivating the memory of a dead person or an event, usually involving homage and a visit to a remembrance space (monuments, cemeteries, remembrance sites).

The motivations referring to the profane of death space include the following:

- cognition motivation – generating the need to learn about history, culture, space or the dead person;
- curiosity motivation – expressed in the desire to see and remember a unique, unusual place or event;
- entertainment motivation – the taboo of death evokes the need for a fictional contact with it through pleasure;
- emotion motivation – results from the global development of pop and mass culture; a need to experience the emotion of fear, horror or repulsion through a fictional contact with suffering and death.

Berbeka (2012) studied the motivations for visiting the Auschwitz-Birkenau Museum and confirmed the research conducted by (Stone, 2010). The majority of respondents, who had four motivations to choose from, indicated interest in the history of the concentration camp (63%), the memory of the victims of the Nazi crimes (49%) and the wish to pay homage to the camp prisoners (45%). However, as many as 40% of the respondents pointed to curiosity as one of the four main motivations to visit the museum. The remaining motivations included search for identity and religion. The results of the study confirm the functions of thanatourism: cognitive, educational and religious (Fig. 4.4).

When analysing the phenomenon of travelling to death space, issues related to death culture should be accurately interpreted. The societies of today want to maintain their 'ontological safety', which results in separating tourists from death space. Curiosity and cognition, as well as entertainment and emotions in extreme cases, play an even greater role in the sphere of the profane, similar to the role empathy, identity, remembrance and religion play in the *sacrum* sphere.

Social processes sometimes lead to a change of human values, morality included. Morality is a particular product of social communication, resulting from individual decisions and interpersonal coexistence. Its most important components are moral values which reflect what is commonly considered to be good or bad, proper or improper, noble or ignoble (Mariański, 2001). Values understood in this way reflect collective awareness and not objective or true values.

Globalization leads to a shaking of traditional moral stances, including tourism. In radically modern societies, tourism, similar to other sectors of life, follows incoherent, sometimes even contradictory values and systems of meanings. Institutionalized values and norms, passed on from generation to generation, lose

Fig. 4.4. Auschwitz-Birkenau Museum, Oświęcim, Poland. (Photo: the author)

their evident cultural character in conditions of change and movement. Death, the deceased, a funeral and suffering have gained new dimensions for the present generation compared to the past. It is virtual and unreal and does not concern an individual: 'death affects others, not me'. Morality perceived from the religious perspective does not allow using death for the purposes of the profane.

A major element in culture is religion. In traditional societies, people accepted what religious law ruled. Specific ways of treating a dead person stemmed from imposed norms. As society was undergoing modernization, the *sacrum* became separate from the profane and religiosity from morality.

Departing from religious norms favours the replacement of the death space function by cognitive, recreation or entertainment functions, not accepted by religious institutions. Understanding and respecting grief and mourning, remembering the dead, cultivating the customs and rites which soothe the pain caused by the death of those close, are all behaviours of this kind. Therefore, learning about customs through travel and visiting death-related places can be defined as an educational and pro-social activity.

Cultural ideologies, beliefs and norms define the emotions which should be experienced and the ways of expressing them. They do not arise until objects or events in a given situation are evaluated when an emotion is aroused (Arnold, 1960). Emotions related to death belong to the group of primary emotions (Turner and Stets, 2005) and include fear, anger, grief and repulsion.

Emotions which appear in the process of building an identity stem from the relationship between expressions of identity and cultural models. If a tourist's behaviour is considered appropriate in the light of the cultural norms of a given community, positive emotions will be the result. On the contrary, when tourists believe that their behaviour based on identity does not meet expectations or does not express the values imposed by the cultural model, negative emotions (a sense of guilt) will be experienced, and other members may feel anger or disappointment. Therefore, if the tourist is properly prepared to perceive a visited space and follows the current cultural model, such a tourist will identify with the place and the social environment, reacting positively in terms of emotions and behaviour. If the tourist perceives death space in a way which does not fit the cultural model, he/she will feel alienated and alone (or, being unaware of the opinions of others, will remain in a state of self-acceptance), and those around will criticize this person's behaviour. In such a situation, emotions show whether the tourist's identity is authentic and acceptable in the visited space. A perfect confirmation of this is the restraints on entering the State Museum of Auschwitz-Birkenau in Oświęcim for tourists with non-normative behaviour; they are not managed by the museum staff like a collective tourist.

Interest in death is usually inspired by cultural difference, a reason why its perception varies as regards the religious, ritual aspect (including the form of cult and commemoration of the dead person), faith (folk culture, the afterlife, ghosts, supernatural powers, phantasms, immortality, reincarnation, legends, the cult of death), the treatment of dead bodies (methods of burying people, cremation, mummification, cannibalism, collecting human remains), as well as ways of expressing emotion in the face of death (seriousness, sadness, fear, respect, celebration).

Death culture may be perceived differently by an aware tourist, usually someone who is well prepared to perceive higher culture. Death culture will be treated as an element of cultural heritage, tradition, and contact with the artefacts of death will be of an educational nature. An unaware tourist, unprepared to interpret the culture of death, will perceive it as a modern cultural (tourist) product. This perception may (though not necessarily) be limited to a simple popular perception characterized by an element of entertainment (Fig. 4.5).

Summary

Tourism in death space is about the coexistence and interpenetration of the *sacrum* and the profane. Due to the transitory character of cultural contact during travel, tourism may lead to social conflicts, resulting not from the clash of the *sacrum* and the profane themselves, but from tourists' attitudes to specific values. A tourist's interest in death space, without proper knowledge and mental readiness, may lead to a mistaken interpretation. This is particularly noticeable at places which are popular tourist attractions (e.g. the Cave of Skulls in Czermna; Paris catacombs), which are usually assessed through pop-cultural theatricalization or historical *kitsch*.

The evaluation of thanatourist journeys is not simple and depends on motivation. A

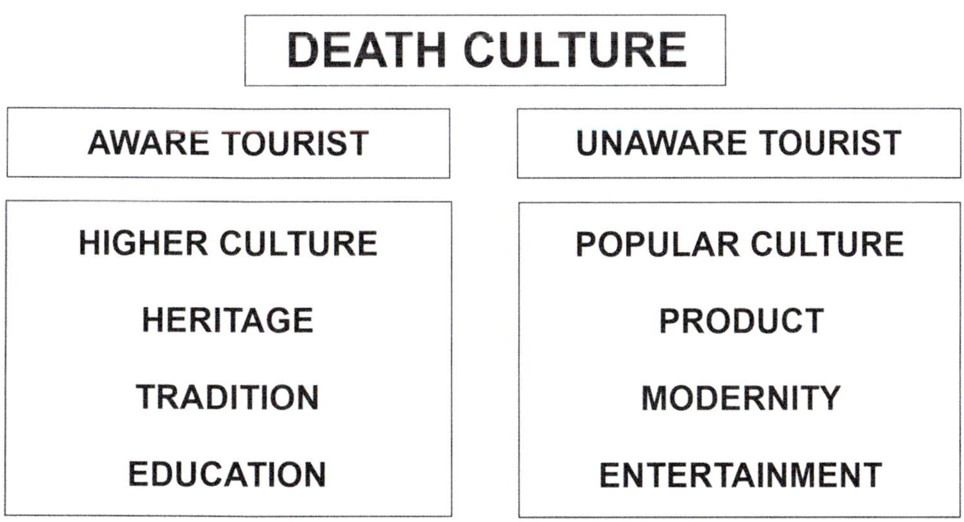

Fig. 4.5. Perception of death culture by an 'aware' and an 'unaware' tourist. (From Tanaś, 2013a)

negative opinion about thanatourism may stem from the cultural distance between domestic and imported culture, as well as from social differences and ethnocentrism, which are extremely important in interpreting new or unknown phenomena.

Geographical diversity and the historical evolution of cultures are reasons why a contemporary tourist takes an interest in cultural heritage, including the heritage of death. This, in turn, makes cultural tourism a market commodity, an example of which is thanatourism with its specific motivations to travel. Thanatourism has been developing in the western world since the 19th century, although journeys that could be classified as thanatourist had been documented much earlier. In the second half of the 20th century, the development of pop culture and the mass media was followed by a rapid growth of tourist products based on the culture of death, together with a controversial and morally doubtful need for direct or indirect contact. Social and cultural changes lead to a change in travelling preferences, including a tourist interest in originally non-tourist spaces, which now gradually acquire the characteristics of an individual or developed tourist space. It may be concluded that the links between death space and tourism space may be more permanent, and from an economic point of view, bring the development of tourist products.

The author has tried to demonstrate changes in motivations for journeys to death space, ranging from religious pilgrimages to graves, through the historical interest in places commemorating death, including educational motivation, to the commercialization of death space and the entertainment motivation, which encourage travellers to visit contemporary tourist attractions based on the death-culture heritage.

References

Aries, P. (1985) *Images of Man and Death*. Harvard University Press, Cambridge.

Arnold, M. (1960) *Emotion and Personality: Psychological Aspects*. Columbia University Press, New York.

Berbeka, J. (ed.) (2012) *Turystyka martyrologiczna w Polsce na przykładzie Państwowego Muzeum Auschwitz-Birkenau*. Kraków, Proksenia.

Farrington, K. (2000) *History of Punishment: A Journey Through the Dark Side of Justice*. Hamlyn History.

Jackowski, A. (2003) *Święta przestrzeń świata. Podstawy geografii religii*. Wydawnictwo Uniwersytetu Jagiellońskiego, Kraków.

Mariański, J. (2001) *Kryzys moralny czy transformacja wartości? Studium socjologiczne*. Towarzystwo Naukowe KUL, Lublin.

Martin, D. (2010) Medical travel and the sale of human biological materials: suggestions for ethical policy development. *Global Social Policy: An Interdisciplinary Journal of Public Policy and Social Development* 10(3), 377–395. DOI: 10.1177/1468018110379979.

Safyan, A.R. (2011) A call for international regulation of the thriving 'industry' of death tourism. *Loyola of Los Angeles International and Comparative Law Review* 33(2), 287–319.

Seaton, A.V. (1996) Guided by the dark: from thanatopsis to thanatourism. *International Journal of Heritage Studies* 2(4), 234–244. DOI: 10.1080/13527259608722178.

Seaton, A.V. (2002) Thanatourism's final frontiers? Visits to cemeteries, churchyards and funerary sites as sacred and secular pilgrimage. *Tourism Recreation Research* 27(2), 73–82. DOI: 10.1080/02508281.2002.11081223.

Srinivas, R.V. (2009) Exploring the potential for American death tourism. *Michigan State University College of Law Journal of Medicine and Law* 13(1), 92–123.

Stone, P.R. (2009) Dark tourism: morality and new moral spaces. In: Sharpley, R. and Stone, P.R. (eds) *The Darker Side of Travel: The Theory and Practice of Dark Tourism*. Aspect of Tourism Series, Channel View Publications, Clevedon, UK, pp. 56–72.

Stone, P.R. (2010) Death, dying and dark tourism in contemporary Society: a theoretical and empirical analysis. PhD, University of Central Lancashire, Preston.

Stone, P.R. and Sharpley, R. (2008) Consuming dark tourism: a thanatological perspective. *Annals of Tourism Research* 35(2), 574–595. DOI: 10.1016/j.annals.2008.02.003.

Tanaś, S. (2006) The meaning of death space in cultural tourism. *Turyzm/Tourism* 16(2), 145–151.

Tanaś, S. (2008) *Przestrzeń turystyczna cmentarzy. Wstęp do tanatoturystyki (Cemetery Tourism Space Introduction to Thanatourism*. Wydawnictwo Uniwersytetu Łódzkiego, Łódź.
Tanaś, S. (2013a) *Tanatoturystyka. Od przestrzeni śmierci do przestrzeni turystycznej (Thanatourism From Death Space to Tourism Space)*. Wydawnictwo Uniwersytetu Łódzkiego, Łódź.
Tanaś, S. (2013b) Tourism death space and Thanatourism in Poland. *Current Issues of Tourism Research* 3(1), 22–27.
Turner, J.H. and Stets, J.E. (2005) *The Sociology of Emotions*. Cambridge University Press.

5 Pilgrims as Strangers

Larry Russell*
Retired from Hofstra University, Hempstead, New York, USA

> But suppose that instead of trying to rise above our perception of things we were to plunge into it for the purpose of deepening and widening it. Suppose that we were to insert our will into it, and that this will, expanding, were to expand our vision of things. (Bergson, 1946b, p. 111)

Introduction

For me, pilgrimage begins at the carved doors of a Roman Catholic shrine of healing in New Mexico and extends for more than a decade as I participate in a hundred-mile pilgrimage each June. In my search for a robust methodology that makes sense of the mystery at the core of my experience, I enlist Edmund Husserl's philosophic techniques of phenomenology – his pioneering studies of consciousness and the objects of direct experience (Husserl, 1970). The first step is to explore scenes that show how pilgrimage variously presents itself (through a road, desert, meditation or song). To do this, I engage in 'a peculiar kind of reflective attentiveness that distances itself from everyday life in order to get to the root of that very life' (Steinbock, 2007). Because my goal is not an academic depiction, I follow Steinbock's examples of first-person descriptions, 'stirring in us the lived experience he or she is trying to awaken' (2007, p. 27). When I want to demonstrate how a spark from pilgrimage elicits a response from my identity, I choose narratives that display an arc of recognition between my body and the environment, my spirit and the community – deep calling unto deep (Psalm 42: 7, KJV). I post signs along the way as I discover the conditions shaping my pilgrim experience, but the reader must judge whether 'the situation itself seems to call the action out of them' (Dreyfus and Kelly, 2011, p. 3). These conditions shape the dynamics at the heart of my pilgrim practice (Jongmeewasin, 2015). Another philosopher, Henri Bergson, suggests a study like this is possible only by 'placing oneself within it instead of adopting points of view toward it, of having the intuition of it instead of making the analysis of it' (Bergson, 1946c, p. 136). I learn pilgrimage as I would learn a foreign language, through immersion. When anthropologist Clifford Geertz describes the goal of 'being there', he urges me 'to sound like a pilgrim and a cartographer at the same time' (Geertz, 1988, p. 10). I lay out a map to trace our route, but New Mexicans insist that the destination is not our goal. Our destination may be a joyous reunion, but we live pilgrimage one day at a time. Though I sense it has a connection with my identity, pilgrimage isn't heroic or virtuous. I'm trying to find the way home, but I share that way with a community of pilgrims. Ritual connects the two as my identity is challenged within broken borders that admit *communitas* (Turner and Turner, 1978, pp. 13-14). This isn't Bunyan's triumphant progress of a pilgrim, but an experience closer to Piers Plowman in a 'difficult dance with steps forward and back, jerkily aimed toward a salvation one can only intermittently grasp while puzzling out the actions

*l.russell.pilgrim@gmail.com

Fig. 5.1. El Santuario. (Photo: the author)

and intentions of one's own soul' (Morrison, 2016, p. 4). My narrative demonstrates that the practice of pilgrimage nurtures discrete identities within a source flowing through all of us.

A Journey Toward Pilgrimage

When people ask about my first pilgrimage, I begin by telling them about El Santuario, an adobe shrine in northern New Mexico that's been a pilgrimage site since the early 19th century (Fig. 5.1). It's likely that Native Americans came to Chimayó for healing mud long before the hot springs dried up and the Spanish arrived. There are legends of a farmer who found a large wooden crucifix like the one of Nuestro Señor de Esquipulas in Guatemala (Graziano, 2019, p. 5). In both places, the earth is considered a healing agent. The farmer belonged to a Franciscan group of laymen, and after he built a small hermitage on the spot, he petitioned a bishop for permission to build a chapel in 1816. His lay brothers carved *santos* and painted *retablos* for the shrine and for many old adobe churches of New Mexico. Karl Jung (1955) would call it a 'meaningful coincidence' that the brothers, *hermanos*, are visiting the shrine to sing their haunting *alabados* during Holy Week when I first walk through the door. Their melodies can break your heart. I enter sacred space as an agnostic, and my lapsed Protestant credentials don't include pilgrimage. So I spend two years reading books and recording interviews before I acknowledge I must make the voyage in order to understand.

The particular pilgrimage I join is organized in four groups that set out from the cardinal directions (following a Native American tradition that orients life from the north, south, east and west). We walk 100 miles in five days before meeting on the sixth at the shrine that has been a treasure of this homeland for two centuries. Each pilgrim group slowly cuts a swath across their portion of northern New Mexico. Villagers know we're coming, and they greet us on the road and bring us to their churches where they feed us before entrusting us with scraps of paper that ask us to pray about their lives and their struggle with illness, violence and hardship.

The Conditions that Shape Pilgrimage

To tell this story, I need a research model based on experience. Phenomenology reclaims everyday experience once lost in a sea of academic terms and theories. Husserl, Heidegger, Merleau-Ponty, Sartre and Beauvoir were the first phenomenologists – if you don't count Kierkegaard and Dostoevsky – who were fascinated by the way we participate in the world. The phenomena of the world are not neutral, as we can guess from the Greek root *phainómenon* – 'that which appears, shining forth' (Dreyfus and Kelly, 2011, pp. 88-89). The world (in this instance New Mexico) appears in various forms – geologic, cultural, meteorological, religious. My senses are so quickly configured by what I anticipate that I miss the subtle sparkle inviting me to respond. Immersed in the sensory fireworks of the high desert, there's a delicate pattern of four or five stimuli that summon something below the surface of my consciousness. I've never been inside a structure like the shrine, yet there's something familiar. The figures speak to me and make sense. It's an intriguing process – not making meaning but drawing meaning out of me. Before I can articulate what's happening, I know that I want something from this. From college, I remember Rudolf Otto's description of a numinous experience (Otto, 1958, p. 10). There's a pang of emptiness I've been waiting to fulfil all my life. Pilgrims speak to me of human suffering, compassion and camaraderie in a language I've been waiting to hear. Phenomenology charts my response to this world – a road leading to an adobe church, the shallow blue bowl of sky and the songs of *hermanos* who understand my yearning. The portal for my experience is surprisingly physical. My body responds, even before my mind. I'm interested in the lives of pilgrims, but I have no idea that my body's relationship to the road will be crucial. The evidence in this chapter of my transformation isn't a snapshot but rather the narrative of an emerging scene. It won't hold still for my camera. Pilgrimage is a frequent metaphor for the movement of our lives. Bergson writes, 'Instead of states contiguous to states, which become words in juxtaposition to words, we have here the indivisible and therefore substantial continuity of the flow of the inner life' (Bergson, 1946a, p. 20). The sensuality of this movement is significant. I catch myself holding my breath and waiting. For what? The strong colours, sounds and smells invite physical pleasure, but it's the subtle animation of sparkling details that reveal its spiritual vitality. Something flows from inside life and then through me.

I'm getting ahead of myself because the story unfolds so quickly. My first impression of New Mexico is actually a rush of images while crossing the high desert in a rental car – a steering wheel and tyres extending the illusory boundaries of my body. Speed and power distract me. Even mountains disappear as I hurry past. When I'm warm, there's air conditioning, and I control the bright sun with dark glasses. I'm a middle-aged (and middle-class) graduate student funding my research on a student loan and a credit card. In a rush to begin research, I ignore the delicate movement of scenery while I focus on the next stop – eight miles to Chimayó and the pilgrim shrine. I turn a dial for music to calm the anxiety of my speeding body, and I lean into the next curve of the road. Suddenly, I'm within sight of a village. As I veer left, it appears. There are four cars in the dirt lot where a wooden sign tells the history of the site in one paragraph. In my excitement, I haven't taken it in – the terrain, the Sangre de Cristo mountains, the clumps of green bushes dotting a ruddy landscape. My five senses don't yet realize how pilgrimage will transform this research question into a quest.

Strangeness

When I enter, I'm girded like a tourist, but haste and constant analysis won't prepare me. I look up the whitewashed adobe walls to the massive roof timbers – *vigas*. Everything comes from the earth. The bright colours of painted holy figures – *santos* – are collected from minerals and herbs nearby. These saints have the faces of Spanish farmers, weavers, shepherds and pottery makers. Their stories are not idealized or sentimentalized. When I stop interpreting them, their faces and gestures announce my strangeness in their world. If the root of the word pilgrim suggests a foreigner, someone from abroad, then I arrive as a spiritual geek, my head full of academic information about the village and chapel. What matters are the conditions that make pilgrimage possible, like my strangeness in their world. To continue, I must leave the comfort of what I know and become a stranger as I head toward a mysterious quality of life for which we reserve the term 'holy'.

Without the security of home and assumptions about who I am, I'm a stranger to myself – this uncomfortable space between me-as-pilgrim and me-as-a-tourist. Discomfort sharpens my senses. I don't know how to pray, but I can kneel. This jolts my lapsed Protestant spirit as I learn to bend – not before a hierarchy but in wonder. What's happening before I tell myself the story of what's happening? Without a familiar context, I stumble from one experience to the next. There are no familiar terms, and I examine my prejudices about kneeling, praying and religion. A year later, I've signed up, packed my bags and returned to a church in Albuquerque that is 100 miles from Chimayó. I join a pilgrim group for whom these roads and towns are their homeland. Their rituals are Roman Catholic in a northern New Mexican vernacular based on their history. Most of the group have walked the 100-mile trek at least once, yet they don't treat me as an outsider.

My journey begins in a parish hall with the clank of a bar on a metal door swinging out into the pre-dawn darkness. A rectangular shaft of light on cement creates a portal through which I pass from my old life onto an uncertain path. Once beyond the building, there's only the crunching sound of shoes on gravel. I look back at the light for a moment as we walk from the church. When I turn to the road, my eyes can't see where I'm going, but my feet find the way. Our band walks from dark to dark, passing under pools of light from the lamp-posts. My body adjusts to the slant of the berm along the road. It's difficult to find a gait that works in the shifting line of pilgrims. Without thinking, I surrender to our pedestrian rhythm. To be a pilgrim is to leave where I belong and head toward a new life.

Solitude

While I'm still absorbing the lesson about strangeness, I'm confronted by a second condition: solitude. Fifty-seven men walk in silence before dawn, each alone with himself, a rosary wrapped around his hand. We're on the outskirts now of Albuquerque, and the lights of the city behind us emphasize the immense basin we cross. We read the scraps of paper in our pockets, collected in parish after parish. I'm unsure about God, but if she's listening, I sense she's aware of the suffering. The folded papers nudge me to grasp the lives of their authors, and I read their pleas. One man's wife has breast cancer. A grandmother worries about the rage of her son-in-law. A nephew needs a job to support his family. I look at my rosary and pray their concerns. My fingers follow their lives through a human story that begins with a young Hebrew girl who senses a sparkling possibility that feels like an angel reaching out to her. Her story is ours. It's wonderful, but in the end we give up the thing we love most. It's so important to let the story sink in that we repeat each scene of the story ten times. The first scenes are filled with joy. She shares her discovery with her cousin who is also full of spirit. An old priest recognizes the possibilities this child offers. Then the harsh world rejects their understanding of vulnerable love. The scene of unbearable suffering leads to a mystery that goes beyond death. The intimacy of these prayers weaves a thread between me and the people who need someone else to care about them. I remember Caravaggio's painting of Mary's death – the bare peasant feet extending out toward the viewer – poor, vulnerable. After a few miles, my feet feel the weight of their troubles. We press on, but it's hard to keep the pace. Is this the purpose of prayer – to share the load? Sometimes I stumble as I pray the joy and agonizing loss replicated in their lives. Gravity drags my body down the slanted gravel berm of a highway in the middle of nowhere. Speeding cars take no notice of us, alone with our lives and the pleas for prayer. Does the parishioner who asked for help know I'm here on the road praying about his wife's cancer?

The condition of my body is not encouraging – raw feet from practice walks and knees that could use a little oil. My middle-aged chassis must sustain this pace for 20 miles today – one lone body in an immense world. If it weren't for pilgrimage, I'd be whizzing past this scene in a car and focused on a destination beyond the difficult present. Our quiet expedition frames my noisy, frenetic life. Alone, yet walking with others, I grasp the intimacy of human suffering. It was Viktor Frankl in a German concentration camp who described how 'each prisoner craved to be alone with himself and his thoughts. He yearned for privacy and for solitude because this made him an individual with an identity' (Frankl, 1959, p. 51). The intensity of solitude is a space where I might grow a soul. There's almost no sound, but the world doesn't feel silent. I'm unprepared for impulses that flicker through my consciousness: anger, fear, regret. They make a jumble of everything, tossing me about. The repetition of prayers in the Rosary draws my attention to lives for whom I pray – their issues and their suffering. My journey is expanding to include other lives while solitude locates me where I need to be.

Then in the red and gold streaks of dawn, my journey becomes radiant. We pause to pray the announcement in the Angeles and then continue down the road. I wince when the brilliant morning light exposes my unresolved issues. To be alone is to be present in the quiet soundscape. With no buzz and bustle to protect me, the deep turmoil of my life rumbles to the surface. Pilgrimage removes the cushion between me and the life I avoid. The pace of our walk in a boundless desert offers the possibility – not of running away but of moving through. My journey becomes a dogged persistence along a never-ending road. It's no longer a highway, this pilgrim path, this way to go through life.

Momentum

As I look at the highway I wonder about the direction I've chosen. No arrow points the way, yet something is beckoning me. The rigour of our journey pares down my life to necessities. I come from a culture based on an accumulation of goods, but here less is more. Less diversion and more awareness. When there's no clutter, each pilgrim has to face his life story, and then we tell it to each other. Elsner emphasizes that narrative is as important to pilgrimage as mobility because it can locate the sacred within experience (Elsner, 2020, p. 270). Who is this other me that observes my life as I live it (Grotstein, 2000)? Who authors my story? Time slows to a desert crawl as I progress along a vast landscape. In the glare of sunlight, I still face the dark. The only way to illuminate the darkness in life is to keep going. All I have is the experience that pilgrimage draws out of me; one mile and another; one day and another. Though I stumble, I discover the momentum of pilgrimage, which is the third condition.

Other people seem aligned with their bodies, but mine is uncoordinated and clumsy. I want to run for joy, but I hobble until I force myself to walk through the pain. When I reach my limits late in the afternoon, the world shrinks, and all I see is the next 100 yards. I manage one patch of highway and then another and another, neither doubting nor believing I can make it to the end. A small space of unknowing is all I can muster. Then, there is a change. Though my mind doesn't realize what's happening, my body continues. It's more of a subtle flow than a second wind. I don't know if this gradual transformation is the way a miracle is supposed to feel. Carl Jung describes 'a wholeness which embraces the very depths of Nature...the strongest, the most ineluctable urge in every being, namely the urge to realize itself' (Jung, 1949, p. 170). The strength is not familiar, yet I claim it. David Kalsched asks (Kalsched, 2013, p. 170), 'What if there were an instinct for spiritual development in the personality? How would that be satisfied?' My life is under the aegis of something beyond me. My only proof is the mystery of momentum and the surprise of my survival when this is repeated each day.

Our pilgrimage follows a pattern that traces a road through pueblos and over a mountain range. My spiritual freefall begins while walking in prayer as I learn to let go. Instead of resisting, I release fear and grief because I can't carry them any further. That makes room for something else. Is this what they mean by faith? Kierkegaard met it with fear and trembling. When there's no place to hide, I listen to the silent flow within me. All I have is an intuitive path that is little more than a spark or a nudge. This vital connection between body and road is a new experience (Feldman, 2017, p. 75). I walk on concrete or asphalt, but my support is a passionate spark of interest that sustains me each afternoon. How can a spiritual dimension in pilgrimage take my body at its weakest and fill me with the wine of gratitude? I'd laugh at my rescue, but I haven't the strength. Though I'm out of my league, spiritually, pilgrimage gives me an appetite for life. Agnostic and atheist friends later say to me, 'I wish I had your faith'. It's not that simple. We all wish we knew how life persists, but we never do. Faith is continuing to walk when there's nothing left.

Turbulence

In each day's walk I negotiate the fragile balance of my existence – feet, knees, strength, spirit. It's up to me to integrate these experiences. If the instrument of my body is in tune, I breathe more deeply and unclench my hands. To complete pilgrimage is to abandon the way I was. That way is no longer possible in the turbulence of this next condition. My old solutions slide out of control, and the pieces dangle. I look around and realize that no pilgrim is exempt. Our common bond is our admission that none of us is that tough. Each walks through the turbulence of his failure, loneliness, guilt and loss. My assignment is to initiate an inquiry into how I looked at life, or didn't look. Life demands everything, and the best I can do is to walk beyond hope. How can I trust an unprepared life that I bring to this task? All I have of my six decades are the images I carry, like yellowed photographs. I plod through the day, scuffing through a heap of memories I can't forget. Next is a pile of unresolved troubles my family bequeathed me. And finally, there are the testaments of suffering in a handful of pleas for help. It isn't much, but this is what I have. When I slow my walk to catch the details, when I attend to life, it is luminous, calling me to love what shines forth in the world. The memories and troubles and pleas are my vocation and my identity. Simone Weil teaches me about purpose (1942)(Weil, 2013, p. 136):

> There is an instant when, for an infinitesimal fraction of time, the pure, naked truth, certain and eternal, enters the soul. For each person it consists of a series of acts and events that are strictly personal to them, and those who pass such obligations by miss the goal.

Burdens

This is my opportunity to not pass such obligations by. Once I admit what I have of life, I make my pilgrim way through turbulence. When Moses (Ex. 3: 4) and Abraham (Gen. 22: 1) are summoned, they answer, 'Here I am'. Little Samuel answers (I Sam. 1: 20), and his name means 'God heard'. Call and response. We're called to forge identities when we show up for pilgrimage. There's a unique luminescence animating each set of pilgrim shoes. Steinbock defines a calling as the discovery of 'the good-in-itself-for-me as a way of aligning my will to the "will of God"' (2007, p. 196). I cannot reach my identity except through brokenness. When I admit that, my treasure is the pile of burdens I carry. Their images speak with greater power than words – an 8-year-old boy dying of leukemia on the sixth floor of Coney Island Hospital (his father reaching under the oxygen tent to gather him in his arms); a college professor molesting me (as I crouch in shame against the locked door of his Cadillac); a U-Hall truck departing from a brown, clapboard parsonage (after I'm excommunicated); my mother refusing to look at a photograph of David (whose name she never utters); and me at a street corner in Manhattan (instinctively reaching for the hand of my five-year-old daughter who sleeps on the other side of the river). Pilgrimage has room for grief, fear, anguish and despair. To leave them behind is to deny who I am. So I walk with them because pain isn't the end of the story.

In late afternoon, we climb a steep road up the mountain ridge. Our bodies and memories are burdens for legs and spirits to support. Charlie tells me this may be as far as he can go – tough, lean Charlie, a seasoned pilgrim who is my age. My body is too weary for hope, and my heart is flooded with every memory from which I've fled. There is nothing left to offer but my weakness. I don't find the emotional strength; it finds me. Near the end of Thursday's journey, I realize the source of my strength – a connection between the perseverance in my limbs and the compassion in my prayer. They're not so different.

Twenty years of grief flow through a crack in my armour. No one notices my tears – each man in our line of pilgrims sounding the particular depth of his history. All I can do is to make my way through the wreckage of years when I didn't mourn. Pain thrusts my stories up through the cracks of the road, but they no longer haunt me. Pilgrimage is its own teacher and its own reward as 'landscapes are woven into life, and lives are woven into landscapes' (Tilley, 1994, p. 30). Our movement is unremarkable, yet our bodies absorb a lesson in fortitude. This practice of a walking prayer strikes a tiny spark at the core of life, and I don't turn away. It demands nothing while asking for everything: participation.

Yielding

At the limits of my strength, I finally yield with a sigh. The parable of my life begins with an observation that 'there's something more when nothing is left'. It belongs to the mysteries of this world. First I find a depth of strength in failure. Then I see that my protective shell is just a vessel, and I allow the stream of experience to flow through. And finally, there's room for other lives to intersect with mine – other pilgrims and their stories, slips of paper that ask me to pray, villagers who feed me, my family passing on what they have. I don't ask for a spiritual epidermis to protect me from the suffering of the world or for tough skin to protect my feet. Yet when I muster the courage to yield, I want for nothing. This understanding is unacceptable in a culture driven by commodities. The psalmist sings of a

Fig. 5.2. Pilgrims. (Photo: the author)

pilgrimage that leads me beside still waters to restore my soul. I hear this song as my strangeness moves through solitude toward the spiritual momentum of pilgrimage. When memories are turbulent, I keep walking until I can embrace the burdens. Physical conditions draw these spiritual lessons out of me, and I pay attention to the vehicle of my journey – a body infused with spirit (Fig. 5.2).

In a long and arduous pilgrimage, there's time and space, not just to think but to change. At times, it's repetitive and boring – today's road like yesterday's. The quiet miles and empty hours provoke a silent investigation. It's the same life in New Mexico that I left in Illinois – I just listen and observe more carefully now. The integrity of my life isn't an achievement but a gift – a body's wisdom and a prayer's strength coming together. My blistered feet, the slanting berm and a blazing sun are part of a spiritual world that is bloody eloquent.

It's not easy – 20 miles a day with the increasing weight of meaning in each lesson. Late on Friday we face the incline of another road. We're Catholic, so someone says, 'We need a litany to get up that hill.' The leader shouts a description of God...and the group responds:

Seeker of Silence, Walk with us.

Giver of Hope, Walk with us.

O Compassionate One, Walk with us.

Parent of the Poor, Walk with us.

Somehow the rhythm of prayer braids our struggle with our aspirations until we make it to the crest. The flow between physical effort and compassionate prayer won't let me partition my life. This is what Roy Rappaport means when he claims that 'the self-referential messages occurring in ritual do not merely say something about the state of the performer. They do something about it' (1999, p. 107). Prayer crosses a border separating me from the needs of other people. Our litany teaches me how to incorporate shattered lives. Is that what Isaiah means? 'They that wait upon the Lord shall renew their strength' (40: 31, KJV). My life is waiting for this moment. I don't come to New Mexico with faith, yet something shines through the life I once ignored. Climbing this road demands something more than piety. By the top of the hill, my thinking, speaking and walking are no longer separate. I become a pilgrim to learn this – this refusal to divide my life into compartments.

If *inspiration* comes from breath and *incarnation* materializes the flow of spirit through my body and *enliven* restores vitality, then the etymology of my language also embodies spirituality. Though the body's wisdom is ancient, my culture ignores the power in rites of passage. Weddings, graduations and funerals no longer enable us to sense our transformations. My preparation in graduate school is a litany of Van Gennep's (1908), Turners', (1978, pp. 2-3), and Rappaport's (1999, pp. 216-219) wisdom about ritual thresholds between stages of life. They help me grow into my life. My intuition and yearning are strengthened by mentors I meet during pilgrimage. A monk asks me if I ever considered the priesthood; a nun sings a joyous song of deliverance; a deacon welcomes me to participate in the Eucharist because I'm a pilgrim. I was hungry, and they fed me, thirsty and they gave me drink, a stranger and they took me in. They help me find the narrative I've been waiting to tell.

Frontier

Neither the crest of a hill nor an adobe shrine are the end of our walk. It's a holy site, but perhaps no more holy than the road or the land. A better word than destination in speaking of pilgrimage is 'frontier' because we cross into unexplored territory. I find my gait and the rhythm of my days to carry burdens (Oviedo *et al.*, 2014). It takes 80 miles in a walking meditative prayer for me to consider the spiritual possibilities of a new world. That leaves the 20 miles of our last day as a rehearsal for the rest of my life. Until this week, secular life was my endgame, but pilgrimage, like a puff of wind, scatters the cards. This is my portion in life, these Christian symbols and stories I once thought were inadequate. They shine through people – a mostly working-class, Hispanic, Catholic culture outside the borders of my secular world. My life is renewed through a vital community of pilgrims who live the parables and proverbs: Blessed are the merciful, the pure in heart, the poor. I need captions and subtitles to follow the story because American culture doesn't see anything below the surface. My culture's irony is pallid among these bright New Mexican colours. When I choose a painful struggle of pilgrimage over the jittery pace of post-modernism, all I can offer are wounded feet, aching knees and an awareness – here, now, this.

Risk

As my psyche catches up with my body, I forsake my little hoard of comfort to care about people and suffering. Risk is the final condition. The mystery of love leaves me in a precarious state, so I risk everything for the meaning that sparkles through pilgrimage. When I get too serious, it winks and makes me laugh. Spirituality demands a lithe stance and resilience. I take a ritual plunge into life, risking a performance of symbols that means something more than the repeated words (Rappaport, 1999). I risk what I thought was certain about my identity. I risk old distractions for a new solitude – alone and yet walking in a line of pilgrims. I risk my familiar emptiness to accept the burdens of my history; and risk propels me toward the frontiers of a life where my particular spark finds its mirror in everything around me. My inner life erupts like an artesian spring in this desert. The journey calls to me, and I respond to the images we sing and pray against a background of the life stories of pilgrims.

Six days of walking is a brief episode measured against a whole life. It's a beginning, but hark, the herald angels sing of transformation. Once, I thought I answered a call to be a minister, but I missed the point and hid myself from life. It was easier to be a nice person than a gay male. Those were precious years, the ones I wasted in a closet. When I finally came out, I had to break the habit of hiding. Now I break the habit of secularizing all experience, and I'm able to recognize the religious vocation of artists and teachers in my culture, 'devoted, working at something, something that is very precious to them – some calling or vocation in the old sense, the priestly sense. They are working at something which fate has called them to, somehow, and which they work at and which they love' (Maslow, 1972, p. 42).

In the stretch before Chimayó, the conditions of my pilgrimage reveal a pattern, and the journey is worth it – all 57 years. This fulsome experience recapitulates the course of my life in a new framework. Life has always been a pilgrimage, and I laugh when I realize I'm hearing the melody and mimicking the gait of John Travolta in the first scene of *Saturday Night Fever* to the beat of 'Staying Alive'. And I walk with T.S. Eliot in the darkness of World War II: 'We shall not cease from exploration, and the end of all our exploring will be to arrive where we started and know the place for the first time' (Eliot, 1942, p. 208). And I hear communion in an audience singing, 'Amazing Grace' with Joan Baez after the murder of a gay politician (Baez, 2014). And I resonate with Leonard Cohen's experience as he sings 'Hallelujah' (Cohen, 1984) in his gravelly voice: 'There's a blaze of light in every word. It doesn't matter which you heard –the holy or the broken hallelujah.'

All my life, a phenomenology of pilgrimage was also illuminating music, film and literature. It anointed me for this journey. William Faulkner was speaking of something like pilgrimage in his Nobel speech (Faulkner, 2006): 'I believe that man will not merely endure, he will prevail. He is immortal, not because he alone among creatures has an inexhaustible voice but because he has a soul, a spirit capable of compassion and sacrifice and endurance.' Faulkner's burdens were alcoholism and a Jim Crow society – his burdens and his people's burdens. His Nobel could have been a victory lap around the track, but he made it an opportunity to carry the burden another mile of the pilgrimage.

My pilgrimage is only 100 miles, and I wonder about other pilgrimages. What do they learn as they approach Santiago de Compostela at the end of the long walk in Spain? Is there a ritual – a mass, a song, a toast? They tell me that some pilgrims travel another 50 miles to Finisterre to look out on the Atlantic as they pare down their lives once more, burning something and throwing something into the ocean. David Whyte wrote:

> The road in the end taking the path the sun had taken,
> into the western sea, and the moon rising behind you
> as you stood where ground turned to ocean: no way
> to your future now but the way your shadow could take,
> walking before you across water, going where shadows go,
> no way to make sense of a world that wouldn't let you pass
> except to call an end to the way you had come,
> to take out each frayed letter you had brought
> and light their illumined corners; and to read
> them as they drifted on the late western light;
> to empty your bags; to sort this and to leave that;
> to promise what you needed to promise all along,
> and to abandon the shoes that brought you here
> right at the water's edge, not because you had given up
> but because now, you would find a different way to tread,
> and because, through it all, part of you would still walk on,
> no matter how, over the waves.
> (Whyte, 2014, p. 24)

References

Baez, J. (2014) *Amazing Grace*-11/3/1991Golden Gate Park (Official) on YouTube. Available at: https://www.youtube.com/watch?v=3ZZtai8poo0 (accessed 27 December 2019).

Bergson, H. (1946a) The perception of change. In: *The Creative Mind: An Introduction to Metaphysics*, Trans. M. Andison. Dover Publications, Mineola, New York, pp. 107–132.

Bergson, H. (1946b) Introduction to metaphysics. In: *The Creative Mind: An Introduction to Metaphysics*, Trans. M. Andison. Dover Publications, Mineola, New York, pp. 133–169.

Bergson, H. (1946c) Introduction (Part II): Stating of the problems. In: *The Creative Mind: An Introduction to Metaphysics*, Trans. M. Andison. Dover Publications, Mineola, New York, pp. 18–72.

Cohen, L. (1984) *Hallelujah: Leonard Cohen* on Genius.com. Available at: https://genius.com/Leonard-cohen-hallelujah-lyrics#note-2031942 (accessed 7 January 2020).

Dreyfus, H. and Kelly, D.K. (2011) *All Things Shining: Reading the Western Classics to Find Meaning in a Secular Age*. Free Press, London.

Eliot, T.S. (1942) Little Gidding. In: Eliot, T.S. (ed.) *Collected Poems:1909–1962*. Harcourt, Brace & Company, New York and London, pp. 173–210.

Elsner, J. (2020) Excavating pilgrimage. In: Kristensen, T.M. and Friese, W. (eds) *Excavating Pilgrimage: Archeological Approaches to Travel in the Ancient World*. Routledge, London, pp. 265–274.

Faulkner, W. (2006) William Faulkner's Nobel Prize acceptance speech. *Southern Cultures* 12(1), 71. DOI: 10.1353/scu.2006.0005.

Feldman, J. (2017) Key figure of mobility: the pilgrim. *Social Anthropology* 25(1), 69–82. DOI: 10.1111/1469-8676.12378.

Frankl, V.E. (1959) *Man's Search for Meaning*. Boston, Massachusetts.

Geertz, C. (1988) *Works and Lives*. Stanford University Press, Redwood City, California.

Graziano, F. (2019) *Historic Churches of New Mexico Today*. Oxford University Press, New York.

Grotstein, J. (2000) *Who Is the Dreamer, Who Dreams the Dream? A Study of Psychic Presences. The*. Analytic Press, Hillsdale, New Jersey.

Husserl, E. (1970) *The Crisis of European Sciences and Transcendental Phenomenology: An Introduction to Phenomenological Philosophy. Trans. D. Carr*. Northwestern University Press, Evanston, Illinois.

Jongmeewasin, S. (2015) Religious tourism, pilgrimage, and cultural tourism. Available at: http://www.academia.edu/12252237/Religious_Tourism_Pilgrimage_and_Cultural_Tourism (accessed 4 November 2019).

Jung, C.G. (1949) *The psychology of the child archetype. Collected Works. 9/1*. Princeton University Press, Princeton, New Jersey, pp. 151–181.

Jung, C.G. (1955) *Synchronicity: An Acausal Connecting Principle*. Routledge, London/New York.

Kalsched, D. (2013) *Trauma and the Soul: A Psycho-spiritual Approach to Human Development and its Interruption*. Routledge, London and New York.

Maslow, A. (1972) *The Farther Reaches of Human Nature*. Penguin Books, New York.

Morrison, S.S. (2016) Walking the walk: experiential learning, pilgrimage, and 'kynde knowynge'. *The Once and Future Classroom. XII.2*. Available at: https://once-and-future-classroom.org/XII/2/12-2morrison.pdf (accessed 18 November 2019).

Otto, R. (1958) *The Idea of the Holy*, Trans. J. Harvey. Oxford University Press, New York.

Oviedo, L., de Courcier, S. and Farias, M. (2014) Rise of pilgrims on the Camino to Santiago: sign of change or religious revival? *Review of Religious Research* 56(3), 433–442. DOI. 10.1007/s13644-013-0131-4.

Rappaport, R.A. (1999) *Ritual and Religion in the Making of Humanity*. Cambridge University Press, Cambridge, UK.

Steinbock, A.J. (2007) *Phenomenology and Mysticism: The Verticality of Religious Experience*. Indiana University Press, Bloomington, Indiana.

Tilley, C. (1994) *A Phenomenology of Landscape: Places, Paths, and Monument*. Berg, Oxford, UK..

Turner, V. and Turner, E. (1978) *Image and Pilgrimage in Christian Culture*. Columbia University Press, New York.

Van Gennep, A. (1908) *The Rites of Passage*, 2nd edn, trans. M. Vizedom and G. Cafee (1960). University of Chicago Press, Chicago, Illinois.

Weil, S. (2013) *Awaiting God: A New Translation of Attente de Dieu and Lettre á un Religieux*, Trans. B. Jersak. Fresh Wind Press, Abbotsford, British Columbia.

Whyte, D. (2014) *Pilgrim Finisterre*, 2nd edn.Many Rivers Press, Langley, Washington.

6 The Timeline of The Way of St James: Motivations and Impacts of the Pilgrimage on Personal Lives

Luis Alfonso Escudero Gómez[1], Lucrezia Lopez[2*], María Ángeles Piñeiro Antelo[2] and Rubén Camilo Lois González[2]
[1]*University of Castilla-La Mancha, Spain;* [2]*University of Santiago, Spain*

Introduction

The Way of St James[1] is a network of interconnected pilgrimage routes which begin in various places throughout Europe (Caucci, 1993), crossing the Iberian Peninsula to reach their final destination: the tomb of St James the Great in Santiago de Compostela, a city of approximately 100,000 inhabitants located in northwestern Spain and the capital city of the region of Galicia. The Way was included on the list of UNESCO World Heritage Sites in 1993, and this heritage listing was extended to include several additional routes in 2015. In 1985, the historical centre of Santiago de Compostela was also designated a World Heritage Site. It has been a meeting place for pilgrims since its beginnings some 11 centuries ago, representing a commercial hub and an area for spreading and sharing knowledge. It is an area in constant evolution that today holds a rich cultural heritage, both tangible (churches, hospitals, hostels, monasteries, bridges) and intangible, represented by the stunning natural landscapes that are to be seen along The Way (UNESCO, 2020). At present, walkers of The Way perform unique rituals and traditions, giving the walking experience a symbolic significance and emotionally connecting the pilgrim to the route. Thus, The Way exemplifies values, practices, rites and social projects that represent particular spatial dimensions. In addition, it also provides pilgrims with the opportunity to access renovated sites and, at the same time, socialize with other tourists and local people.

The main aim of this chapter is to analyse the motivations and impacts of the pilgrimage on pilgrims' personal lives. The research will address the importance of 'mobility' in this sacred and spiritual space in order to analyse its impacts on personal lives. It provides a fresh and original contribution to the concept of the pilgrimage, as well as a case study, The Way, both of which have been widely addressed in academic literature (Frey, 1998; Santos, 2002; Lois González and Somoza Medina, 2003; Tilson, 2005; Maak, 2010; Santos and Lois González, 2011; Lois González, 2013; Lois González *et al.*, 2015, 2016; Lois González and Santos, 2015; Santos, 2016; Lopez *et al.*, 2017; Lois González *et al.*, 2018; Pérez Guilarte and Lois González, 2018; Moscarelli *et al.*, 2020). Using both qualitative and quantitative methodologies, which are explained in the corresponding section, this research explores the multi-faceted nature of the

*Corresponding author: lucrezia.lopez@usc.es

pilgrims following The Way, who are inspired by a complex series of motivations. It is clear that the pilgrimage experience is currently associated with a multi-cultural and multi-religious content (Lois González, 2013). The examination concludes with an overview on how the pilgrimage experience survives and permeates in a phase that we can call 'The Post-Way', which can be considered as the key to its post-contemporary success. In other words, the pilgrimage does not end once the pilgrim has returned home, but rather it continues and revives.

Literature Review

A pilgrimage is a movement and a journey of people and ideas, maintaining the sacred value of the place alive and creating spatial relationships (Lois *et al.*, 2015). Going on pilgrimage is one of the oldest practices of humanity and is associated with a wide range of religious, spiritual and secular traditions. The sacred spaces which are the final destination of pilgrims, defined as venerated geographical locations with deep-rooted spiritual beliefs, practices and observances, are to be found throughout the world. Today the number of visitors to these sacred places is on the rise: more than 330 million people around the world set off on traditional pilgrimages (McIntosh *et al.*, 2018). At the same time, more and more public officers responsible for promoting and managing tourist destinations around the world are including pilgrimage routes in their tourist development plans in order to diversity their offer in an increasingly competitive tourism market (Olsen *et al.*, 2018).

The pilgrimage is an interdisciplinary field of scientific study. We have witnessed a boom in academic literature which attempts to analyse and understand the importance of pilgrimages (Santos and Lois González, 2011). A starting point could be the descriptive geographical study by Stoddard and Morinis (1997), which maps and describes sacred spaces and destinations around the globe. Coleman and Eade (2004) maintain that the pilgrimage is but one of the many forms of cultural mobility in the 21st century. They arrive at this conclusion by analysing the significance of the pilgrimage in Christian, Mormon, Hindu, Islamic and Sufi traditions, as well as in secular contexts, and from there create a new theory of the pilgrimage as a form of voluntary movement or migration. This voluntary movement creates cultural significance in a world that participates in an ongoing journey (Coleman and Eade, 2004).

Papers on pilgrimage have adjusted their focus to address today's spiritual trends. Of interest in this sense is Ivakhiv (2003) evaluating the pilgrimage phenomenon as part of a New Age spirituality, including the notions of sacred space, nature and the self. This author compares pilgrimage practices and sensory interactions with traditional tourist and merchandization practices. His case study is based in Sedona, in the state of Arizona (USA), but the theory is easily applied to The Way of St James, where pilgrimage practices exist alongside an aggressively promoted (Escudero Gómez, 2013) and merchandized (Lois González and Somoza Medina, 2003) tourist product. Ivakhiv's (2003) conclusion is that each pilgrim is searching for a personal expression of spirituality, and for this reason both the motivation for doing the pilgrimage and the impact that this has on the pilgrim's personal life are very heterogeneous.

Blom *et al.* (2016) and Lopez *et al.* (2017) also examine different ways of addressing the concept of pilgrimage in today's world. Today we can talk about post-secular pilgrims, who are fully contemporary individuals differing greatly from the past pilgrims as far as their values and perceptions are concerned. The post-secular situation is characterized by a significant shift from organized and normative religion towards subjective and experience-based spirituality. The motivations of post-secular pilgrims are varied, although returning to the sacred place and the desire to walk and reassert oneself in one's surroundings are essential for forging both personality and effort (Lorimer, 2011).

Studies from different scientific disciplines that make recurring reference to the attitudes, behaviours, motivations, opinions and perceptions of visitors to different places are quite frequent and are particularly salient in tourism-related research (Poria *et al.*, 2013). Several authors have analysed the visitors travelling to a particular tourist destination and studied their motives, including cultural and social concerns, as we see in Glasson *et al.* (1992) in the historical city of Oxford (UK), and Rudsari and Gharibi

(2019) in the Iranian city of Chalus. This type of work may also focus on the international visitors to a given country, as we see in Henkel *et al.* (2006) in their research on Thailand.

For decades, a clear relation has been described between pilgrimage and tourism (Cohen, 1974; Nolan and Nolan, 1989; Smith, 1992). The pilgrimage is generally seen as a precursor of modern tourism (Kark, 2001), creating a natural link between pilgrimage and tourism. Churches, cathedrals, religious routes and ways, and religious festivals and ceremonies are easily accessible cultural assets with tangible and intangible qualities, meaning that they are easily incorporated into the tourist market (Olsen *et al.*, 2018). Collins-Kreiner (2010) concludes that 'pilgrims cannot be differentiated from tourists, both kinds may be motivated to undergo an experience which will add more meaning to their lives' (p. 446). Notwithstanding, while the behaviour of pilgrims reflects the characteristics of contemporary tourists, they do indeed form part of a clearly differentiated group (Urry and Larsen, 2011; Harman, 2014). Given the strong link between pilgrimage and tourism today, it has been common practice in tourism studies to address and heed the opinions and perceptions of pilgrims. We can find relevant studies covering a wide variety of pilgrimage destinations, such as China (Wong *et al.*, 2013), India (Bagri *et al.*, 2009; Rana and Singh, 2011) and Europe (Eade, 1992). Similar studies exist in relation to visitors to sacred places such as cathedrals (Williams *et al.*, 2007; Hughes *et al.*, 2013).

Previous studies have indeed applied this methodology of analysing the opinions and perceptions of visitors on The Way. Frey (1998) did an anthropological study of The Way by conducting multiple interviews. Through this technique, the author arrives at the conclusion that the physical and mental journey of The Way offers pilgrims a closer-knit community, greater personal introspection and links to the past and to nature. Frey (1998) discovers that pilgrims often reflect extensively on their lives, and some institute profound changes. Others repeat the pilgrimage or join a pilgrims' association to keep their connection to The Way alive in their daily lives. Lois González (2013) has also analysed pilgrim responses from the perspective of geography. In this case, the primary focus was to establish the top reasons for travelling. According to the author's findings, these reasons include the search for unusual experiences, slow travel, contemplation of the landscape, learning about history and culture, and meeting new people (Lois González, 2013).

The Case Study: The Way of St James

The Way is one of the most important and significant Christian pilgrimages in the world (Lois *et al.*, 2015). Its origin goes back to the ninth century and follows the discovery of the remains of the body of the Apostle St James. It is an event known as the *inventio*, i.e. the discovery of the relics of a martyr in a place where there is no previous tradition of their existence (Santos, 2002; Baliñas Pérez, 2017). Thousands of people from all over Europe travel along the different routes to Santiago de Compostela. The route to Santiago is a linear way. It stretches along more than 700 km of northern Spain – nearly 2000 km if its four stems in France are to be included. It crosses many rural communities, villages and cities that flourished during the Middle Ages (Lois González and Somoza Medina, 2003).

From its beginnings in the Middle Ages it has held an unparalleled prominence in the transformation of Europe, with moments of great religious fervour and floods of pilgrims – in particular, the period between the 11th and the 14th centuries (Garton, 2018), and others in which its importance was less prominent, with pilgrims representing a small testimony of a very different past (Santos and Lois González, 2011). So despite the fact that pilgrimage to Santiago never ceased, a downturn was recorded beginning in the 15th century (Lois González, 2013). In our days, however, The Way has witnessed a profound transformation which began during the second half of the 20th century, when the strictly religious phenomenon gave way to a tourism product. Of great importance for this transformation are the Años Xacobeo, or jubilee years, which occur when the main holy day devoted to the apostle St James (25 July) falls on a Sunday. This happens at intervals of six, five, six and eleven years, with the next jubilee year occurring in 2021. The Way has always had spiritual connotations that embody a spirituality which has evolved over time, adapting to historical and social events (Santos, 2016). During the jubilee years of 1954, 1965 and 1971, The Way was officially promoted as both a spiritual and tourist route, offering numerous features of cultural appeal.

Since then, The Way has transformed into a pilgrimage route, a way for setting forth on ambulatory journeys with spiritual, religious, social and tourist connotations that attracts hundreds of thousands of people from around the globe each year (Lopez *et al.*, 2017). The spectacular increase in the number of pilgrims to Santiago over the last few decades has prompted the model of The Way to be cultivated in other contexts and countries where it was not present until only a few years ago. This has occurred with the Via Francigena that connects Canterbury to Rome, and the St Olav's Way between Sweden and the Norwegian city of Trondheim (Lois González *et al.*, 2018). Other routes in Spain have been created as well, drawing on the success of The Way, such as the Don Quixote route in Castilla-La Mancha, which stretches some 2500 km and offers ten different routes based on scenes from the adventures of Cervantes' beloved hero. This increase in visitors has, however, generated issues of sustainability, prompting the need to manage the number of visitors arriving at the final destination of The Way (Pérez Guilarte and Lois González, 2018). It has also called for management of the pilgrimage itself, both to Santiago and to other urban areas along The Way such as Leon and Astorga (Lois González and Somoza Medina, 2003). This growing risk of overcrowding and saturation can clearly be seen on The Way, and is especially patent in jubilee years, during high season (e.g. the month of August) and among Spanish pilgrims. The accumulation of people in certain sections of the route, in the Cathedral of Santiago and in the surrounding areas inside the historical city centre, has become a serious problem for those pilgrims who are primarily motivated by spirituality, finding time to think or to contemplate unique places (Lois González *et al.*, 2018).

The transformation process and the success of The Way have been so intense that it has been deemed 'the renaissance of The Way of St James' (Lois González, 2013) – a material renaissance, exemplified by the beautification and renovation of cultural heritage elements, the improvement of infrastructures and the installation of necessary conveniences such as a network of hostels and lodging, as well as a renaissance in the number of pilgrims, which has grown exponentially following decades of stagnation. As a result, the original motivation of faith has given way to a concept of living cultural heritage thanks to the restoration of routes, monuments and landscapes (Moscarelli *et al.*, 2020), and to the thousands of pilgrims who walk The Way inspired by any number of motives (Fig. 6.1).

The turning point was the jubilee year 1993, when a combination of various circumstances (explained in Santos and Lois González, 2011) and the common link of significant public investment (Lois González and Somoza Medina, 2003; Escudero Gómez, 2013) led both The Way and Santiago de Compostela to become a highly successful and original tourist phenomenon (Lois González and Somoza Medina, 2003; Tilson, 2005). Santos and Lois have described the elements that characterize The Way of St James, making it distinct, original and unique:

- slow movement;
- its markedly experience-based character;
- recovery of settings of sociability;
- the prominence of the landscape and of nature;
- the importance of ethnographic and gastronomic elements;
- the plural character of religion and culture;
- a low-cost tourist journey.

The Way, likewise, represents the monumentalization of sacred values. Sacred buildings and structures dedicated to the apostle St James (churches, monasteries and cemeteries), as well as other artistic, historical, human and cultural elements found along The Way, are clear examples of the monumentalization that sanctifies and symbolizes the space (Lopez *et al.*, 2017). This cultural and religious heritage forms part of its identity, constantly evoking the historical origins of The Way and the religious ideology that inspired it.

Notwithstanding, the rebirth of The Way has been focused primarily as a tourist destination (Lois González and Santos, 2015). The Way has a clear and final purpose and destination: the city of Santiago de Compostela, and, more specifically, the shrine of the apostle St James located inside the cathedral. Upon their arrival in the city, pilgrims perform other rituals as

Fig. 6.1. Pilgrims walking along Rúa de San Pedro, Santiago de Compostela. (Photo: the authors)

well, such as obtaining the Compostela, which is a certificate that shows that the holder has completed at least 100 km of The Way on foot or on horseback, or 200 km by bike. Pilgrims may also attend mass at the cathedral, purchase souvenirs and enjoy a hearty meal to recover their strength. The pilgrim therefore becomes a tourist visiting the city. In fact, once they reach Santiago, many pilgrims who have walked The Way no longer walk, they no longer sleep in hostels or travel in groups with other pilgrims, as they did along The Way, but rather begin to sleep in hotels and drive from place to place, often dedicating several days to conventional tourism in Galicia (Lois González and Santos, 2015). Alternatively, they may visit the entire north-western region of the Iberian Peninsula, including northern Portugal. There are also many people who choose to continue walking to Finisterre (Blom *et al.*, 2016), one of the westernmost points of continental Europe, where they can contemplate the immensity of the Atlantic Ocean, enjoy Land's End or simply gaze at the magical sunsets (Lois González and Santos, 2015).

Sources and Methodology

From a methodological standpoint, we combine theoretical and empirical methodologies in order to explore the motivations and impacts of the pilgrimage on pilgrims' personal lives. Based on these methodological guidelines, we have also used different sources which respond both to the need to take stock of the current situation and to the question posed in the introduction regarding the 'Post-Way' concept.

In the first phase, we reviewed the scientific literature related to the broader concept of pilgrimage and the way in which it has been addressed from the point of view of different disciplines, focusing particularly on its relationship with tourism. Regarding The Way in particular, we analysed research related to the evolution of the different routes and the impact on the city of destination, and, very specifically, on those studies focusing on pilgrims' rites, on the individual significance given to the pilgrimage, and on its power to transform the pilgrim's daily life.

During the second phase we used a qualitative methodology based on 39 semi-structured interviews conducted by the authors in Santiago de Compostela during the months of January and February 2020. This methodology was chosen because our goal was to discover and highlight information obtained through inductive reasoning. We also considered the importance of identifying and analysing cases with common characteristics in order to ensure that the results of successive interpretive processes could be compared in order to best reflect reality (Kurrant, 2019). The following criteria were used to select the pilgrims for interview: (i) origin: as all of the subjects interviewed were of Spanish origin, this allowed us to adopt a common interpretative approach; (ii) diversity of routes: we attempted to select subjects that had followed different routes in order to explore diverse perceptions of the pilgrim experience. We should point out that having done The Way more than once was not a criterion for exclusion, and among the pilgrims interviewed, 43.6% had done The Way more than once, adding value to their experience as well as to their perception of the route.

The questions were structured in three blocks: the first related to motivations, the second related to experiences during the pilgrimage, and the third related to determining whether having done The Way had brought about any changes in the pilgrim's daily life. For those who responded that they had indeed made changes, we determined what these were, and to what extent they had been implemented. The results were treated both quantitatively and qualitatively. In the first case, we quantified the results in order to obtain a global view of the field study sample, and from there to provide an overall assessment of the findings. In the second case, we focused on analysing and interpreting the pilgrims' responses using their exact words, as we felt that this was the clearest and most direct way of transmitting the feelings they wished to express. As we will see below, these responses have been coded alphanumerically, with 'P' referring to *pilgrim*, followed by the number of the interview.

Analysis and Results

In the study, 39 pilgrims were included, and the socio-demographic factors were as follows: 17 men and 22 women with ages ranging from 23 to 72. All resided in Spain, and all have done The Way at least once in the last 15 years (between 2006 and 2019). Just over half (54%) have done The Way only once (22 people), and of these, more than 68% preferred The French Way. In terms of motivation, the many different reasons reflect the inclusive dimension of The Way, as it attracts pilgrims driven by very diverse motivating factors. The following list shows some of the categories the pilgrims' responses fell into: (i) sport (hiking, trekking etc.); (ii) personal (promise, friendship, retirement), including the need to 'find time for myself and to discover my place in the world' (P36); (iii) cultural; (iv) religious; (v) spiritual; (vi) adventure and landscape. By and large, the responses were not exclusive, and pilgrims often gave more than one reason from various categories. Of the responses given, we would like to highlight the following, which are relevant for understanding the motivational inclusiveness of The Way (Moscarelli *et al.*, 2020): 'My reasons for doing The Way were sport, nature and meditation' (P15); 'My reasons for doing The Way were hiking and trekking with friends while spending time in nature. A feeling of serenity' (P28).

Nearly 70% of those interviewed stated that doing The Way had prompted subsequent changes in their daily lives. These changes were tangible even among pilgrims who had done The Way only once, as 72% said their lives changed in some way. This aspect holds the greatest interest for us, as the aim of our research is to determine whether a 'Post-Way' dimension exists, and if so, to describe its characteristics. We should point out here that there were also multiple responses regarding how doing The Way contributed to transforming the lives of the interviewees. Pilgrims referred to changes in their daily lives, such as learning to 'appreciate the little things' (P13). This feeling also led to a new outlook on life, as doing The Way helped them to 'see things from a different perspective; it opened up a range of values that I didn't consider so much before' (P1). Having time is an advantage for those pilgrims who, driven by personal motives, consider The Way as an opportunity for self-discovery, as expressed by one of the pilgrims who said that The Way helped to 'discover yourself, both the good things and the bad' (P4). This cathartic process, representing an 'opportunity for self-improvement and rebirth' (P38), is also an act of self-reconciliation. If our frantic lifestyle does not allow us time for reflection, doing The Way allows us to 'meditate more and think more about life' (P14). This experience-based reassessment helps to create a new set of values based primarily on 'discovering that we don't need so many things, learning to prioritize, to see religion from a different perspective...' (P27). While this pilgrim justified his response from a religious standpoint, the rekindling of the desire for a simpler life is quite common, as the experience of the pilgrimage requires us to 'learn to appreciate the little details in life and in our relationship with others' (P28).

It is interesting to point out that for some of the people interviewed, The Way has 'led to a greater appreciation of the effort made, both from an emotional and a motivational standpoint, as it teaches you to set personal goals' (P11), including from a physical point of view. Others discover their spiritual side, as 'The Way has become an activity that allows me to "reboot" after the typical stress of my daily life. I can stop rushing and enjoy a few days of silence, solitude, no phone, no media.... This has become my annual spiritual retreat, and I feel that it is necessary' (P39). 'It helps me to be aware of the spirituality that I already had, and to rest from the accumulated stress of life. It also helped me to find places where there were other people who shared my same values' (P15). This pilgrim also mentioned other values associated with The Way: the therapeutic value and sociability (Lopez *et al.*, 2017). In the first case, it helped him to value his own 'personal strength, be more positive, and less self-centred. I learned that anything can be done alone, and that I can take important decisions in my life' (P36). In the second case, it taught 'greater respect for people and for the environment' (P12), as well as 'how to appreciate the little details in life and in our relationship with others' (P28). In fact, improving social relations was also a motivating factor, as we can see from the following response: 'I did The Way for enjoyment and to meet people' (E5). The dichotomy of 'solitude and community' is nothing new in the experience of the pilgrim, nor are the two concepts mutually exclusive; rather, they are complementary and relational (Fig. 6.2).

Pilgrims' responses also pointed to the connection between nature and spirituality (Lopez *et al.*, 2017); in this case, being outdoors nurtured spirituality: 'The Way reinforced my spirituality, which derived in a greater importance and enjoyment of nature' (P28). This relationship is also evident when asking about motivation, as is clearly shown in the following response: 'I did The Way by bike with the aim of listening to the feelings that arise from doing The Way' (P4), which reinforces the idea that doing the pilgrimage for sport facilitates contact with nature, allowing the pilgrim to explore new feelings.

During the interviews, pilgrims were encouraged to reflect on the changes they perceived after doing The Way. One person interviewed said that he quit his job to do what he really enjoyed; other people talked about greater maturity, a stronger appreciation for spirituality, patience and respect for others, and friendship. As has been the case for centuries, The Way continues to be compared to the path a person follows through life: 'It helped me to redefine my priorities .... It reminded me that The Way is a metaphor of life itself, with moments in which we are walking through a lush, cool forest, moments in which we are struggling uphill and our strength is flagging, moments to relax and enjoy the little things, moments in which blisters and tendinitis remind us of our fragility... but we always arrive at the Portico of Glory' (P39).

Of those who said that The Way had not changed their life, half had done it just once, while the other half had walked it several times, up to seven times.

The majority of those interviewed stated that they did not maintain the personal relations they had made during The Way. Of the 13 pilgrims who do still maintain these friendships, the great majority have maintained them for one, two or three years, up to the present. Two respondents indicated that they had maintained the new friendships only for a brief period of one or two months.

With the aim of going even deeper into 'Post-Way experiences', we asked the pilgrims about their interest in reliving the experience,

Fig. 6.2. Pilgrims in Plaza do Obradoiro, Santiago de Compostela. (Photo: the authors)

and 28 people stated that they do not do any special activities in this sense. Those who do participate in activities related to their experience cited meeting up with people they met along The Way, membership in associations of friends of The Way, annual meals, hiking along The Way, and in two cases the pilgrims had got tattoos with Jacobean designs.

Conclusions

In conclusion, The Way is the main historical pilgrimage route in Europe (Lois González *et al.*, 2018). From the original religious character it held during medieval times, it has transformed into the cultural, tourist, monumental, spiritual and even sporting route that it is today, as pilgrims can follow the route on foot, bicycle or horseback. It is a polysemic route, an inclusive and complex space that has been transfigured from a purely religious route into a multifaceted tourist reality (Moscarelli *et al.*, 2020). The combination of these spatial and motivational transformations responds to the factors characteristic of the post-secular pilgrimage, such as slow movement, liminality and a sense of community. The Way discloses new worlds and new habits as pilgrims learn to make time for others. Among others, one of the main aspects is that the pilgrimage experience continues after the physical journey. As pilgrimage is a cathartic experience, it transforms pilgrims' lives and priorities. Indeed, as the field work has confirmed, The Way arouses an intense series of feelings and sensations, intensifying sociability among individuals who are generally caught up in a frantic model of urban life and encouraging pilgrims to return (Santos and Lois González, 2011). The significance of The Way is based on the close relationship between cultural heritage and individual will, making each journey or itinerary unique (Lopez *et al.*, 2017). It is characterized by being notably experience-based, preparing pilgrims for a change in their lives (Santos and Lois González, 2011). In this sense, it can be defined by its liminality, i.e. representing an intense experience that supposes a shift from one stage of life to another (Turner, 1974; Possamai, 2003).

One of the common questions when studying The Way is to try to quantify the number of days spent along the route. However, the duration of a pilgrim's experience can hardly be quantified and limited in time (Lopez, 2013). As our fieldwork shows, the pilgrimage is an experience of personal identity, the duration of which cannot be quantified if we assume that it impacts on people's lives even after they have finished their journey. As a consequence, it is also difficult to fix a timeline of the experience, as it is present and effective each time the pilgrims reflect on what they have learned. Considering the responses provided in the previous section, and if we recognize the didactic dimension of The Way as a set of lessons learned by the pilgrims, the next step is to conceptualize a 'Post-Way'. This is a phase which is chronologically subsequent to The Way and is characterized by its relation to the experience of doing The Way and the impact it has on daily behaviour and relations.

Although this topic has been widely addressed in research, we feel that The Way still has much to offer to sociological research, precisely because it has become a phenomenon of international reference. The study looks closely at the personal and subjective changes and transformations perceived by the pilgrims themselves, as well as the impact that the pilgrimage has had on their daily lives. We should focus more closely on the personal and spiritual evolution among pilgrims in order to expand on the impact that The Way has on people's daily lives. Understanding these aspects is of great value both to researchers in this field and for those who plan and manage pilgrimage routes. Comprehending the motivations for doing The Way and its impact on the personal lives of the pilgrims is fundamental for successfully planning and managing a sustainable pilgrimage route and making the experience as enriching as possible for those who decide to do it.

Note

[1] Henceforth we will use the abbreviation 'The Way' to refer to The Way of St James.

References

Bagri, S., Gupta, B. and George, B. (2009) Environmental orientation and ecotourism awareness among pilgrims, adventure tourists, and leisure tourists. *Tourism: An International Interdisciplinary Journal* 57(1), 55–68.

Baliñas Pérez, C. (2017) The origins of the Inventio sancti Iacobi and the making of a Kingdom. In: Pazos, A.M. (ed.) *Translating the Relics of St James. From Jerusalem to Compostela*. Routledge, London, pp. 75–87.

Blom, T., Nilsson, M. and Santos, X. (2016) The way to Santiago beyond Santiago. Fisterra and the pilgrimage's post-secular meaning. *European Journal of Tourism Research* 12, 133–146.

Caucci, P. (1993) *Santiago: la Europa del peregrinaje*. Lunwerg, Barcelona, Spain.

Cohen, E. (1974) Who is a tourist?: a conceptual clarification. *The Sociological Review* 22(4), 527–555. DOI: 10.1111/j.1467-954X.1974.tb00507.x.

Coleman, S. and Eade, J. (2004) *Reframing Pilgrimage: Cultures in Motion*. Routledge, London.

Collins-Kreiner, N. (2010) Geographers and pilgrimages: changing concepts in pilgrimage tourism research. *Tijdschrift voor economische en sociale geografie* 101(4), 437–448. DOI: 10.1111/j.1467-9663.2009.00561.x.

Eade, J. (1992) Pilgrimage and tourism at Lourdes, France. *Annals of Tourism Research* 19(1), 18–32. DOI: 10.1016/0160-7383(92)90104-W.

Escudero Gómez, L.A. (2013) La imagen Urbana de Santiago de Compostela (España), un estudio de su representación pública, mediática, promocional Y artística. *Boletín de la Asociación de Geógrafos Españoles* 62, 265–294. DOI: 10.21138/bage.1578.

Frey, N. (1998) *Pilgrim Stories: On and Off the Road to Santiago*. University of California Press, Berkeley, California.

Garton, T. (2018) The experience of medieval pilgrims on the route to Santiago de Compostela, Spain: Evidence from the 12th-centruy *Pilgrim's Guide*. In: McIntosh, I., Quinn, E.M. and Keely, V. (eds) *Pilgrimage in Practice. Narration, Reclamation and Healing*. CABI, Wallingford, UK, pp. 1–16.

Glasson, J., Thomas, J., Chadwick, A., Elwin, J., Crawley, R. *et al.* (1992) *Oxford Visitor Study*. Oxford Brookes University, Oxford, UK.

Harman, L.D. (2014) *A Sociology of Pilgrimage. Embodiment, Identity, Transformation*. Urus, London, Canada.

Henkel, R., Henkel, P., Agrusa, W., Agrusa, J. and Tanner, J. (2006) Thailand as a tourist destination: perceptions of international visitors and Thai residents. *Asia Pacific Journal of Tourism Research* 11(3), 269–287. DOI: 10.1080/10941660600753299.

Hughes, K., Bond, N. and Ballantyne, R. (2013) Designing and managing interpretive experiences at religious sites: visitors' perceptions of Canterbury Cathedral. *Tourism Management* 36, 210–220. DOI: 10.1016/j.tourman.2012.11.022.

Ivakhiv, A. (2003) Nature and self in new age pilgrimage. *Culture and Religion* 4(1), 93–118. DOI: 10.1080/01438300302812.

Kark, R. (2001) From pilgrimage to budding tourism: the role of Thomas Cook in the rediscovery of the Holy Land in the nineteenth century. In: Wagstaff, M. and Searight, S. (eds) *Travellers in the Levant: Voyagers and Visionaries*. ASTEN, London, UK, pp. 115–174.

Kurrant, C. (2019) Biographical motivations of pilgrims on the Camino de Santiago. *International Journal of Religious Tourism and Pilgrimage* 7(2), 11–18.

Lois González, R.C. (2013) The *Camino de Santiago* and its contemporary renewal: pilgrims, tourists and territorial identities. *Culture and Religion* 14(1), 8–22. DOI: 10.1080/14755610.2012.756406.

Lois González, R.C. and Santos, X.M. (2015) Tourists and pilgrims on their Way to Santiago. Motives, Caminos and final destinations. *Journal of Tourism and Cultural Change* 13(2), 149–164. DOI: 10.1080/14766825.2014.918985.

Lois González, R.C. and Somoza Medina, J. (2003) Cultural tourism and urban management in northwestern Spain: the pilgrimage to Santiago de Compostela. *Tourism Geographies* 5(4), 446–460. DOI: 10.1080/1461668032000129164.

Lois González, R.C., Paül, V., Pazos-Otón, M. and Santos, X.M. (2015) The Way of Saint James: a contemporary geographical analysis. In: Brunn, D. (ed.) *The Changing World Religion Map*. Springer, New York, pp. 709–732.

Lois González, R.C., Castro Fernández, B.M. and Lopez, L. (2016) From sacred place to monumental space: mobility along the Way to St. James. *Mobilities* 11(5), 770–788. DOI: 10.1080/17450101.2015.1080528.

Lois González, R.C., Santos, X.M. and Taboada de Zúñiga Romero, P. (2018) The Camino de Santiago de Compostela: the most important historic pilgrimage way in Europe. In: Olsen, D.H. and Trono, A. (eds) *Religious Pilgrimage Routes and Trails*. CABI, Wallingford, UK, pp. 72–86.

Lopez, L. (2013) How long does the pilgrimage tourism experience to Santiago de Compostela last? *International Journal of Religious Tourism and Pilgrimage* 1, 1–15.

Lopez, L., Lois González, R.C. and Castro Fernández, B.M. (2017) Spiritual tourism on the Way of Saint James the current situation. *Tourism Management Perspectives* 24, 225–234. DOI: 10.1016/j.tmp.2017.07.015.

Lorimer, H. (2011) Walking: new forms and spaces for studies of pedestrian. In: Cresswell, T. and Merriman, P. (eds) *Geographies of Mobilities: Practices, Spaces, Subjects*. Ashgate, Farham, UK, pp. 19–35.

Maak, K. (2010) *Der Jakobsweg als Faktor Touristischer Entwicklung in Ländlichen Regionen*. Verlag, Hamburg, Germany.

McIntosh, I., Quinn, E.M. and Keely, V. (2018) Introductions: pilgrimage in practice-narration, reclamation and healing. In: McIntosh, I., Quinn, E.M. and Keely, V. (eds) *Pilgrimage in Practice. Narration, Reclamation and Healing*. CABI, Wallingford, UK, pp. ix–0.

Moscarelli, R., Lopez, L. and Lois González, R.C. (2020) Who is Interested in developing the Way of Saint James? The pilgrimage from faith to tourism. *Religions* 11(1), 24. DOI: 10.3390/rel11010024.

Nolan, M. and Nolan, S. (1989) *Christian Pilgrimage in Modern Western Europe*. University of North Carolina Press, Chapel Hill, North Carolina.

Olsen, D.H., Trono, A. and Fidgeon, P.R. (2018) Pilgrimage trails and routes: the journey from the past to the present. In: Olsen, D.H. and Trono, A. (eds) *Religious Pilgrimage Routes and Trails*. CABI, Wallingford, UK, pp. 1–14.

Possamai, A. (2003) Alternative Spiritualities and the cultural logic of late capitalism. *Culture and Religion* 4(1), 31–45. DOI: 10.1080/01438300302807.
Pérez Guilarte, Y. and Lois González, R.C. (2018) Sustainability and visit management in tourist historic cities: the case of Santiago de Compostela, Spain. *Journal of Heritage Tourism* 12(6), 489–505.
Poria, Y., Reichel, A. and Cohen, R. (2013) Tourists perceptions of World Heritage Sites and its designation. *Tourism Management* 35, 272–274. DOI: 10.1016/j.tourman.2012.02.011.
Rana, P.B. and Singh, R.P.B. (2011) Perceptions and images of tourists and pilgrims in Banaras. In: Singh, R.P.B. (ed.) *Holy Places & Pilgrimages: Essays on India*. Shubhi Publications, New Delhi, India, pp. 165–206.
Rudsari, S.M.M. and Gharibi, N. (2019) Host-guest attitudes toward socio-cultural carrying capacity of urban tourism in Chalus, Mazandaran. *Journal of Tourism & Hospitality Research* 6(2), 31–47.
Santos, X.M. (2002) Pilgrimage and tourism at Santiago de Compostela. *Tourism Recreation Research* 27(2), 41–50. DOI: 10.1080/02508281.2002.11081219.
Santos, X.M. (2016) The Way of Saint James as an event: politics and nation. *Journal of Policy Research in Tourism, Leisure and Events* 8(3), 233–248. DOI: 10.1080/19407963.2016.1214960.
Santos, X.M. and Lois González, R.C. (2011) El Camino de Santiago en El contexto de Los nuevos turismos. *Estudios Turísticos* 189, 87–110.
Smith, V.L. (1992) Introduction. The quest in guest. *Annals of Tourism Research* 19(1), 1–17.
Stoddard, R. and Morinis, A. (eds) (1997) *Sacred Places, Sacred Spaces: The Geography of Pilgrimage*. Louisiana State University, Baton Rouge, Louisiana.
Tilson, D.J. (2005) Religious-spiritual tourism and promotional campaigning: a Church-state partnership for St. James and Spain. *Journal of Hospitality & Leisure Marketing* 12(1-2), 9–40. DOI: 10.1300/J150v12n01_03.
Turner, V. (1974) *Dramas, Fields and Metaphors: Symbolic Action in Human Society*. Cornell University Press, Ithaca, New York.
UNESCO (2020) Routes of Santiago de Compostela: *Camino Francés* and Routes of Northern Spain. Available at: https://whc.unesco.org/en/list/669/ (accessed 27 February 2020).
Urry, J. and Larsen, J. (2011) *The Tourist Gaze 3.0*. Sage, London.
Williams, E., Francis, L., Robbins, M. and Annis, J. (2007) Visitor experiences of St Davids Cathedral: the two worlds of pilgrims and secular tourists. *Rural Theology* 5(2), 111–123. DOI: 10.1179/rut_2007_5_2_004.
Wong, C.U.I., McIntosh, A. and Ryan, C. (2013) Buddhism and tourism: perceptions of the monastic community at Pu-Tuo-Shan, China. *Annals of Tourism Research* 40, 213–234.

7 The Multiple Views on the Values and Identity of the Pilgrimage to Santiago

Xosé M. Santos*
Department of Geography, University of Santiago de Compostela

Introduction

On the Pilgrim's Way to Santiago, as on other pilgrimage routes, there is a debate about the motivations of pilgrims and the frontiers between tourism and religion (Cohen, 1992; Maclean, 2009; Collins-Kreiner, 2010, 2018; Terzidou *et al.*, 2018; Timothy and Olsen, 2006). The literature on the Pilgrim's Way to Santiago de Compostela is extensive (Murray and Graham, 1997; Graham and Murray, 1997; Frey, 1998; Magry, 2015; Lois-González and Santos, 2015; Sanchez, 2016; Kim *et al.*, 2016). Research has addressed numerous aspects of the Jacobean phenomenon in an attempt to understand the reasons for its success. Post-Secularism is one of the central axes on which some of the studies are focused (Lois-González and Santos, 2015; Nilsson and Tesfahuney, 2016; Blom *et al.*, 2016).

The enormous popularity that the Way has achieved, with more than 300,000 people moved by all kinds of reasons, and arriving every year in Santiago to receive the certificate of completion, the Compostela, has raised doubts about the actual concept of pilgrim. The traditional definition refers to a journey of a religious nature, regardless of the motivations that complement it, or if the religiosity is equally regarded as spirituality. Barber's (1993, p. 1) classical approach says that it is 'a journey resulting from religious causes, externally to a holy site, and internally for spiritual purposes and internal understanding'. In this sense, the Chapter of the Cathedral of Santiago, the body in charge of the handing over of the compostelas, states that to get this certificate, among other things, the applicant must 'make the pilgrimage for religious or spiritual reasons, or at least an attitude of search'.[1]

The statistics collected by the Pilgrim Reception Office[2] list very few options when it comes to selecting the motivations: Religious and others; Religious; Non-religious. That is, in all of them the central axis is religion. In the case of those people who declare non-religious reasons, a different certificate to the Compostela is issued. In any case, the volume of this last group ranges between 5% in a holy year and slightly less than 10% in the other years. However, they are accounted for by the Church in Santiago in the total amount of pilgrims.

There are other works, outside the scope of the Church, that, without elaborating on obtaining the Compostela, depend on the motivations of the people who do the Way (Alvarez, 1999; CETUR and Xacobeo, 2007–2010; Amaro *et al.*, 2018). They usually draw a more complex picture in which religion is much less significant than spirituality, contact with nature, art or even reasons related to sports or simple fun. It

*xosemanuel.santos@usc.es

is also noted that, frequently, motivations intersect, change and evolve, which makes it difficult to reduce them to a single answer.

The Governance of the Way

Despite its relevance, this is a subject very little studied (Santos, 2017). The motivations stated by people who undertake the ways to Santiago are a reflection of the agents that manage the different routes to Compostela. In summary, they can be reduced to three: international agencies, public administrations, and the Church. In all these cases there are relationships and contradictions that draw a complex picture in which the pilgrim is a subject repeatedly trapped in a maze of motivations.

The international agencies

Much of the international recognition of the Way of Saint James has its background in the UNESCO declaration of the city of Santiago in 1985, and the Way in 1993, as World Heritage Sites, as well as the recognition of the Way as the First European Cultural Route in 1987 by the Council of Europe. These are relevant facts for at least two reasons. First, because they link the Way with culture and not exclusively with religion. Secondly, because it opened a new heritage category for cultural routes, and that UNESCO incorporated in 2005 into its Operational Guidelines for the Implementation of the World Heritage Convention[3].

In both 1987 and 1993, the Way of Saint James was beginning to take its first steps towards its popularization as a pilgrimage route. The actions that had been carried out in the 1980s had a strong religious bias, but also cultural and European ones. The best example of this last connection was the visit of John Paul II to Santiago in 1982 when his speech of 9 November was seen as a 'European event'[4] in which the Jacobean route was extolled as a religious route that had contributed to the creation of the identity of Europe. The Pope of Rome was dressed in traditional clothing and symbols that identify the pilgrims with Compostela.

The 1987 declaration of the Council of Europe classified the Way of Saint James as a cultural route, resulting in the creation of the European Institute of Cultural Routes in 1998, and opening the door to the recognition of many other routes – up to 38 as of December 2019. The value attributed to these routes has to do with the European identity, with the existence of a plural and diverse common past that currently contributes to improving the quality of life and intercultural dialogue. They are regarded as educational tools of Europe's memory and history that help draw a better future.[5]

But if the declaration of the Council of Europe was important, even more so was the recognition by UNESCO in 1993, the year of the boom of the Way, with almost 100,000 pilgrims. Regardless of the media relevance, we are interested in pointing out the cultural values that are associated with the Way. It is important to indicate that from that date a debate began on the relevance of cultural routes as a new category of heritage that was not a sum of elements, but represented a unit with an intangible historical significance. The joint work of UNESCO and ICOMOS, in which an International Scientific Committee of Cultural Routes was created, derived from the Icomos Charter on Cultural Routes,[6] which served as a reference for the definition of cultural routes.

Despite the differences that exist between UNESCO and the Council of Europe on cultural routes, the truth is that the Way of Saint James was, in both cases, the starting point for the recognition of many other routes. This declaration, still valuing the religious origin of the Way, highlighted the cultural value that the route has had from the Middle Ages to the present, as a meeting place for people from all over the continent and, consequently, for exchange in its broadest sense (Fig. 7.1).

Public administrations

If international agencies have highlighted the cultural values of the Way of Saint James, public administrations have focused more on purely economic aspects, mainly on tourism. Because the various routes cross several countries and regions, we will focus mainly on the Spanish case

Fig. 7.1. Cathedral of Santiago de Compostela. (Photo: D. Liutikas)

and, more specifically, on Galicia, the region in which the city of Santiago is located and where all the routes converge. In Spain, tourism and cultural policy is managed by the Autonomous Regions and therefore everything concerning the Way of Saint James depends on the regional administration of Galicia.

Although the official stance of the Galician authorities with regards to the Way as a cultural route serves to position Galicia in a relevant place in Europe, in practice it points in a different direction – that is tourism. There are several arguments that support this statement. First, the leap forward in 1993 was based on tourist pillars. The Spanish political decentralization of the 1980s allowed the autonomous regions to design their own political strategies that, in the Galician case, were marked by efforts to differentiate themselves from the Spanish sun-and-beach offer. The Europeanism of those years, as well as the celebration of the Olympic Games in Barcelona and the Universal Exhibition of Seville in 1992, facilitated the organization of the Holy Year as an event (Santos, 2016). The creation of a brand (Xacobeo) and a mascot (Pelegrin) evidenced this approach.

Secondly, the successive governments of Galicia have used the Way as an effective form of tourism promotion. The creation in 1991 of the Management Company of the Jacobean Plan, initially linked to the Department of Institutional Relations, showed that, despite everything, tourism and the Way were different things. However, over the years there has been a gradual integration of the Jacobean department into tourism. Although the public company Sociedad Anónima de Gestión del Plan Xacobeo currently continues to exist, organically it depends on the Galician Tourism Agency, and its functions have been very limited.

Thirdly, and giving continuity to what is stated in the previous paragraph, the Way is a fundamental piece of the Galician government tourism policy. In the 2020 Tourism Strategy, the Way of Saint James appears as one of the four differentiating features and is mentioned, directly or indirectly, throughout the entire document.[7] Through the different routes, the objective is to contribute to the internationalization of tourism in Galicia and to favour the growth not only of Santiago, as the final destination, but also of the territories, mostly rural, crossed by the route.

However, as already mentioned, the discourse continues to maintain an important cultural bias. It can be verified in the Galician 2015–2021[8] Directing and Strategic Plan of the Way of Saint James, in which the conservation of the roads, the preservation and enhancement of heritage values or the quality of the pilgrim's customer service are some of the priority lines. However, in that same document there is a forecast of 464,000 pilgrims by 2021 and actions with a clear tourism-based approach are proposed.

The Church

Due to its religious origin, the Catholic Church, especially the Church of Santiago, has always been very involved in all the issues related to the pilgrimage to Compostela. Its vision, as expected, is orientated to aspects of a religious nature. At the practical level, it is the institution that sets the standards to certify the pilgrim status through the regulation of elements such as the distance travelled, the motivation, or the use of the official credential. In addition, many of the rites related to the pilgrimage are directly connected to the Church, such as the *botafumeiro* or the pilgrim's mass. In the same way, a very important part of the heritage that is found along the different routes is religious and therefore depends on this institution.

In regards to the discourse, there is a continuing concern for the preservation of the Christian values of the Way, both religiously and spiritually. The Pilgrim Reception Office[9] is the place where the Compostela is handed out. It is also responsible for distributing or, where appropriate, certifying the pilgrim's credential. There is comprehensive information on its website about the meaning of the pilgrimage. The Church also has the role of criticizing the excessive touristization of the route. The Cardinal of Santiago did that in 1975, although without rejecting the practice of tourism (Rodríguez, 2004), and it was repeated in December 2019 in the presentation of the 2021 Jubilee Year by the Archbishop of Santiago. In this latter presentation there was a pastoral letter[10] in which the Archbishop spoke of the Christian sense of the pilgrimage, above values such as landscape charm or historical heritage.

Despite this discourse, the Church clearly benefits from the economic income derived from the pilgrimage. The credential, the Compostela, an increasing number of diplomas (distance, welcome, visit) or the multiple souvenirs that can be purchased in the Church official stores show an active participation in the tourism business. In addition, there is the entry to different dependencies of the cathedral, some of them without entry restrictions until a few years ago, or the many accommodation establishments that run along the Way and in the city of Santiago.

The Friends of the Way Associations

We have so far seen three hegemonic actors in the governance of the Way that have transversal discourses, sometimes coincidental, but also different. Although the discourse does not always converge with action, each of these actors represents cultural, tourist and religious values. This same multiplicity is what can be observed in the pilgrims whose motivations, as we have said, are diverse and changing in space and time.

Together with these three key actors, there is another, rarely visible, one which largely reflects the visions of the previous ones and combines them in a kind of tradition-based pragmatism. It is about the associations of Friends of the Way of Saint James. Their origin dates back to the year 1950 when the first of them was born in Paris, thus resuming a medieval tradition of support for the pilgrim (Rodríguez, 2018). Later, in the 1960s, others appeared until its definitive expansion in the 1990s. Initially, their creation had an obvious historicist content in which the recovery of the Jacobean tradition was at the centre of their work. In this sense, the participation of researchers specializing in the Middle Ages, or the organization of pilgrimages recreating those of several centuries ago, were significant.

Gradually these associations were undertaking a more relevant role, with actions such as the identification of traditional routes, many of which had been lost in the middle of urbanization processes, the recovery of identification documents for pilgrims or support for pilgrims under different formulas, from the information given to the reception of the pilgrim. The latter is going to be linked in a very special way to the network of shelters that have been appearing and that currently constitutes one of the most unique elements of the entire route, based on non-commercial accommodation principles.

There are more than 300 associations scattered throughout the world although with very different characteristics. In Spain, for example, there is a great atomization compared to other countries such as the USA, Canada, Australia or the UK that have a very small number, one or two, but which are larger and have greater capacity for organizing activities. Based on this, their functions are different, although they are also conditioned by their location. For example, many of the European organizations are responsible for the conservation and maintenance of the routes. Other larger ones, generally from outside continental Europe, develop volunteer programmes in the shelters. Historical research is another point that differentiates them.

However, they all have elements in common that refer to the distribution of the pilgrim's credentials or giving information about the Way. This last support activity is very important since the members of the associations, often elderly people, start an experience that consists of crossing, in many cases, depopulated rural territories with their own traditions or even a language they do not know. Less common is the help to return to everyday life. This action is offered only by some associations, but it is of great relevance as it contributes to facilitating the return to everyday life, by incorporating the values acquired or reinforced along the pilgrimage route (Lopez, 2013).

In this sense, the associations of Friends of the Way represent a different view from that of international agencies, public administrations or the Church. Their main objective is to spread the Jacobean culture and help maintain the traditional sense of pilgrimage. In no case is this associated with religion but rather with spirituality. This allows them to maintain their independence with respect to the Church but also generates certain misgivings from the Church. It is true that we often find religious people in associations, and were even, in some cases, promoted by priests, although not always Catholic.

Also the objectives of the associations are not coincident with those of public administrations. In fact, one of their main complaints is the excessive commercialization of the route that generates massification, banalization and tourism processes. These problems are evidenced mainly in the last 100 km, which is the minimum length required to obtain the Compostela, according to the regulations set by the Church. There is a certain disappointment among them in regards to the idea of the Way becoming a tourist product, although they mostly still believe it is a pilgrimage route (Fig. 7.2).

The key tools for the dissemination of these traditional values are actions such as talks or general information provided to their members. In some cases, especially in those countries furthest from the final destination, the number of credentials handed out by the associations may represent up to 50% of the pilgrims from that country. This implies that many pilgrims begin the route knowing the historical and cultural significance of the route; that is, they understand that the Way of Saint James is not a tourist product, or at least not exclusively, and that it offers opportunities to find oneself, to meet other people and to be part of a community that shares values, at least while the journey lasts.

Another of the tools used by some associations, in particular the larger ones, is the volunteer *hospitaleros* programme. After a small training course, these people volunteer mainly in their own shelters, public ones or those run by other associations or institutions. Among the support actions that they carry out, their work is mainly aimed at transmitting those traditional values that define the route, contributing to reinforce a vision that is neither religious nor purely tourist. The role of these volunteers is very important for the economic sustainability of many shelters that are based on non-commercial pilgrim reception, but especially for the dissemination of the Way's identity and unique values.

The associations of Friends of the Way are therefore in a different position from the other

Fig. 7.2. Pilgrims and Tourists in front of Cathedral of Santiago de Compostela. (Photo: D. Liutikas)

institutions. Their support for the Jacobean culture and tradition could make us think of their excessively orthodox character in terms of their idea of the Way and of the pilgrimage. However, their perspective is compatible with a pragmatic view of the pilgrimage, bringing associations closer to the complex reality of the pilgrimage. Also, their work alongside the Way means that they have a much closer knowledge of the pilgrim than the Church or the public administrations may have.

A reflection of that pragmatism resides in their opinion about some more or less controversial aspects of the Way. One of them is the 100 km necessary to acquire the Compostela. This measure, implemented at the beginning of the pilgrim boom to prevent anyone from requesting Compostela, has become a problem since more than 40% of pilgrims start the route around the 100 km mark, especially in the summer months, thus contributing to the overcrowding of some routes and services in certain stages and days of the year. Although the associations have identified this problem, they are also aware that the time and physical conditions necessary for longer journeys are not always available. Being aware that more days on the Way and fewer people can help to better soak up the values of the route, they also state that these values do not depend exclusively on the number of days or the volume of people, but that they reside inside each person.

In the field of motivation, this pragmatism is also observed. Normally, the statistics, both the official ones and those analysed in different investigations, are quite limited in terms of the number of responses. The associations, far from a religious and orthodox vision of the Way, adopt a much more enriching perspective where the pilgrimage experience includes a constant flow of emotions that change, evolve or transform, just as in daily life. The Jacobean culture and being a pilgrim does not only imply a religious (or rather spiritual) motivation, but it can also be lived together without contradictions or less *deep* reasons, such as sports or having a good time. This flexible look serves to appreciate the potential that everyone has to be a pilgrim at some point during the journey. Hence they condemn the concept of false pilgrim.

In academic literature there is a debate about the false and the true pilgrim (Dunn *et al.*, 2016; Olsen and Wilkinson, 2016). The majority of associations agree that this dichotomy is not real. In fact, it is common that alongside the route, pilgrims who are linked to the associations use, according to their needs and abilities, standardized commercial accommodation or services such as the transfer of backpacks. In their opinion, these actions do not turn them or anyone else into false pilgrims. The concept of pilgrim is much more intimate. Even the fact that they support non-commercial pilgrim reception does not mean that by resting in a hotel they are no longer pilgrims.

Values and Identity on the Way

These aforementioned institutions, to some extent, collect the values and identity of the Way. The purely tourist narrative as opposed to the strictly religious as well as cultural ones, give way to a multiplicity of motivations that overlap and follow each other continuously. To exemplify this complexity, we have selected a newspaper article that was published in 2019 in the newspaper *El País*,[11] the most widely read newspaper in Spain with a strong audience in Latin America. The piece received 354 comments mostly criticizing the author's words.

The title of the article was 'The Antipilgrims' and had the subtitle 'Walk for 6 hours a day, for more than a week, to reach the cathedral of Santiago and fail'. The author recounts, with a sarcastic tone, her experience on the Way in search of what is supposed to represent the pilgrim's main motivation, spirituality. At one point in the text it says, 'I search inside me during the 6 hours of walking each day between mountains, rivers, and also asphalt, but I don't find questions, nor answers. Not the slightest depth of thought or spirituality.' The conclusion she arrives at is that the only thing that the Way gave her were blisters on her feet and a 1.5 kg weight increase. Hence, she talks about failure in relation to what should be expected from this pilgrimage experience.

Readers' reactions, in the form of comments, reflect, in general, a critical attitude towards the content of the text, although there are some nuances that we will try to highlight. It seems normal to us that, in general, there is an attitude contrary to the ideas expressed by the author of the article. We consider that whoever read it did so because he or she was interested in the Way of Saint James. It is therefore easy to notice that many of the comments come from people who have done the pilgrimage route. We have divided the responses to the article into two different categories. First of all there are those who mainly agree with the author. The other comments, the majority, are from people who are critical of the text, although there is a great diversity of opinions that, to a large extent, reflect on the complexity of the Way and even the different views of the institutions involved in the governance of the route.

The comments supporting the text are few but express an interesting opinion about the Way. They highlight that the article provides a different vision of the route, away from the majority focused on its spiritual and cultural value, and thank the author for not taking the subject too seriously. Another comment is nearer to reality, as someone who has done it several times (the author) can confirm. Finally, another comment speaks of the *instagram pilgrimage*.

Although in the previous paragraph we have referred to those comments that explicitly praise the content of the article, there are many others that do so indirectly and that connect with the idea that the Way is, fundamentally, a tourist product. In this sense, they affirm that the business, the economic profit, represents the predominant vision. There are even those who go further in responding to those who point out that the Way is barely hiking, saying that it is not even that because it is closer to mass tourism.

The debate about motivations is thus reached, which is, perhaps, one of the most interesting debates among the people leaving comments. We have seen that in some cases they refer to the concept of tourist, understood in their purely recreational perspective. There are those who identify the Way with hiking, giving them the opportunity to exercise but also incorporate a spiritual value since walking

Fig. 7.3. On the Way to Santiago. (Photo: D. Liutikas)

in nature acts as a tool for personal reflection. This last idea is not always regarded as valid. In fact, one of the comments says that without religious motivation there are no differences with a long distance path, implying that hiking is only a sport.

The religious values identified with the Way are the subject of numerous comments. Its authors are critical of the content of the article and even become quite belligerent with other comments referring to Christian phobia or the ability to pervert the figure of the pilgrim. Sometimes these are comments made from non-believing positions. For example, one says that, 'For those of us who do not have religious motivation, I do not understand what the Way can offer. It would be a valuable experience for someone very religious who really takes it as a pilgrimage, I don't deny that. The rest of us do not belong there.' Another comment insists that without religious motivation, what remains is hiking, and therefore he does not understand those people who do the Way without those religious reasons (Fig. 7.3).

However, most of the time these comments start from religious approaches. Thus, another comment highlights that the specificity of the Way is the religious pilgrimage: 'This is the only pilgrimage that makes sense.... The other walkers participate in an absurd wander.... The cultural pilgrimage is meaningless but it is a good business for many.' Along the same lines, another person expresses that, without its religious value, doing the Way is absurd, recognizing that many of those who do so only seek to leave the house and have a good time. Finally, another entry believes that pilgrims with religious motivations are a minority, while for the majority the reasons that lead them to walk are multiple including tourism, sport or socializing. In any case, there would be more than enough reasons to do the route, if you like to walk.

Finally, the greatest number of comments, generally critical of the author, have an impact on the complexity of the Way, in which there can be multiple reasons for undertaking it, but in which spirituality plays a very special role. Along the same lines of the discourse of the associations of Friends of the Way, it is said that 'On the Way everyone fits', which is 'a summary of what life is' or that 'there are as many ways to make the Way as you want, because the Way are many lives and many personal stories'. The person-centred idea is quite frequent and indicates that the Camino is a way of finding oneself. It is a challenge, and as such depends on the intentionality with which you do it.

One of the comments goes further by stating that whatever the reason for beginning the Way, it ends up being a spiritual experience, because its essence is to be a path into oneself. The word spiritual is frequently cited as the most significant component of the route. On some occasions, doubts are expressed as to whether this interior pilgrimage is religious or not. However, there are also those who want to make it clear that they are atheists or agnostics, and despite this, the Way has been a personal, spiritual, unforgettable experience.

In this debate about the motivations there are those who think that with more than 300,000 pilgrims a year, these reasons are varied, although it clarifies that it is more of a tourist product and that a transformative spirituality cannot be reached in less than a week. The debate of the 100km distance, largely responsible for the overcrowding, is introduced. We close the circle with criticism on the massification, the loss of values and the distortion of the itinerary of the route, which has become 'a cultural sports event'. In line with this historical rupture of the meaning of the Way are the institutions responsible for them. Like the associations of Friends of the Way, these responsibilities are shared between public administrations and the Church.

Public administrations are accused of making the Way pagan, turning it into a profitable spectacle. But it is the Church that receives most of the criticism. The commercialization and the little *religious* interest that it seems to show for the route appear in some comments. For example, the depersonalized handing over of the Compostela is criticized as it follows a number-based queuing system on a screen, as if it were a type of office. But criticism of the Church is more incisive when talking about closed temples along the Way, of the difficulty of attending mass on the route or the transformation of monasteries into luxury hotels instead of serving for the reception of pilgrims.

Conclusions

Throughout this text we have tried to focus on the complex governance that constructs the routes to Santiago. It is not possible to understand the success of this pilgrimage route and its multiple meanings, if we do not value the role played by the different agents that interact in the public and private management of the Way. We have focused on what we consider to be the main ones: international agencies, public administrations, Church and associations of Friends of the Way. The discourse of all of them is quite convergent although the responses that derive from their actions are much more divergent. Through the analysis of the comments of a newspaper article, we have seen a rich vision that helps shape a dense network of values that give the Way a unique identity, yet sometimes full of contradictions.

The comments analysed show us the approaches on which the institutions involved in governance focus, the tensions between them and their perspectives, sometimes opposed, on the Way of Saint James. First, there is the tourist approach that turns the Jacobean route, in the best case, into a hiking trail, if not into a more typical type of mass tourism. Also, there is the religious view that exclusively understands the Way in its Christian aspect without which its existence would be meaningless. Thirdly, there is the perspective of cultural dialogue offered fundamentally by international agencies.

Finally, most of the comments are very close to the discourse of the associations of Friends of the Way and show a more enriching and inclusive view. These are comments that insist on the idea that the Way is a route that fits all the motivations that are not exclusive. Personal challenge, spirituality and sense of community are part of the content of its own existence. We should remember that the associations provide a perspective that, even rooted on tradition, is based on the pragmatism of reality.

Although the reading of the comments does show the existence of certain conflicts between these different perspectives, the institutional reality shows a greater degree of collaboration. The criticisms in some entries of the Church and public administrations are proof of how many walkers see the conflict between these two institutions. It does not seem good that the Way becomes a mass tourism product or a route in which the Catholic religion acts as the main engine. In both cases the identity and the values that define this route would be lost.

The post-Secularism that some authors (Lois-González and Santos, 2015; Nilsson and Tesfahuney, 2016; Blom *et al.*, 2016) have indicated as the basis for understanding the contemporary success of the Way of Saint James is perfectly represented by those comments that we have identified as closer to the general discourse of associations of Friends of the Way. This means that it would be necessary to strengthen the role of these associations without implying a loss of their independence. Like many other pilgrimage routes, that of Santiago is a metaphor for our wandering the earth (Lois-González and Santos, 2015), with its own physical difficulties and metaphysical uncertainties. It is in this context that it seems that the associations of Friends of the Way place their emphasis on the spiritual depth of the human being together with the enjoyment of daily life in the form of culture, sport or fun.

Notes

[1] https://oficinadelperegrino.com/en/pilgrimage/the-compostela/ (accessed 17 December 2019)
[2] https://oficinadelperegrino.com/en/# (accessed 17 December 2019)
[3] https://whc.unesco.org/archive/opguide12-en.pdf (accessed 17 December 2019)
[4] http://www.vatican.va/content/john-paul-ii/es/speeches/1982/november/documents/hf_jp-ii_spe_19821109_atto-europeistico.html (accessed 17 December 2019)
[5] https://www.coe.int/en/web/cultural-routes/about (accessed 17 December 2019)
[6] https://www.icomos.org/images/DOCUMENTS/Charters/culturalroutes_e.pdf (accessed 17 December 2019)
[7] https://www.turismo.gal/canle-institucional/turismo-de-galicia/a-axencia/estratexia-do-turismo-de-galicia-2020 (accessed 18 December 2019)
[8] https://www.turismo.gal/canle-profesional/plans-e-proxectos/plan-director-Way-de-santiago?langId=en_US (accessed 18 December 2019)
[9] https://oficinadelperegrino.com/en/ (accessed 18 December 2019)
[10] https://anosantocompostelano.org/wp-content/uploads/2019/12/Carta-PastoralA%C3%B1oSanto21.pdf (accessed 12/18/19)
[11] https://elpais.com/ccaa/2019/09/27/catalunya/1569590522_694152.html (accessed 26 February 2020)

References

Alvarez, A. (ed.) (1999) *Homo peregrinus*. Xerais, Vigo/Spain.

Amaro, S., Antunes, A. and Henriques, C. (2018) A closer look at Santiago de Compostela's pilgrims through the lens of motivations. *Tourism Management* 64, 271–280. DOI: 10.1016/j.tourman.2017.09.007.

Barber, R. (1993) *Pilgrimages*. The Boydell Press, Woodbridge, UK.

Blom, T., Nilsson, M. and Santos, X. (2016) The Way to Santiago beyond Santiago. Fisterra and the pilgrimage's post-secular meaning. *European Journal of Tourism Research* 12, 133–146.

CETUR and Xacobeo (2007–2010) *Observatorio estatístico do Camiño de Santiago 2007, 2008, 2009 e 2010*. Xunta de Galicia and University of Santiago, Santiago de Compostela/Spain.

Cohen, E. (1992) Pilgrimage and tourism: convergence and divergence. In: Morinis, A. (ed.) *Sacred Journeys: The Anthropology of Pilgrimage*. Greenwood Press, Westport, Connecticut, pp. 47–.61.

Collins-Kreiner, N. (2010) Researching the pilgrimage: continuity and transformations. *Annals of Tourism Research* 37, 440–456.

Collins-Kreiner, N. (2018) Pilgrimage-tourism: common themes in different religions. *International Journal of Religious Tourism and Pilgrimage* 6, 8–17.

Dunn, M. and Sanchez, S. (2016) Historical and modern signs of 'real' pilgrims on the road to Santiago de Compostela. In: Sanchez, S. and Hesp, A. (eds) *The Way of Saint James in the 21st Century*. Routledge, New York, pp. 13–35.

Frey, N. (1998) *Pilgrim Stories: On and Off the Road to Santiago*. University of California Press, Berkeley, California.

Graham, B. and Murray, M. (1997) The spiritual and the profane: the pilgrimage to Santiago de Compostela. *Ecumene* 4, 389–409. DOI: 10.1177/147447409700400402.

Kim, B., Kim, S.S. and King, B. (2016) The sacred and the profane: identifying pilgrim traveler value orientations using means-end theory. *Tourism Management* 56, 142–155. DOI: 10.1016/j.tourman.2016.04.003.

Lois-González, R.C. and Santos, X.M. (2015) Tourists and pilgrims on their way to Santiago: motives, Caminos and final destinations. *Journal of Tourism and Cultural Change* 13(2), 149–164. DOI: 10.1080/14766825.2014.918985.

Lopez, L. (2013) How long does the pilgrimage tourism experience to Santiago de Compostela last? *International Journal of Religious Tourism and Pilgrimage* 1, 1–14.

Maclean, K. (2009) Seeing, being seen, and not being seen: pilgrimage, tourism, and layers of looking at the Kumbh Mela. *CrossCurrents* 59(3), 319–341. DOI: 10.1111/j.1939-3881.2009.00082.x.

Magry, P.J. (2015) To be or not to be. . a pilgrim. Spiritual pluralism along the Way Finisterre and the urge for the end. In: Sánchez-Carretero, C. (ed.) *Heritage Pilgrimage and the Way to Finisterre*. Springer, Heidelberg/Germany, pp. 175–211.
Murray, M. and Graham, B. (1997) Exploring the dialects of route-based tourism: the way of Saint James. *Tourism Management* 18, 513–524.
Nilsson, M. and Tesfahuney, M. (2016) Performing the 'post-secular' in Santiago de Compostela. *Annals of Tourism Research* 57, 18–30. DOI: 10.1016/j.annals.2015.11.001.
Olsen, D.H. and Wilkinson, G. (2016) Are fast pilgrims true pilgrims? the Shikoku pilgrimage. *Annals of Tourism Research* 61, 228–230. DOI: 10.1016/j.annals.2016.08.004.
Rodríguez, M. (2004) *Los años santos compostelanos del siglo XX*. Xunta de Galicia, Santiago de Compostela/Spain.
Rodríguez, M.F. (2018) Función de las asociaciones de amigos del Camino. Sus conexiones con las antiguas cofradías de peregrinos a Santiago. *Proceedings of the VI International Conference of Christian Welcome in the Way to Santiago*. Council of the Cathedral of Santiago, pp. 163–173. Available at: http://archicofradia.org/wp-content/uploads/2019/07/6-d-MANUEL-F.-RODRIGUEZ-2.pdfhttp://archicofradia.org/wp-content/uploads/2019/07/6-d-MANUEL-F.-RODRIGUEZ-2.pdf
Sanchez, S. (2016) *The Way of Saint James in the 21st Century: Interdisciplinary Perspectives and Global Views*. Routledge, New York.
Santos, X.M. (2016) The way of Saint James as an event: politics and nation. *Journal of Policy Research in Tourism, Leisure and Events* 8(3), 233–248. DOI: 10.1080/19407963.2016.1214960.
Santos, X.M. (2017) La construction touristique des chemins de Saint-Jacques-de-Compostelle en Espagne : acteurs, regards et contradictions. *Sud-Ouest Européen* 43, 7–19. DOI: 10.4000/soe.2522.
Terzidou, M., Scarles, C. and Saunders, M.N.K. (2018) The complexities of religious tourism motivations: sacred places, vows and visions. *Annals of Tourism Research* 70, 54–65. DOI: 10.1016/j.annals.2018.02.011.
Timothy, D. and Olsen, D. (2006) *Tourism, Religion and Spiritual Journeys*. Routledge, London.

8 Commensality, Communitas and Contestation on the Camino Francés: Making Pilgrims through Meals

Augusta X. Thomson*
New York University

Introduction

In *Remembrance of Repasts: An Ethnography of Food and Memory*, David Sutton (2001) points to the ways that collective Orthodox Catholic memory informs a sense of community defined by commensality on the Greek island of Kalymnos. He looks specifically at the tradition of preparing and consuming special bread used to celebrate the Eucharist on Saints' Days. The collective consumption (and to a certain extent, preparation) of this bread binds the Kalymnian community through a history of commensality linked to personal, cultural and religious memory. Eucharistic commensality creates and affirms Kalymnian kinship relations, informed by an overarching ontology of kinship: that of the Catholic family, bound to Christ and God ('Our Father') through communion.

In the following chapter, I draw from Sutton's discussion of Eucharistic commensality to argue that communal meals given to and shared by pilgrims (at albergues, or pilgrim hostels, and a few restaurants) along the Camino Francés, a 500-mile Christian pilgrimage, serve as historically embedded instances of *communitas* aimed at shaping and framing pilgrims' identities and motivations and connecting pilgrims to each other and to the Camino's 1000-year-old religious history. For the sake of this chapter, I define 'pilgrims' according to the criteria established by self-described 'pilgrims' undertaking the Camino Francés: while not required to walk the route in its entirety, pilgrims differentiate themselves from tourists by their commitment to walking the route and personally transporting their packs along the way. Pilgrims travel to the Camino Francés from diverse regions, such as wider Europe, Australia, New Zealand, Brazil, the USA, Canada, Japan and Korea,[1] with varying religious and secular motivations and values, and yet pilgrim recognition relies less on the uniformity of these values and motivations than on the collective experience of the walk and its resultant physical, mental and emotional impacts. A pilgrim is 'made' along the way.

Likewise, communal meals and conversations shared between pilgrims during mealtimes contribute to ideologies about the pilgrimage route and pilgrim identity, informing pilgrim values, subjectivities, motivations and behaviour. Despite the increasingly secular nature of the pilgrimage route, the close connection of these free, communal meals to the Eucharist, and thus an overarching Catholic family, provides a subtle backdrop to contemporary Camino meals, through which pilgrims come to

*axt203@nyu.edu

view themselves as connected to each other and to the pilgrimage route.

In order to make the above argument, I engage with several inter-related theories. I begin with an introduction to Turner's (1969) concept of communitas, followed by a brief discussion of Eade and Sallnow's (1991) critique of communitas and corresponding emphasis on contestation. I turn to Carreño's (2014) elaboration of the coexistence of communitas and contestation to emphasize how communitas and contestation intertwine and intersect. In addition, Sahlins's (2011) theory of kinship, as 'mutuality of being', in addition to Munn (1986) notion of gift exchange, as a food-mediated 'transaction', provide important theoretical frameworks for understanding the role of commensality in aggregating and informing an overarching pilgrim community. I engage with Asad (2003) theories of secularism, briefly drawing on Hirschkind's (2011) discussion of the 'secular body' to clarify what is meant by secularism along the Camino. Finally, I draw from Connerton's (1989) discussion of social memory to emphasize the role that visually mediated memories of mealtimes play in sustaining the bonds among pilgrims. Using these theories as a backdrop, I provide a brief survey of some ethnographies of pilgrimage that address food.

By exploring the history of commensality to Christianity through assorted literature on the Eucharist, I strive to connect the above theoretical perspectives to the Camino France's historical past. This chapter's capsule history of the Camino de Santiago sets the tone for a more detailed foray into an ethnographic case study developed from personal ethnographic research undertaken along the Camino de Santiago in the summer of 2017. This example particularly addresses the mnemonic significance of food and the intersection of food with pilgrim values and identities, as presented visually, and drawing from Connerton's text, *How Societies Remember*. While commensality has been tangentially explored within anthropological pilgrimage literature it remains less explored than themes of space and place, material culture, and embodiment, especially with regard to Christian pilgrimage. In this way, this paper also points to a lacuna in anthropological studies focused on the relationship of food to Christian pilgrimage.

Theorizing Pilgrimage, Theorizing Commensality

The anthropology of pilgrimage is largely attributed to Victor Turner, who first developed the notion of communitas in his text *The Ritual Process: Structure and Anti-Structure*. Turner (1969) defines communitas as 'society as an unstructured or rudimentarily structured and relatively undifferentiated *comitatus*, community, or even communion of equal individuals who submit together to the general authority of the ritual elders' (Turner, 1969, p. 96). He argues that communitas and structure are mutually exclusive, yet symbiotic. They must exist in a cyclical balance during which structure is temporally suspended as participants engage in bonding rituals. Turner (1969) posits that separation from everyday activities and social relations coincides with personal and/or societal crises, in which an individual or group suspends involvement in daily life. The result of this separation is liminality – a halt in normal social life, during which ritual participants re-enact the crises underlying the ritual, in order to both affirm and challenge the structures of social life (Turner, 1969, p. 96-97).

While Turner's theory of communitas dominated anthropological studies of pilgrimage throughout the seventies and eighties, it was problematized by Eade and Sallnow (1991), who introduced the notion of contestation. They argue that pilgrimages, rather than dissolving social boundaries and distinctions, often serve to reinforce those boundaries. In this way, pilgrimage might be more productively viewed as 'a realm of competing discourses' (Eade and Sallnow, 1991, p. 5). Preferring not to discount communitas altogether, they suggest that communitas, while not an 'empirical description' of pilgrimage, might 'coexist or compete with alternative discourses' (Eade and Sallnow, 1991, p. 5).

This is echoed by Carreño (2014), who note the simultaneous presence of coexistence and the reproduction of 'semiotic diversity' at the Quyllurit'i pilgrimage in the southern Peruvian Andes. During the pilgrimage, celebrated the week before the feast of Corpus Christi, in May, the shrine, located at the base of a glacier, is flocked by a plethora of pilgrims – foreign New

Age pilgrims, researchers, film crews and 'all the sociocultural diversity present in the city of Cuzo' (Carreño, 2014, p. S188). While differences exist to mark and distinguish pilgrims, these differences are simultaneously transcended through 'the strong emotional experiences of the pilgrimage,' in which 'collective effervescence' – particularly that associated with the dancers in the procession – contributes to 'an overwhelming experience' that 'can be seen as socially constructed by the attendance of each and every person at the pilgrimage' (Carreño, 2014, p. S193). In this way, despite pilgrims' varying interpretations of the pilgrimage and its ritual sequences (and the racial-ethnic labels and hierarchies associated with control of the shrine and participation in these rituals), these diverse interpretations do not lead to open contestation. Rather, these diverse interpretations contribute to the unique melding of communitas and contestation that heightens 'collective effervescence' (Durkheim, 1912; Carreño, 2014).

The notion of communitas, as one of a range of 'coexisting or competing discourses', is central to the argument of this paper, in which I focus on the centrality of commensality to pilgrim relations. Drawing from and expanding on Turner, I define communitas as 'a feeling of collectivity based on the assumption of shared experiences among co-participants in a highly routinized and geographically determined experience of physical, mental and emotional endurance'. The ties of commensality to Christian 'kinship' are central to the communitas fostered through meal-sharing on the Camino de Santiago; these mealtimes succeed in bonding pilgrims to one another (despite cultural, socio-economic and religious differences) and forging pilgrim subjectivities through their simultaneous incorporation of two 'discourses' that, arguably, compete against communitas – secularism and individuality.

Sahlins (2011) defines kinship as 'a manifold of intersubjective participations, founded on mutualities of being', in which 'mutuality of being' suggests 'persons who are members of one another, who participate intrinsically in each other's existence' (Sahlins, 2011, p. 1, 10). In this way, Sahlins' kinship analogue incorporates social construction. He includes 'shared food' within his 'catalogue of commonplace post-natal means of kinship formation' (Sahlins, 2011, p. 5). Sahlins also points to the ways that 'shared life conditions and shared memories' inform kin relations.

Much of the effectiveness of commensality in binding kin derives from its connection to gift-giving. Munn (1986) notes how Gawa men first chew the food they feed to their babies in order to produce a fosterage relation. According to Munn, 'this transaction is a paradigmatic instance of food-giving as the separation of food from one's own body for incorporation by another' (Munn, 1986, p. 50). Like the above, the act of sharing a free, communal meal, prepared by unpaid volunteers and often financed by the Church (even if this is commonly ignored by pilgrims), in turn comprises an 'instance of food-giving as separation' from the body – in this case, from the body of Christ. Thus, Sahlins' and Munn's arguments are relevant to the idea that certain communal mealtimes along the Camino de Santiago, symbolically tied to the Eucharist and Christian kinship through Christ, bind pilgrims with diverse values and motivations through commensality – incorporating exchanges of food and memories.

As stated briefly above, the Camino has become increasingly 'secularized' as mainstream marketing and tourism initiatives popularize the journey. Frey (1998) notes, regarding this phenomenon, that this is central to the effectiveness of communal meals in bonding pilgrims from diverse backgrounds – many of whom are not Christian, or even religious. Asad (2003) defines the secular as 'the concept that brings together certain behaviors, knowledges, and sensibilities in modern life' (Asad, 2003, p. 25). For Asad, secular practices are not secular because they are non-religious, but rather because they owe their recognition to the discourse of secularism (distinguished from the political realm), associated with the moral, social and political transformations that have informed global modernity. In this sense, 'the secular marks a relational dynamic more than an identity' (Hirschkind, 2011, p. 641).

Hirschkind (2011) emphasizes that 'the boundaries' of religious and secular categories are 'continuously determined and reciprocally redefined' through an ongoing power struggle that, in turn, informs the identification of certain practices as secular versus religious. According to Hirschkind, 'every secular practice is accompanied by a religious shadow', such that 'the secular will always be subject to a certain indeterminacy or instability' (Hirschkind, 2011, p. 643). This is especially true of the Camino de Santiago, where the secular and the religious are imbricated; while the Camino itself exists as a World Heritage Site and a secular, nationalist project, it passes through towns with historic religious architecture. It is thus difficult to fully disentangle the religious from the secular. Because communal mealtimes on the Camino de Santiago (hosted by parochial albergues or served in religious contexts) often incorporate secular rituals – personal introductions and/or narratives, musical performances and diverse multicultural foods – the religious history of the Camino de Santiago is carefully backgrounded to include pilgrims from diverse social and cultural backgrounds. This 'religious history' persists through memories of the pilgrimage route, as contained in continuous and repeated rituals undertaken at heritage sites, landmarks and over meals (Halbwachs, 1980) – rituals that, in turn, inform pilgrims' identities. It pervades the various guidebooks written about the pilgrimage route, most notably Brierley's *A Pilgrim's Guide to the Camino de Santiago* (Brierley, 2019); and it presents one important aspect of the commensality engendered by collective meals, even if unconsciously.

In *How Societies Remember*, Connerton (1989) writes that 'we preserve versions of the past by representing it to ourselves in words and images' (Connerton, 1989, p. 72). For many pilgrims, the Camino de Santiago is highly mediated; pilgrims take regular photographs of the landscape and community – memorializing the experience and cementing bonds formed along the Way. Many of these photographs wind up in personal and commercial albums, on blogs and/or Facebook pages, and in Instagram archives (Geismar, 2017). The subjects of these photographs often overlap: historic buildings, churches, signs and mealtimes. Overlaps in scenery and context suggest, as emphasized by Halbwachs (1980), that individual memory and social memory are indeed imbricated (Halbwachs, 1980, p. 37). Photographs taken during meals, and even of meals, along the pilgrimage route serve to prolong and, arguably, intensify the experience of communitas, one that simultaneously allows for contestation, as and through diversely mediated impressions. Parallel images inform pilgrim identities by marking them part of an overarching pilgrim community defined via overlapping experiences. In this way, memories of the pilgrimage and participation in the pilgrimage are 'assured' by this act of 'orienting', of becoming emplaced through the sharing of space, meals and memories (which are then re-memorialized through photographs) (Ricoeur, 2006, p. 41).

While commensality has been addressed in some ethnographies of pilgrimage, more attention has focused on themes of space, place, embodiment, material culture artifacts and tourism. Toomey's (1994) ethnography of food ritual and feasting during the Krishna pilgrimage to Govardhan, Mathura district, Uttar Pradesh, comprises one of the few ethnographies focused primarily on food rituals. Toomey (1994) notes how devotees offer foods to Krishna associated with feelings of well-being, joy and auspiciousness. Pilgrims believe that Krishna consumes this food, metonymically transforming it into 'more love-laden *prasad*', which, in turn, leads to emotional transformations among pilgrims (Toomey, 1994, p. 54). Werbner (2010) describes the way that commensality bonds Sufi women during the pilgrimage to Ghamkol Sharif in Pakistan. Based on research conducted at the lodge of the living saint Zindapir, in 1989, Werbner documents how women bond through cooking chapatis and cleaning cups and dishes. After meeting with the living saint, these women (from diverse backgrounds and geographic locations) 'share a communal meal and listen to each other's "sorrow stories" – connecting over issues of marriage, family, betrayal, infertility, bereavement, abandonment, illness, and suffering' (Toomey, 1994, p. 378). Similarly, during the most renowned and attended Roman Catholic feast of St Anne in Bukit Mertajam, Malaysia, pilgrims from Thailand and Singapore are fed free meals by volunteers, which are thought to be blessed by St Anne, thus possessing 'trans-valued properties' (Yeoh, 2006, p. 21). And yet, despite the above studies on the relationship between food and pilgrimage, surprisingly little research has been undertaken on the relationship of commensality to pilgrimage journeys, beyond the consumption of officially blessed and ritually significant food.

Christianity, Commensality and the Camino Context

The history of the Eucharist in the Catholic tradition is central to contemporary practices of serving free, communal meals to pilgrims along the Camino de Santiago. While this practice extends the historical reliance of medieval pilgrims on the hospitality and alms of those permanently settled along the route, it derives from the Catholic ideal of communion. This eucharistic history is tied to feasting and fasting traditions that, in turn, informed many of the festivities and rituals enjoyed by medieval communities, and that helped to construct them. Rubin (1991) notes that the Eucharist – a 'ritual which turned bread into flesh – a fragile, small, wheaten disc into God' – was at the core of the whole religious system of the Middle Ages (Rubin, 1991, p. 1). In the Middle Ages, social relations were informed by religion, which provided a cosmic order. From approximately 1100, this cosmic order was mediated by a sacrament linked to the Eucharist. According to Bossy (1983), the mass served as a catalyst for social fusion among those receiving communion. This social fusion was emphasized during the feast of Corpus Christi, a summer feast that aggregated medieval communities through public processions.

Burke (2009) highlights the traditional emphasis on festivals within European popular culture. Community festivals celebrating the feast of the patron saint of a town or parish, along with annual festivals marking the religious holidays, such as Easter, Advent, Carnival and Epiphany, comprised 'special occasions when people stopped work and ate, drank and spent whatever they had' (Burke, 2009, p. 179). Prior to Lent, during Catholic Carnival, local peasants swelled the streets, partaking in the largely food-centred festivities and/or feasts. Pilgrimages to holy places also tended to occur on the occasion of major festivals, suggesting the centrality of food and, particularly, feasting to pilgrimage cults.

Bynum (1985) notes that in the medieval period, the possibility of over-eating and donating food was only available to those of aristocratic or patrician status. In this way, demonstrations of charity, as acts of feeding the poor and ill, were central to religious orders. 'Food', Bynum writes, 'was not only a fundamental material concern to medieval people; food practices – fasting and eating – were at the very heart of the Christian tradition' (Bynum, 1985, p. 2). Christians were required to fast on certain days and receive communion at least once a year. Food (the Eucharist), as the most direct way of encountering God, appeared as a prominent theme throughout Christian poetry, devotional literature and theology. According to Bynum, hundreds of years before transubstantiation was defined as a doctrine, most Christians believed that the sacrament *was* Christ's body and blood. Thus food, 'as a practice and symbol', was central to medieval spirituality (Bynum, 1985, p. 2).

The complex entanglement of medieval Christian history with the Camino de Santiago is often the primary attraction for pilgrims, regardless of their religious orientation. The Camino de Santiago, frequently abbreviated to the Camino, is now largely associated with the Camino Francés, the most popular pilgrimage route for historic and contemporary pilgrims. It comprises a 500-mile medieval pilgrimage route, beginning in France and entering Spain at two mountain passes – Roncesvalles and Somport – which connect in Puente la Reína, a small town in the Basque region. The Camino played an important role in the Christian repopulation of the peninsula encouraged by the kings of Navarre, Castile and Galicia, all seeking to control the lands and expand reformed monastic orders south of the Pyrenees. The repopulation drove commerce in the area, attracting merchants and artisans (especially those of Frankish origin), and leading to the construction of Romanesque and Gothic churches and the development of far-reaching infrastructures of villages, bridges and roads.

Frey (1998) notes that while the Camino de Santiago flourished from the 11th to the 13th centuries, largely due to papal support from Rome, travel along the route declined in the 16th century. The Spanish Inquisition led to lower numbers of visitors, as did the publication of writings critical of the pilgrimage route. Due to various historical developments, this decline continued into the 20th century until after

World War II, when attention began to shift to rebuilding Europe through a 'collective past' (Frey, 1998, p. 27).

In the 1980s, the pilgrimage experienced a marked reanimation (that continues to today). This reanimation was part of a European social phenomenon aimed at collapsing political and social borders. The increased mobility of middle-class Europeans seeking leisure travel opportunities, combined with the 'transnational publicity of the Camino', led to an increased pan-European interest in the Camino as a tourist destination (Frey, 1998, p. 27). Despite the drop in church attendance across Europe following the 1960s, the dwindling membership in religious orders and the widespread secularization of civil society (or public life), an increased interest in alternative spiritual and religious movements also informed the Camino's reanimation. This reanimation was spurred by the grassroots efforts of Friends of the Camino associations, clever politicians, scholars of Medieval Studies, returning pilgrims and eager adventurers (Frey, 1998, p. 27). Frey (1998) points to the 'ethos of political and cultural nostalgia' that motivated many of these pilgrims, eager to rediscover 'a direct link to the past, an authenticity based on sacrifice, endurance, and austerity' thought to have characterized the journeys of medieval pilgrims. While there are many varieties of pilgrims, multiple ways of reaching Santiago and myriad ways to interpret what it means to be a pilgrim, the majority of contemporary pilgrims hope to connect to the historical heritage of the pilgrimage route and its ties to a community of like-minded seekers (*ibid.*, 30).

Commensality Along the Camino

It is no coincidence that much of the community bonding that happens on the Camino de Santiago occurs over meals. On the one hand, the eucharistic goal of the religious pilgrims is to participate in a eucharistic meal at the Cathedral of Santiago de Compostela. On the other hand, because the pilgrimage involves long periods of solitary walking that end in social meals shared with other pilgrims, these meals stand out in contrast to the lone physical exertion of the daily walk as times for bonding and sharing.

While pilgrims spend differing amounts of time undertaking the pilgrimage, for many the pilgrimage lasts at least a month. Much of the day is spent walking. Depending on the time of year and weather patterns, pilgrims typically set out at 6 am. They walk an average of 20 km over the course of seven to eight hours. Many pilgrims stop in cafés along the way to purchase food and consume coffee; others, however, prefer to eat breakfast before they leave and to wait to eat lunch when they arrive at their daily destination. Because of the individual aims of pilgrims, many of whom arrive on the pilgrimage route with deep existential questions necessitating daily contemplation, walking can be a solitary experience. After several hours spent with minimal caloric intake and little social stimulation, pilgrims typically arrive at albergues eager to share their daily reflections and connect with fellow pilgrims. After eating lunch, pilgrims wash clothes and nap before preparing for a communal dinner. Commensality, together with the wine often served with the meal, prompts conversations among those who have shared in this daily walk. The food, drink and convivial conversation are especially highlighted, with little likelihood of conflict, since pilgrims, as fellow travellers, may not be able to avoid seeing one another again, and indeed rely on one another for aid if crisis arises. In this way, pilgrims are 'made' through mutual recognition and even concern.

As noted above, the history of the Camino is entangled in the medieval history of Christianity, and thus a history of feasts, festivals and food consumption. The following case study, taken from a personal experience along the Camino de Santiago this past summer, 2017, examines how the free, communal dinners prepared by two *hospitaleros* volunteering at a small albergue, San Anton, on the outskirts of Castrojeritz, a town about 30 km outside Burgos, functions to connect pilgrims to the Camino's Catholic heritage, and through that heritage to a wider pilgrim community defined by specific values, identities and rituals.

The albergue of San Anton sits in the ruins of the Monastery of San Anton, dating to the 14th century (Fig. 8.1). It is a *donativo* pilgrim hostel; pilgrims do not have to pay to stay in or eat at the hostel, but are, rather, encouraged to make a donation in gratitude for the services

Fig. 8.1. Albergue of San Anton. (Photo: the author)

provided to them. The albergue is smaller than most, with only 12 beds, no electricity and no hot water. Pilgrims who wish to wash must take cold showers and, occasionally, when the water isn't functioning, wash themselves with water from an irrigation pipe belonging to the farm next door. I stayed in San Anton for three nights during the summer of 2017 (one night in late June and two others in July), while undertaking research along the Camino. This gave me the chance to observe and participate in the communal meals, to note how they shifted with different volunteers. Unsurprisingly, the rituals surrounding the evening meal remained largely the same.

Mealtimes are particularly central to the San Anton albergue experience. During the evenings, pilgrims can choose to help prepare the meal if they are not too tired. While I was there in June, one woman asked to chop vegetables for the salad, while another helped to prepare the lentil stew with one of the *hospitaleros*. During my stays in July, no one offered to help; instead, guests, overwhelmed by the heat of the hottest month in that region, rested and slept. Again, unsurprisingly, the same meal was served at all three suppers. This meal consisted of white bread, a lentil stew, a simple salad, cheap red wine and yoghurt with fruit for dessert. Before eating, each pilgrim introduced himself/herself with a story or song – sharing his/her name, home town, occupation and reasons for walking the Camino.

The *hospitaleros* (of which there were always two) also introduced themselves, and expressed their reasons for returning to 'serve'. Each time (in June and during both evenings in July), the volunteers cited their love of the Camino, and particularly the pilgrims. *Hospitaleros* are required to walk the full Camino at least once prior to volunteering, and at each meal, both volunteers recalled the intimate connections they had made walking with pilgrims in the past. One of the *hospitaleros* volunteering during my second visit in July had returned, for the past five years, to spend a portion of each summer serving in San Anton; this *hospitalera* had developed close bonds with some of the pilgrims passing through the albergue, not to mention locals living in Castrojeriz.

While the religious implications of the meal were inescapable (given that we ate amidst the ruins of an old Catholic monastery), the dinner was cleverly crafted to include pilgrims of varying cultural and spiritual persuasions. In line with the increasing secularization of the Camino by the Spanish government (evident in the investment in tourism infrastructures), the meal evoked an inclusive family supper. There was no Christian blessing or prayer; rather, the loosely structured introductions engendered a warm and neutral atmosphere. During the June evening, one Korean pilgrim, Deborah, felt moved to tears, as she opened up about the stresses of her life in Seoul, where she had previously been a schoolteacher serving underprivileged youth. At the close of all three meals, volunteers collectively cleaned the dishes – washing, drying and sharing stories. Each evening concluded with communal tea-drinking in the kitchen, at which point (and on all three occasions) the pilgrims staying in the albergue appeared to have bonded.

In this way, the act of eating – a secularized ritual performance, embedded within the Camino's Catholic past – provided the context for communitas, with room for contestation. The intentional avoidance of any overt Catholic ritual (such as saying grace) mirrored the increasing secularization of the pilgrimage route. Pilgrims were encouraged to share stories in any way they desired; they could opt not to share if they so chose. Similarly, pilgrims were not required to eat dinner; it was simply offered as a complimentary 'gift' and symbol of the hospitality extended to pilgrims in the past. Despite distinctive religious views and cultural backgrounds, the inclusive atmosphere of the meal served to create a sense of mutuality, of pilgrim camaraderie centred around corresponding values of inclusivity, care and intention. The Camino's undeniable religious history lingered beneath each dinner's familial ambience, implying sustained connections to the Eucharist and the Church's role as benefactor, extending charity to pilgrims, as God's children, even as the meal unfolded as a secular pilgrimage ritual.

For Turner (1969), communitas relates to co-religious sharing of sacramental goals. The hierarchical rules of ordinary life, including the potential for conflict, are set aside while all engage in a shared ritualistic activity. The resulting communitas is almost utopian, with little room for conflict or confrontation. As noted earlier, Eade and Sallnow (1991) critique this conception of communitas by drawing from ethnographic contexts in which collective pilgrimage experiences actually produce, and in some instances reproduce, social boundaries and distinctions. This is an important consideration. But it is strengthened by Eade and Sallnow's fur ther suggestion that communitas does exist at pilgrimage sites, where it 'coexists or competes with alternative discourses' (Eade and Sallnow, 1991, p. 5). Carreño (2014) expand on this notion of communitas-in-contestation through their discussion of the Quyllurit'i pilgrimage in the southern Peruvian Andes. As noted above, during this pilgrimage, the assumption of shared values/interpretations prevents conflict over diverse perspectives and values. The goal-directedness of pilgrims partaking in the journey and intent on completing the 500-mile journey on foot, the strong personal motivations underlying each individual pilgrimage despite distinct belief systems, and respect for others experiencing similar rigours, discourage disagreement, or contestation, even as differences are noted, acknowledged and accepted.

On the Camino de Santiago, cultural differences and socio-economic distinctions between pilgrims contribute to the collective experience – expanding tolerance and acceptance along the route. One might argue that communitas along the pilgrimage route not only incorporates contestation, but also is heightened by pilgrims' respect for contestation. All pilgrims must share in the rigours of the pilgrimage route and time spent in conversation provides relief from a day of solitary walking; thus, unspoken rules, affirmed at festive, communal mealtimes, support conviviality and discourage conflict. The sense of collectively working towards a mutual goal (the completion of the pilgrimage route and arrival in Santiago de Compostela), while simultaneously meeting fellow pilgrims from diverse areas of the world and thus gaining entrance to an increasingly global community of pilgrims, contributes to a sense of 'collective effervescence', bolstered and reinforced by a form of communitas-in-contestation. This 'collective effervescence', in turn, informs imaginaries about the ideal Camino pilgrim, as inclusive, accepting and dedicated – qualities that shape pilgrims' (often still emergent) motivations, subjectivities and values.

Remembering (The Restaurant)

Memory is central to commensality along the pilgrimage route. Yet it is also interwoven with the Camino's history. Halbwachs (1980) makes the distinction between memory and history: he argues that history refers to a past 'no longer included within the sphere of thought of existing groups'. Collective memory, in contrast, can stretch back into various pasts, but is most preoccupied with events within living memory (Halbwachs, 1980, p. 106). Along the Camino, memory and history are imbricated. Memories of mealtimes gather significance when contextualized historically. As noted above, the pilgrimage route's ingrained religious history, directly informing memories made in the present, contributes to the power of commensality. The pilgrimage's history of hospitality, including collective mealtimes (particularly in parochial albergues), generates an aura around contemporary shared mealtimes. This extends to restaurants and bars, locales where pilgrims often gather after hiking to eat, drink and reflect on the day.

Returning to the example of the albergue of San Anton – the communal dinner and breakfast were memorialized via photographs and videos. The table spread was photographed multiple times before we sat down to dinner – capturing the rugged tableware and dripping candles. Deborah, the Korean schoolteacher, walked around the table as we ate, taking photographs of us mid-bite, or, if we were lucky, posing with our arms around each other. At the end of the meal, Carla, a biologist from northern California, took a group selfie as we huddled around her jostling for room in the frame.

In the morning, the same photographic ritual occurred over a candlelit breakfast. While in parallel circumstances most people would have felt irritated to be the subjects of a photoshoot at 6.30 in the morning, we all smiled for the photographs, recognizing the significance of the memories we had shared (and would continue sharing via Facebook). These photographs, while not about the food, *per se*, *were* about the collective experience of commensality. They were also about mutual participation in an overarching and inclusive pilgrim community – one defined by recognized values and collective engagement with the pilgrimage route that, in turn, came to inform a collective pilgrim identity. While many of the images later found their way onto social media platforms (where they remain to remind us of the evening and the pilgrimage journey), participating in the photographs, as photographers and/or subjects, was mnemonically significant in and of itself; the moments we shared for photographic documentation contributed to our overall sense of community and inclusive pilgrim identity. In the above instance, commensality provided the context for these mediated expressions of communitas, defined by mutuality.

Ricoeur (2006) writes that 'the transition from corporeal memory to the memory of places is assured by acts as important as orienting oneself, moving from place to place, and above all inhabiting' (Ricoeur, 2006, p. 41). Perhaps the visual mediation of commensality serves as a form of orientation – of emplacement, through the almost performative recognition of food and community in space and place. Furthermore, as mentioned above, the underlying historically embedded religious significance of communal meals on the pilgrimage route is likewise memorialized through visual representations of mealtimes, even if unintentionally. In this way, visually mediated memories of mealtimes serve to connect pilgrims in the moment, to the past (history) and future of the pilgrimage route and overarching pilgrim community; these photographs function because they simultaneously allow for contestation. Pilgrims with diverse motivations and values become part of inclusive performances of mnemonic inscription that function through their simultaneous recognition of difference. While most pilgrims choose to participate in photographs, pilgrims can also easily opt out, or take photographs that prioritize different subjects and scapes. In this way, the multiplicity of visual memories of the pilgrimage route, and specifically of mealtimes, heightens the sense of communitas generated by commensality, while contributing to an underlying pilgrim identity premised on inclusivity.

Conclusion

To come full circle, in his ethnography, *Remembrance of Repasts*, Sutton (2001) looks at the preparation of a special offering bread stamped with a religious seal; this bread, prepared by Orthodox Catholic women in Kalymnos, is then crumbled into the communion wine to serve as the Eucharist on the feast days of the saints for whom family members are named. After the Eucharist, the rest of the bread is distributed to the community, connecting it to 'a sense of community defined by commensality' (Sutton, 2001, p. 34). Sutton's example is central to the aim of this paper, in which I have argued that free, communal meals (and particularly dinner) given to and shared by pilgrims in some albergues and restaurants along the Camino Francés, principally referred to as the Camino de Santiago, function as inclusive rituals to connect pilgrims to each other and to the pilgrimage route. The increasingly 'secular' nature of these dinners heightens their inclusivity; communitas exists alongside the space for 'contestation'.

As Frey (1998) notes, much of the 'Camino's enormous appeal as a symbol is its flexibility to accommodate various interpretations of its spaces and encounters yet still retain its basic structure' (Frey, 1998, p. 219). Still, the slipperiness of 'secularism' suggests that it is not so easily separated from religion (Asad, 2003). The Camino's embedded Catholic history and identity underlies its current transformation into a more secular tourist route. Communal meals offered as gifts within some albergues and restaurants across the Camino, function as Eucharistic analogues, and thus spaces where this foundational religious identity continues to inform the construction of a pilgrim community, bound to an overarching religious 'family' – a family that, in turn, informs pilgrim subjectivities, values and identities. These analogues work largely because of the Camino's history, its ties to 'social memory' – ties that are continually reinforced through textual and/or visual (often digitally disseminated) stories.

In sum, the above chapter has moved from an introduction to Turner's theory of communitas to a brief discussion of Eade and Sallnow's notion of contestation. It has touched on Sahlin's argument that kinship relations can be socially constructed, in addition to Munn's notion of gift exchange as a food-mediated 'transaction'. Through reference to Asad's and Hirschkind's theories of secularism I have emphasized the relatedness of religion and secularism – seemingly intertwined. References to Connerton, Halbwachs and Ricoeur have foregrounded the role that social memory (particularly as visually mediated) plays in perpetuating rituals of communitas along the pilgrimage route. A condensed survey of a few ethnographies of pilgrimage exploring food has allowed me point to some directions future anthropological studies of pilgrimage and food might take, while also problematizing the lack of research on the relationship of food to pilgrimage. By briefly touching on the history of commensality within Christianity through assorted literature on feasting and the Eucharist, I have attempted to background more in-depth case studies involving free, communal meals shared by pilgrims along the Camino de Santiago. Following a condensed history of the 1000-year-old pilgrimage route, this paper has used a case study taken from personal fieldwork, conducted during the summer of 2017, to illustrate that commensality plays a role in informing communitas (with simultaneous room for contestation) along the pilgrimage route. Through the communitas generated by commensality, pilgrims with varying identities, motivations and values become part of an overarching pilgrim community, with room for contestation-in-communitas, which, in turn, actively informs pilgrim values, motivations and identities. While there yet remains a great deal of work to be done on the intersection of food and pilgrimage, particularly with regard to modern Christian pilgrimage, this paper has attempted to take a tentative 'step' in that direction.

Note

[1] http://oficinadelperegrino.com/wp-content/uploads/2016/02/peregrinaciones2019.pdf

References

Asad, T. (2003) *Formations of the Secular: Christianity, Islam, Modernity*. Stanford University Press, Stanford, California.

Bossy, J. (1983) The mass as a social institution 1200–1700. *Past and Present* 100(1), 29–61. DOI: 10.1093/past/100.1.29.

Brierley, J. (2019) *A Pilgrim's Guide to the Camino de Santiago*, 16. Camino Guides, Dyke, UK.

Burke, P. (2009) *Popular Culture in Early Modern Europe*. Ashgate, Farnham, UK.

Bynum, C.W. (1985) Fast, feast, and flesh: the religious significance of food to medieval women. *Representations* 11, 1–25. DOI: 10.2307/2928425.

Carreño, G.S. (2014) The glacier, the rock, the image: emotional experience and semiotic diversity at the Quyllurit'i pilgrimage (Cuzco, Peru). *Signs and Society* 2(S1), S188–S214. DOI: 10.1086/674324.

Connerton, P. (1989) *How Societies Remember*. Cambridge University Press, Cambridge, UK.

Durkheim, E. (1912) *The Elementary Forms of Religious Life*. Oxford University Press, Oxford.

Eade, J. and Sallnow, M. (1991) Introduction. In: Eade, J. and Sallnow, M. (eds) *Contesting the Sacred: The Anthropology of Christian Pilgrimage*. Routledge, New York, pp. 51–76.

Frey, N. (1998) *Pilgrim Stories: On and Off the Road to Santiago*. University of California Press, Berkeley, California.

Geismar, H. (2017) The instant archive. In: Horjit, L. (ed.) *The Routledge Companion to Digital Ethnography*. Routledge, New York, pp. 331–343.

Halbwachs, M. (1980) *The Collective Memory*. Harper & Row, New York.

Hirschkind, C. (2011) Is there a secular body? *Cultural Anthropology* 26(4), 633–647. DOI: 10.1111/j.1548-1360.2011.01116.x.

Munn, N.D. (1986) *The Fame of Gawa: A Symbolic Study of Value Transformation*. Massim Society. Duke University Press, Durham, North Carolina.

Ricoeur, P. (2006) *Memory, History, Forgetting*. University of Chicago Press, Chicago, Illinois.

Rubin, M. (1991) *Corpus Christi: The Eucharist in Late Medieval Culture*. Cambridge University Press, Cambridge, UK.

Sahlins, M. (2011) What kinship is (part one). *Journal of the Royal Anthropological Institute* 17(1), 2–19. DOI: 10.1111/j.1467-9655.2010.01666.x.

Sutton, D. (2001) *Remembrance of Repasts*. Berg Publishers, Oxford, UK.

Toomey, P.M. (1994) *Food from the Mouth of Krishna: Feasts and Festivities in a North Indian Pilgrimage Centre*. Hindustan Publishing Corporation, New Delhi, India.

Turner, V. (1969) *Ritual Process: Structure and Anti-Structure*. Routledge, New York.

Werbner, P. (2010) Beyond division: women, pilgrimage and nation building in South Asian Sufism. *Women's Studies International Forum* 33(4), 374–382. DOI: 10.1016/j.wsif.2010.02.011.

Yeoh, S.-G. (2006) Religious pluralism, kinship and gender in a pilgrimage shrine: the Roman Catholic feast of St. Anne in bukit mertajam, Malaysia. *Material Religion* 2(1), 4–37. DOI: 10.2752/174322006778053834.

9 Identity, Social Practices and Values of the Pilgrims on the Portuguese Inner Way to Santiago de Compostela

Pedro Azevedo* and Xerardo Pereiro
University of Trás-os-Montes and Alto Douro, Portugal

Introduction

Today, the Ways to Santiago de Compostela are characterized by being multi-motivational, not only attending to religious motivations (Lois-González and Santos, 2014; Amaro *et al.*, 2018). In fact, the differences between the pilgrim and the tourist are increasingly smaller, because both travellers are motivated by the experience of realizing the Way and are looking for the same type of services (Amaro *et al.*, 2018; Collins-Kreiner, 2018).

The objective of this chapter is to make a comparison between the identities and values of the Portuguese tourist-pilgrims (Pereiro, 2017, 2019) and the foreign tourist-pilgrims on the Portuguese Inner Way of Santiago de Compostela (PIWSC). The term tourist-pilgrim is perceived as a concept where tourism and pilgrimage are mixed and have no distinctions. There are differences between the motivations of the pilgrims according to their nationality, gender, age, education and other identity variables, including social action, like the option for a less-travelled way and the type of travelling by which pilgrims undertake their journey.

Some of the results obtained are evidence of a motivation for less mass-pilgrimage ways (Pereiro, 2017, 2019a, b; Gomes *et al.*, 2019) – as is the case with PIWSC. The cultural motivations and the combination of various motivations prevail, focused on the search for new experiences, the appreciation of heritage and contact with nature. Religious, social and sports motivations are the least indicated. Articulating with other studies, it is noticeable that the motivations are different according to various factors such as nationality and the way pilgrims undertake their journey.

Portuguese Inner Way of Santiago de Compostela (PIWSC)

The pilgrimage to Santiago de Compostela is a phenomenon that originated in the Middle Ages and has seen progressive changes, where tourism and religion create a symbiosis with each other.

In 2019, the Ways of Saint James reached, and for the third consecutive year set a new record, 350,000 pilgrims arriving in Santiago de Compostela through the various Paths, with emphasis on the French and Portuguese Ways, as shown in Table 9.1, taking into account data provided by Oficina del Peregrino of Cathedral of Santiago de Compostela (2019).

*Corresponding author: pedrodosrc@gmail.com

Table 9.1. Official statistics of pilgrims to Santiago de Compostela and distribution by path in 2018 and 2019.

	2018		2019	
Ways of Saint James	Number of pilgrims	Percentage	Number of pilgrims	Percentage
The French Way	186.199	56.88	189.937	54.65
The Portuguese Way	67.822	20.72	72.357	20.82
The Northen Way	19.040	5.82	22.292	6.41
The Primitive Way	15.038	4,59	19.019	5.47
The English Way	14.150	4.32	15.780	4.54
Vía da Prata (The Silver Way)	13.841	4.23	15.715	4.52
The Portuguese Way of the Coast	9.127	2.79	9.201	2.65
The Fisterra and Muxía Way	1.131	0.35	1.548	0.45
The Winter Way	703	0.21	1.035	0.3
Other Ways	326	0.10	694	0.2

Source: Pilgrimage Office of Cathedral de Santiago de Compostela. Available at: https://oficinadelperegrino.com/

It should be noted that the various Ways of Saint James show a tendency of growth in terms of the number of pilgrims who travel them. It is important to mention that there are about 2500–3000 pilgrims leaving Portugal, but the Pilgrimage Office of the Cathedral of Santiago de Compostela does not insert them in any specific statistics of the Paths. Also, many pilgrims make their way freely, not using official credentials, and there is no registration of them in the Pilgrimage Office, so they cannot receive the Compostela.

In recent years, the PIWSC (Fig. 9.1) has witnessed a profound revitalization. It presents itself as a Path slightly travelled by pilgrims – 338 in 2017; 308 in 2018 – with no data yet for 2019. This situation demonstrates a reduced massification and cultural intimacy that this pilgrimage experience provides.

The PIWSC is a path located in the northern part of Portugal that crosses the municipalities of Viseu, Castro Daire, Lamego, Peso da Régua, Santa Marta de Penaguião, Vila Real, Vila Pouca de Aguiar and Chaves (PIWSC, 2019), with a total length of 205 km in Portuguese territory and 180 km from the border with Galicia (Spain) to Santiago de Compostela, joining the Prata Path or the Mozárabe Path, as shown in Fig. 9.2.

The Path has the particularity of having a double way, i.e. it consists of taking the reverse route to Fátima, the Path being marked by blue arrows. The revitalization of the PIWSC in the 21st century was based on four factors: (i) inspired by a book by Professor Arlindo Magalhães Ribeiro da Cunha (1995) on the history of the Portuguese routes to Santiago; (ii) a walking meeting made on a section of this route in 2000 from Cidadela de Aguiar to Sabroso de Aguiar; (iii) an exhibition of photographs on the Jacobean routes taken in Vila Pouca de Aguiar in 2011; and (iv) the signing, on 7 April 2011, of a co-operation protocol among the municipalities of the PIWSC to homogenize the marking along the route.

The leadership of the project is the responsibility of the Vila Pouca de Aguiar City Council; its politicians, in particular, Ana Rita Dias (Vice-President of the City Council); and its tourism technicians, especially Catarina Chaves. The first step of this regional and transnational collaborative project was to identify, with historical bases, the route's layout. The second step was to create and post-signage along the route. The third was the creation of 'hostels' every 30–35 km to house pilgrims. On 24 April 2012, the first official walk of this new stage of revitalization of the PIWSC was celebrated. It should also be noted that the PIWSC project is part of an international collaborative network of the routes of Santiago (European Federation of Saint James Way, 2020) and the most recent Portuguese Federation of Saint James Way (Federação Portuguesa dos Caminhos de Santiago). The PIWSC is one of the Portuguese Ways to

Fig. 9.1. The PIWSC in Santa Marta de Penaguião. (Photo: the authors')

Santiago applying for UNESCO World Heritage recognition.

In the PIWSC there is a dialectic between different stakeholders such as the Associations of Pilgrims and Walkers, the Catholic Church, political technicians and businessmen, where all intend to develop the Path but not in the same way, because those who travel and those who activate the Path have many different visions. There are different discourses, representations and social practices in the PIWSC, as an arena of dialectical negotiation. In this sense, the Catholic Church of the small town of Vila Real has a very close view of the Church of Santiago de Compostela, as they see the PIWSC as a religious and cultural route and not as a tourist itinerary; therefore, 'It must be travelled along a path of faith and not as a tourist path' (Catholic Church in Vila Real, 4 January 2020). Political leaders and businessmen, on the other hand, want the Way to provide economic returns and, above all, tourism development.

In short, the motivations are influenced by these discourses, visions and different practices, where the dialectics end up changing, as shown in Table 9.2. In view of this, the PIWSC is a multi-faceted and multi-dimensional path.

In this chapter we are going to focus on the tourist-pilgrimages discourses, motivations and experiences. It is important to define the profile of the tourist-pilgrim who travels through CPIS, as well as their motivations and experiences, to better understand the identities and values of the tourist-pilgrimages.

Methodology

The methodology used for the pursuit of this study was essentially focused on conducting open and semi-structured interviews with 100 Portuguese and foreign tourists who walked and cycled the PIWSC. The purpose of these interviews was to identify the motivations and

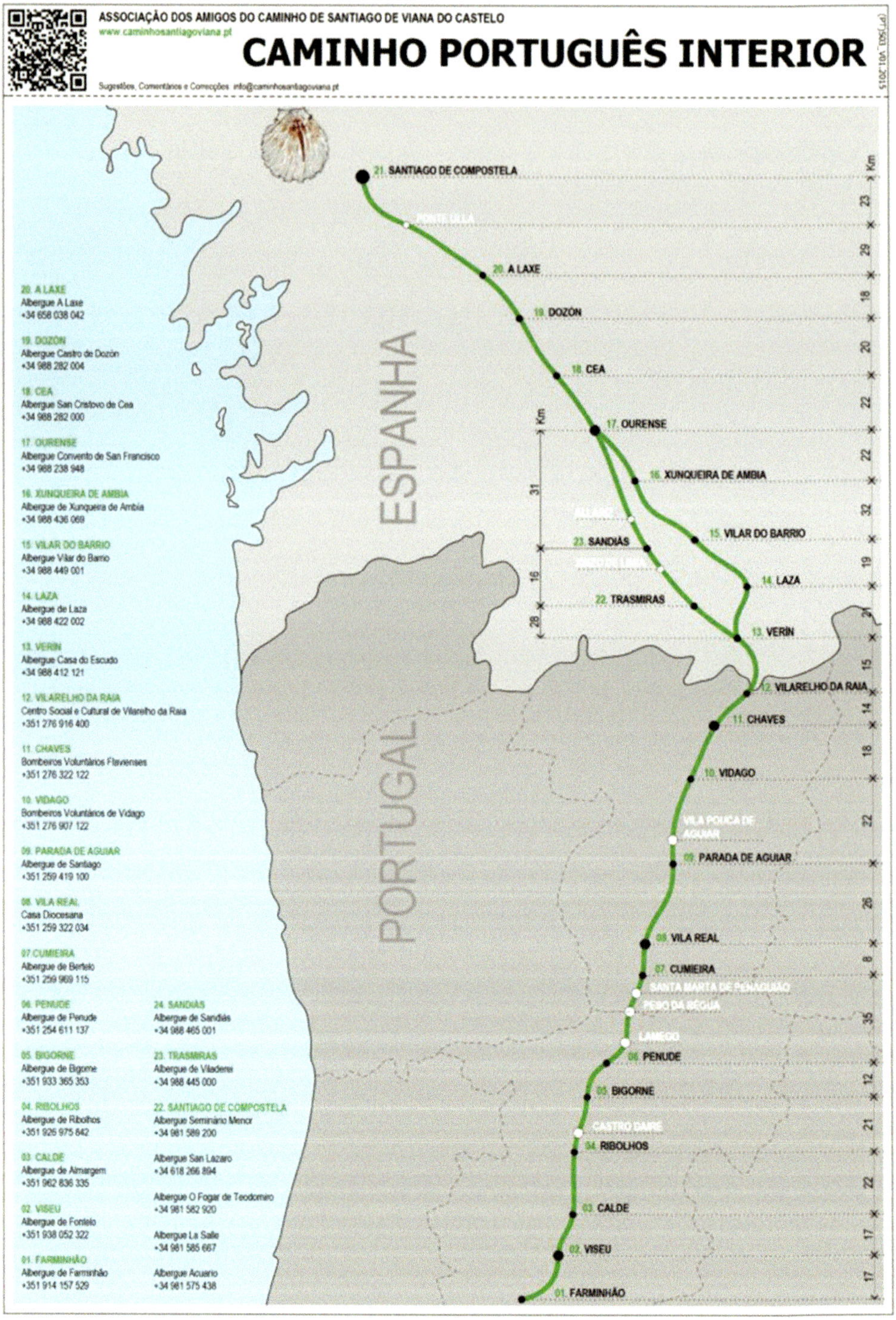

Fig. 9.2. A map of the PIWSC, between Viseu and Santiago de Compostela. (Source: Association of Friends of the Ways of Saint James - Viana do Castelo (2019) http://www.caminhosantiagoviana.pt/caminhointerior.html)

Table 9.2. Summary of the main agents and their discourses on the PIWSC.

Social agents	Discourses: visions and objectives
Pilgrims' associations and groups of walkers (civil society)	Way that combines the sacred, the profane and the spiritual
Catholic Church	Way of faith
Entrepreneurs (hotel agents, merchants, businessmen)	Tourist and economic path
Local authorities: technicians and politics (municipal and local councils)	Tourist and religious path

characterize the tourist-pilgrim experience lived at the PIWSC. At the same time, informal interviews and observant participation were applied to members of pilgrim associations and groups of hikers in the region, especially from Castro Daire, Vila Real and Vila Pouca de Aguiar. Also, we have walked and cycled the entire path and we have produced ethnographies of the walking experience (https://cpis.utad.pt/).

The main questions posed to pilgrims in the interviews were as follows: 1. What are the reasons why you chose to visit the PIWSC?; 2. What are the motivations that lead tourist-pilgrims to go through the PIWSC?; 3. What are the positive and negative perceptions that you obtained in relation to the PIWSC, the differences between Portugal and Spain, even having the Way to Fatima as a comparison?; 4. What are the valuation practices around the PIWSC?; and 5. Is there any feeling of belonging/connection to the Path itself?

The interviews were complemented with fieldwork and participant observation understood as a primary strategy. This combination of research techniques, such as the field diary, audio recording and photographic collection as recording instruments provided an analysis and understanding of the pilgrims' social practices and values, allowing a more interpretative approach to the ethno-biographical nature of the experience (Pereiro and Gómez-Ullate, 2019).

The interviews were conducted between November 2018 and January 2020. Eighty interviews were applied in person, in which the social memories obtained were shared with the researchers (Pereiro and Fernandes, 2018). Many of the interviews were carried out through telephone calls and mail, to overcome the difficulty of interviewing pilgrims in person, many of them of different origin. The interviews were transcribed and analysed according to qualitative content analysis techniques (Bardin, 2003).

Pilgrimages and Ways: Theoretical Perspectives

Pilgrimages are one of the oldest forms of population mobility (Collins-Kreiner, 2010) and are considered the oldest form of tourism (Timothy and Boyd, 2006). Recently, there has been a transition from a traditional pilgrimage to a tourist pilgrimage (Pereiro, 1999, p. 41; Griffin and Raj, 2012). Tourism is considered by some authors as a lay form of pilgrimage (Amirou, 2007, p. 29).

According to Pereiro (2019b), there are four theoretical perspectives of pilgrimages:

a. Functionalist (Halbwachs, 2017) – pilgrimage is seen as a way of social cohesion, of purifying the soul and connecting with the sacred;
b. Symbolic-performative-rituals (Turner, 1969; Turner and Turner, 1978; Pérez, 1995; Clift and Clift, 1996; Pereiro, 1999) – the pilgrimage is understood as a preliminary ritual inversion of everyday social structures, which means they are anti-structural experiences with a connection to tradition and the past, and corporal punishment as a way of salvation and rebirth;
c. Dialectic – pilgrimage is a heterogeneous social field and full of disputes, social divisions, discourses and senses in tension (Murray and Graham, 1997; Coleman and Eade, 2004) – it does not constitute a fixed category and structure; however, it is dynamic, full of confrontations between the local and global, the Catholic faith and the various spiritualities, solidarities and conflicts;

d. Procedural and transformative (Graburn, 1977, 1983, 2001) – the pilgrimage is defined as an ethno-biographical experience of personal spiritual transformation that combines with contemporary tourism, creating multi-dimensional, polysemic and complex experiences (Kurrat, 2018); it is a subculture of global tourism. Thus, the pilgrimage can also be considered as a therapeutic system, suggested by psychiatrists to their patients as a way to rearrange their personal lives and to walk in contact with nature (Ingold and Vergunst, 2008), with others and with themselves on an inner journey.

Peter Margry (2008) states that the pilgrimage, combined with the tourist experience, also becomes a social practice with a more spiritual, polysemic and multi-motivational tendency. The fact of walking involves multiple motivations. Even covering a small section of the Path is still a goal, an objective or an adventure for the walker.

In 1986, officially, there was only one Way – the French Way. Now there is a proliferation of Paths due to the increase in the number of tourists, the reinvention of history, tradition and the Jacobean cultural heritage, and the political and commercial local interests, because people want to 'collect' the Paths travelled.

Currently, several Paths of Saint James are symbols of the local, regional, national, state and European cultural identity. Likewise, they form a model of an international tourist-cultural route that is imitated all over the world in a process that anthropologist Peter Jan Margry (2008) calls 'Caminonization' and that has led to a substantial internationalization of the Ways of Saint James (Murray and Graham, 1997; Pereiro, 2019a). This internationalization is related to different interests of several agents. In the case of the Ways of Saint James, we witnessed four great narratives: (i) those of the Catholic Church: the Way as faith and God; (ii) those of the Policy, based on identity and power; (iii) those of Civil Society, based on cultural and natural heritage; and (iv) those of the Market, which emphasizes tourism, cultural routes and economic development. These four narratives involve and oscillate between the religious, the spiritual and the profane; Catholic and non-Catholic; the religious, the secular and the post-secular (Blom *et al.*, 2015).

In this same thematic line, Joan Prat (2011) defines three versions of the Way:

a. A pagan version ('*pagus*' – land): open, secular and heterodox, full of telluric and spiritual energies, based on an ecosophy with great importance for Mother Nature, the inner change of life and the path as a 'turning point' to the happiness.
b. A Christian and Orthodox religious version, which understands the Way as a meaning of the Christian life, a physical and virtual path to God, of prayer, meditation and bodily pain. The Way is an ascetic, purifying path for the salvation of the soul.
c. A mixed version, religious and spiritual, magic and faith, sacredness and tourism; ritual initiation, Franciscan metamorphosis and interior path.

Pilgrimage today means something different from the past: (i) a purifying physical effort prior to contact with the sacred; (ii) a space for reflection to think about one's feelings and actions; (iii) a return to nature according to the new values in harmony with it; (iv) an important business for local communities, mediators and other social agents; (v) a space for ideological, political and institutional discourses; (vi) a search for inner peace and the meaning of life; and (vii) an antidote against social malaise. Thus the pilgrimage became touristic and more spiritual than religious.

In view of these analysed theoretical perspectives, we can say that the pilgrimage is polysemic and bears multi-motivations.

Experiences and Motivations

Tourist motivations are understood as the reasons that lead a person or a group to travel. Currently, the motivations of tourists have become increasingly complex and diversified (Pereiro and Fernandes, 2018) and assume a central axis in tourism research.

Currently walking the Way to Santiago de Compostela refers to a multi-motivational framework, where there are not only religious motivations (Lois-González and Santos, 2014;

Amaro *et al.*, 2018). In this sense, the differences between the pilgrim and the tourist diminishes, because both travellers are motivated by the experience of perceiving the Way and are looking for the same type of services (Amaro *et al.*, 2018; Lois-González and Santos, 2014). Thus, the concept of tourist-pilgrim arises.

Pereiro and Fernandes (2018) state that the scientific approaches to tourist motivations are the following:

a. economy: the tourist's choice is motivated by price, profit and benefit – therefore it is a matter of cost-opportunity;
b. history: the motivations of the trip have undergone changes throughout history;
c. psychology: its centre was the individual personality and the satisfaction of its desires and needs (intrinsic motivations); and
d. sociology and anthropology: emphasis on society and the social as builders of motivations, more than in the individual; they are extrinsic motivations that are socially and culturally conditioned. Social *anomie* and the search for 'status' are presented by these social sciences as central explanations of tourist motivations, among others.

In summary, tourist motivations obey intrinsic (personal) and extrinsic (structural, social and institutional) factors, as well as factors of internal impulse and external attraction that several authors rank as more or less important determinants or conditions in the tourist experience (Pereiro and Fernandes, 2018).

Each person has their own Way, one in the head and one in the body, and their own particularities.

For Singul (1999), five main motivations are defined for travelling the Way of Saint James: (i) religious/devotional (traditional); (ii) cultural (medieval origin and art history); (iii) ecological (contemplation and appreciation of the landscape, nature and the environment); (iv) spiritual and ecumenical; and (v) personal (meditation on someone's life).

But, based upon our fieldwork research, we could underline that the tourism motivations of pilgrimages are very complex, biographical and with several levels of intimacy. These hidden layers were not expressed in interviews with the researchers.

Results

Notwithstanding the small number of pilgrims who travel through the PIWSC, as we verified earlier, in order to carry out this work we interviewed 100 tourist-pilgrims who had travelled through the PIWSC in recent years, thus presenting themselves as a representative and significant sample and consolidating other studies already carried out (Pereiro, 2017, 2019a, b; Gomes *et al.*, 2019).

The results obtained allow us to perceive that there are distinctions in terms of motivations and experiences.

Profile of Tourist-pilgrims on the PIWSC

The profile of the 100 tourist-pilgrims interviewed is characterized as follows: 79 men and 21 women, aged between 19 and 71. Regarding nationality, 84 are Portuguese, while 16 respondents were foreigners – 3 from Spain, 4 from Brazil, 2 from Germany, 2 from the Czech Republic and 5 from other countries. Therefore, the PIWSC is mostly run by national pilgrims and, to a lesser extent, by foreigners. It should also be noted that 49 pilgrims had received higher education.

In general, the majority of the tourist-pilgrims come from locations around the PIWSC. In this sense, it is important to mention that a large part of the tourist-pilgrims are members of associations of local pilgrims, groups of hikers, mountain bikers and scouts from the region, which helps to reveal their own motivations and personal experiences when travelling the Way, differentiating themselves from other tourist-pilgrims. These associations and groups of walkers are entities that involve the local population and cover themselves with recreational and sporting components.

Most of them made the Way during the spring, and in some cases in the summer, largely due to the better weather conditions and the mild temperatures. The availability of free time on weekends and the enjoyment of holidays leads people to do the Way, providing more introspective and intercultural leisure activities.

Most respondents have repeated the experience on the PIWSC or other Ways of Saint James. This situation refers to the collection of different paths of experience that each pilgrim wants to have.

It should be noted that 87 interviewees followed the Way with more than two people, while only 13 people did it alone and all the lone pilgrims were foreigners. Half of the people revealed that they had met other pilgrims along the Way. These two situations reflect the sociability of this experience and the meaning of the PIWSC as a space of solidarity and social network construction.

The way in which the pilgrims travelled was divided as follows: 79 travelled on foot while 21 travelled on mountain bikes. In this aspect, the time spent walking the Path and carrying out various activities presents differences between walkers and bicycle pilgrims (Cardoso, 2013). Walkers take more time to go through the stages, taking longer to arrive and must rest to regain strength for the next day. Bicycle pilgrims take less time and tire less easily, and when they arrive, earlier, they have more time to get to know the places and points of tourist interest.

Another important indicator is the fact that some tourist-pilgrims do the PIWSC in sections or stages, not reaching Santiago de Compostela, revealing that the interviewees feel fulfilled by doing only part of the Way, where each stage really corresponds to a goal. Foreign pilgrims, in general, travel the Way along its entire length (Viseu-Santiago de Compostela).

Reasons for Choosing the Portuguese Inner Way of Santiago

In this section we have highlighted the reasons why the tourist-pilgrims chose to visit the PIWSC. The motives are essentially related to the landscape, being a unanimous criterion, in the interest of knowing the Way itself and the existing heritage around it: 'little mass route is an asset' (woman, 54 years old, Mondim de Basto); and to 'get to know and travel an international tourist route of this dimension in our region' (male, 24 years old, Cabeceiras de Basto) and the proximity to the place of residence. With less expressiveness, the interviewees stated that they also had sociability and the performance of sports activities, such as walking, running and mountain biking, as reasons.

Motivations

Of the 100 interviews carried out, it was possible to prove that a large portion is part of local associations of pilgrims and groups of hikers, mountain-bike groups and scouts. The people who are part of these pilgrim associations have their own vision of the Way and have the following characteristics:

- connection to one's own path and heritage; referring to a feeling of belonging;
- maintenance of the Way, creating a network of synergies with the local community;
- connection with nature and outdoor activities, articulating with other tourist routes;
- use the PIWSC to perform other sports activities such as mountain biking and trekking;
- allowing coexistence with people in the region; and
- a tendency to walk the Way in stages.

Given these facts, there are several motivations for going through the PIWSC, as shown in Fig. 9.3. Cultural motivations predominate (36%), followed by the combination of several motivations (17%) and adventure motivations (16%). On the opposite side, only 7 pilgrims report that they travel the Way for religious reasons. Lastly, there are sports motivations (8%). These are framed in the motivations proposed by Singul (1999).

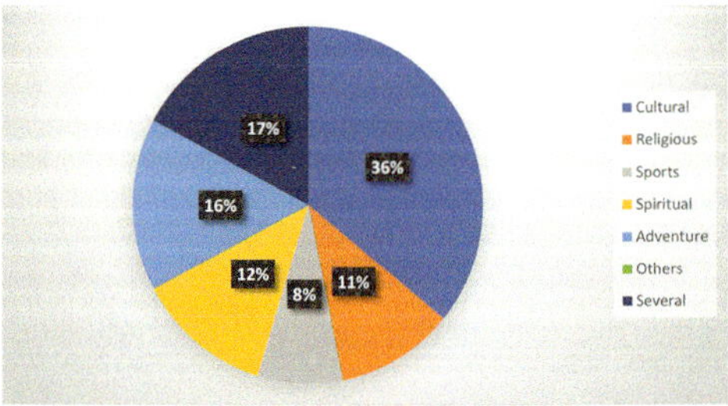

Fig. 9.3. Main motivations for going through the PIWSC.

Meeting what was pointed out by Christian Kurrat (2018), we consider that the motivations of pilgrims must be contextualized in their biographical paths to better understand them. That is why each pilgrim has their own motivations for making the journey. We can assume that they are biographical motivations (Kurrat, 2018) and psycho- and socio-cultural, governed by the absence of guidelines and norms. We speak of tourist-pilgrims' motivations because, in our view, the main motivations link traditional and ancient pilgrim motivations with contemporary tourist motivations, and because the tourist motivations of the pilgrimage are multi-dimensional.

Comparing the data obtained in other studies (Gomes *et al.*, 2019), we found that foreign pilgrims walk the Path alone. We find here a clear difference between Portuguese pilgrims and foreigners. On the one hand, all the Portuguese pilgrims made this journey as a group, but on the other hand, among foreigners, this no longer happens. The motivation of walking together in the Portuguese pilgrimages is a key force to reinforce the social relations in local communities (*i.e.* neighbourhood, friendship, family links, kinship, and so on).

Another relevant fact is related to the journey in stages, fragmented throughout the year. The small towns of Viseu, Vila Real and Chaves are the main points of departure for these pilgrims on their Way to Santiago.

Experiences and Meanings

Another issue that we tried to clarify with the interviewees is related to how they classified the experience as tourism, pilgrimage or both. The testimonies obtained show that 57% of tourist-pilgrims considered that travel through the PIWSC, combined tourism and pilgrimage. Nevertheless, 37% considered that they took the Way for tourism and only 6% claimed it to be a true pilgrimage. In this sense, it is noted that the number of people who travel through the PIWSC with a truly religious feeling is reduced. In view of this, visiting the PIWSC is a polysemic activity that links tourism with the religious and the spiritual (Fig. 9.4).

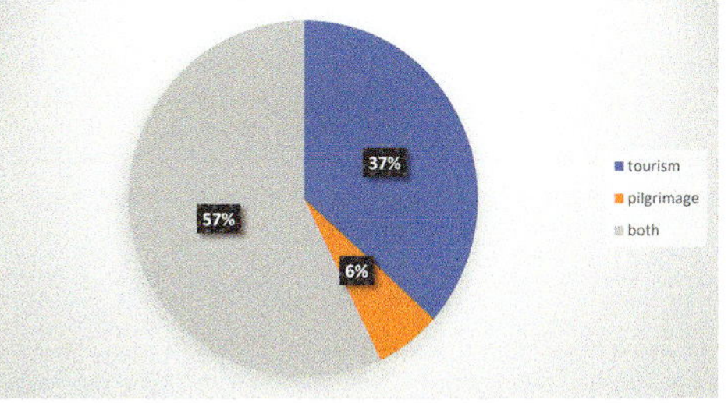

Fig. 9.4. Typology of motives.

We obtained positive and negative perceptions regarding the PIWSC. As a large percentage of the interviewees are members of associations, it is evident that they have a greater knowledge of the positive and negative aspects existing in the course. Positive perceptions are related to the landscape, religious heritage, towns along the way, the encounter with other pilgrims, hospitality and security. Regarding negative perceptions, these are related to the lack of maintenance of the Path in some sections, the reduced offer of hostels, and that much of the religious heritage is closed and the signage is insufficient.

Going through the PIWSC provides an experience of a certain cultural intimacy (Herzfeld, 2005). It allows interaction with local communities, with pilgrims and tourists, knowledge of heritage, involvement with nature and, above all, with oneself.

The first idea that we had after our analysis was the multiplicity and multi-dimensionality of the pilgrim experiences to Santiago de Compostela: namely, each pilgrim has his or her own motivations as well as different ways of articulating the different dimensions of the motivations.

The experience obtained allowed for spiritual growth, cultural enrichment, contact with nature, adventure and a feeling of autonomy. Ninety per cent of the interviewees also mentioned that taking the Path stimulated personal reflection, because in the Path 'faith and values such as tranquillity that allow self-reflection are adjacent' (man, 36 years old, Cabeceiras de Basto). The PIWSC represents a transforming and invigorating human experience for pilgrims. Thus, walking the Path translates and shapes changes in the way of being.

Half of the tourist-pilgrims reveal that they have previous experience of traversing other Paths, such as the Way to Fatima, different Ways of Saint James or other more regional pilgrimage Paths. Nineteen have already travelled to Fátima, 20 have already made other Ways of Saint James, while four have only made pilgrimages to regional places of worship. 'The Fátima Way only has a religious aspect, while the Way of Saint James is free from religions and religious practices, involving other aspects such as the landscape, contact with the population, among others' (leader of a group of hikers in Vila Real; interview held on 16 January 2020).

Most of the interviewed tourist-pilgrims establish a feeling of belonging with other pilgrims as well as with other communities of pilgrims, groups of hikers and mountain bikers in the region. In that sense, the Ways to Santiago are a mechanism of the construction of a sense of global humanity.

As for the differences detected in the PIWSC between the sides of the border between Portugal and Galicia (Spain), the tourist-pilgrims mention the following aspects:

- The PIWSC in Portuguese territory is more attractive, having a unique and diverse landscape.
- The inhabitants are hospitable and the heritage is varied.
- The Path has sections of greater difficulty and poor marking.
- The PIWSC in Galicia has better marking, sections with better conditions, greater offers of hostels and more pilgrims.
- The sign at the entrance to Verín is poor.

Conclusions

The results obtained allow us to understand the multi-motivations, social practices and meanings of the pilgrim experience in travelling through the PIWSC. We present a multiplicity of motivations and lived experiences.

As earlier verified, the PIWSC is no longer an exclusive path of the Catholic faith, but a path that follows the three aspects previously exposed by Joan Prat (2011): secularism, religiosity and spirituality.

The main inventoried motivations that lead pilgrims to travel the Way are as follows:

a. proximity to the place of residence and feeling of identification with the Path;
b. The PIWSC has elements, such as the landscape, that make it distinctive in relation to other roads in Santiago;
c. possibility of socializing;
d. opportunity to practice sports and perform outdoor and nature activities.

In short, the following are true of the PIWSC:

- It is a path in the northern inland of Portugal, presenting itself as a 'resource' that belongs to the community and that values the territory. Little by little, locals feel the Path to be 'their own', valuing it and getting involved with feeling and dedication.
- It presents a multi-dimensionality of motivations.
- It could be a sub-tourism experience, that is no mass tourism and a quiet destination for personal reflection.
- The motivations of foreign and Portuguese tourist-pilgrims differ: pilgrims travel in groups while foreigners travel alone.
- The groups of hikers promote journeys on the Way and reuse it monthly and annually, carrying out various activities, giving it a more playful and sociable aspect; this promotes the coexistence of tourists, pilgrims and the local community (leader of a group of hikers in Vila Real, interview held on 16 January 2020).

Although other Ways of Saint James work thanks to the visitors and tourist-pilgrims who travel through them, the PIWSC has the particularity of being the place to develop, largely due to the role played by associations of pilgrims and groups of hikers.

From a phenomenological perspective (Graburn, 1977, 1983, 2001; Cohen, 1979, 1988), we consider the pilgrimage to Santiago de Compostela as the search for a new spiritual and existential centre. More specifically, we interpret the PIWSC as a metaphor of life, for life and a liminal space, in which there is always a dialectic and tension between social cohesion, the sense of community, communitas, and social division and distinction as complex human practices.

From a dialectical perspective, the PIWSC is a social space where pilgrims and social agents are constructing different versions of social and cultural identities and using different values to create different meanings of the pilgrimage from the point of view of the local offer. Also, there is a symbolic confrontation of discourses and values that are going to influence the future development of this route.

Acknowledgements

This work is carried out within the scope of the research grant with the reference SFRH / BD / 136459/2018 attributed by FCT – Portugal and the Sabbatical License Research Grant with the reference SFRH / BSAB / 150447/2019 attributed by FCT – Portugal. The Center for Transdisciplinary Studies for Development (CETRAD) is financed by national funds through FCT – Foundation for Science and Technology, I.P., under the project UIDB / 04011/2020.

References

Amaro, S., Antunes, A. and Henriques, C. (2018) A closer look at Santiago de Compostela's pilgrims through the lens of motivations. *Tourism Management* 64, 271–280. DOI: 10.1016/j.tourman.2017.09.007.

Amirou, R. (2007) *Imaginário turístico e sociabilidades de viagem*. Estratégias Criativas – APTUR, Vila Nova de Gaia.

Association of Friends of the Ways of Saint James - Viana do Castelo (2019). Available at: http://www.caminhosantiagoviana.pt/caminhointerior.html

Bardin, L. (2003) *Análise de conteúdo*. Edições 70, Lisboa.

Blom, T., Nilsson, M. and Santos, X. (2015) The way to Santiago beyond Santiago. Fisterra and the pilgrimage's post-secular meaning. *European Journal of Tourism Research* 12, 133–146.

Cardoso, F. (2013) *Bom Caminho – Um convite viagem e reflexão*. Planeta, Lisboa.

Clift, J.D. and Clift, W.B. (1996) *The Archetype of Pilgrimage: Outer Action with Inner Meaning*. Paulist Press, New York.

Cohen, E. (1979) A phenomenology of tourist experiences. *Sociology* 13(2), 179–201. DOI. 10.1177/003803857901300203.

Cohen, E. (1988) Authenticity and commoditization in tourism. *Annals of Tourism Research* 15(3), 371–386. DOI: 10.1016/0160-7383(88)90028-X.

Coleman, S. and Eade, J. (2004) *Reframing Pilgrimage: Cultures in Motion*. Routledge, London.

Collins-Kreiner, N. (2010) Researching pilgrimage. *Annals of Tourism Research* 37(2), 440–456. DOI: 10.1016/j.annals.2009.10.016.

Collins-Kreiner, N. (2018) Pilgrimage-tourism: common themes in different religions. *International Journal of Religious Tourism and Pilgrimage* 6(1), 8–17.

Cunha, A. (1995) *Caminhos Portugueses de Peregrinação a Compostela. Itinerários portugueses (Identificação dos caminhos e coordenação da investigação)*. Santiago de Compostela/Porto, Xunta de Galicia/Centro Regional de Artes Tradicionais.

European Federation of Saint James Way (2020). Available at: http://www.saintjamesway.eu/en/ (accessed 11 January 2020).

Gomes, C., Losada, N. and Pereiro, X. (2019) Motivations of pilgrims in the Portuguese inner way to Santiago de Compostela. *International Journal of Religious Tourism and Pilgrimage* 7(2), 31–40.

Graburn, N. (1977) Tourism: the sacred journey. In: Smith, V. (ed.) *Hosts and Guests: The Anthropology of Tourism*. University of Pennsylvania Press, Philadelphia, Pennsylvania, pp. 17–31.

Graburn, N. (1983) *To Pray, Pay, and Play: The Cultural Structure of Japanese Domestic Tourism*. Centre des Hautes Études Touristiques, Aix-en-Provence.

Graburn, N. (2001) Secular ritual: a general theory of tourism. In: Smith, V. and Brent, M. (eds) *Hosts and Guests Revisited: Tourism Issues of the 21st Century*. Cognizant Communications, London, pp. 42–50.
Griffin, K. and Raj, R. (2012) *Reflecting on Religious Tourism and Pilgrimage*. Association for Tourism and Leisure Education (ATLAS), Arnhem.
Halbwachs, M. (2017) *Los orígenes del sentimiento religioso según Durkheim. Introducción a la Sociología de la religion de Émile Durkheim*. Ediciones Dado, Madrid.
Herzfeld, M. (2005) *Cultural Intimacy. Social Poetics in the Nation-State*. Routledge.
Ingold, T. and Vergunst, J.L. (2008) *Ways of Walking: Ethnography and Practice on Foot*. Ashgate, Farnham, UK.
Kurrat, C. (2018) Biographical motivations of pilgrims on the Camino de Santiago. In: *Book of Abstracts. 10th Annual International Religious Tourism and Pilgrimage (IRTP) Conference, 27th–30th*. 37. University of Santiago de Compostela, Santiago de Compostela.
Lois-González, R.C. and Santos, X.M. (2014) Tourists and pilgrims on their way to Santiago. motives, Caminos and final destinations. *Journal of Tourism and Cultural Change* 13(2), 149–164. DOI: 10.1080/14766825.2014.918985.
Margry, P.J. (2008) *Shrines and Pilgrimage in the Modern World. New Itineraries into the Sacred*. Amsterdam University, Amsterdam.
Murray, M. and Graham, B. (1997) Exploring the dialectics of route-based tourism: the Camino de Santiago. *Tourism Management* 18(8), 513–524. DOI: 10.1016/S0261-5177(97)00075-7.
Oficina del Peregrino of Cathedral of Santiago de Compostela (2019). Available at: www.oficinadelperegrino.com/ (accessed 12 January 2020).
Pereiro, X. (1999) *Homo peregrinus*. Xerais, Vigo, Spain.
Pereiro, X. (2017) Turiperegrinos portugueses no Caminho Português interior de Santiago de Compostela. *Revista Turismo e Desenvolvimento* 27/28, 413–423.
Pereiro, X. (2019a) *Património cultural jacobeu, turismo e peregrinação*. O Caminho Português Interior de Santiago de Compostela. PASOS Edita, La Laguna.
Pereiro, X. (2019b) Tourism and pilgrimage, two sides of the same currency: the Portuguese inland way of St James. *Cuadernos de Turismo* 43, 613–616.
Pereiro, X. and Fernandes, F. (2018) *Antropologia e Turismo: Teorias, métodos e praxis*. PASOS, Tenerife.
Pereiro, X. and Gómez-Ullate, M. (2019) Pilgrimage tourism and cultural route team Ethnographies in the Iberian Peninsula: a collaborative study. In: Andrews, H., Dixon, L. and Jimura, T. (eds) *Tourism Ethnographies*. Routledge, London, pp. 112–127.
Pérez, N. (1995) Camiño de Santiago, metáfora da vida humana. *Compostellanum* 40(3-4), 465–480.
PIWSC (2019). Available at: https://cpis.utad.pt/mapas-do-cpis-via-da-prata-sig/ (accessed 11 January 2020).
Prat, J. (2011) ¿Por qué caminan? Una mirada antropológica sobre el Camino de Santiago. In: Nogués-Pedregal, A. and Olmos, F. (eds) *La cultura sentida. Homenaje al profesor Salvador Rodríguez Becerra*. Signatura ediciones, Seville, pp. 495–529.
Singul, F. (1999) *O Caminho de Santiago: a peregrinação ocidental na Idade Média*. UERJ, Rio de Janeiro.
Timothy, D.J. and Boyd, S.W. (2006) Heritage tourism in the 21st century: valued traditions and new perspectives. *Journal of Heritage Tourism* 1(1), 1–16. DOI: 10.1080/17438730608668462.
Turner, V. (1969) *The Ritual Process: Structure and Anti-Structure*. Aldine Publishing Company.
Turner, V. and Turner, E. (1978) *Image and Pilgrimage in Christian Culture*. Columbia University Press, New York.

10 Pilgrimage and Tourism in Japan: Shift in Values and Motivation

Ricardo Nicolas Progano* and Kumi Kato
Wakayama University

Introduction

In a context of massive tourism development in the Asia-Pacific region, the Japanese government has prioritized tourism development since the early 2000s, particularly inbound tourism, as a way to revitalize regional economies suffering from low economic activity and an ageing population, and to assist in the recovery of areas affected by the 2011 Tohoku earthquake and tsunami disaster, as stated in different official documents such as the *Tourism Nation Basic Promotion Plan* (Prime Minister of Japan and Cabinet, 2012) and *Tourism Vision for Supporting the Japan of Tomorrow* (Prime Minister of Japan and Cabinet, 2016). Pilgrimage and religious tourism are connected to these national government policies, particularly regional revitalization, because its resources (such as trails and sacred sites) are commonly found, or, more precisely, found to remain intact, in the countryside. This is the case of Koyasan (高野山), Shikoku Henro (四国遍路) and Kumano Kodo (熊野古道), among other pilgrimages. By directing visitors to said sites, not only regional communities can benefit from tourism development, but it can also help to alleviate concerns of overtourism in major destinations. International trends also highlight the importance of religious and pilgrimage tourism, as this type of travel has become a major tourism market (UNWTO (United Nations World Tourism Organization), 2011). Pilgrimage destinations attract a growing number of tourists, who visit these locations even for reasons not related to religious faith such as non-affiliated spirituality, leisure, intercultural understanding or wellness. In this context, pilgrimage sites play an important role as tourism resources that attract both domestic and international visitors. However, visitor motivations have not been fully investigated – What kind of motivations drive them? In order to shed light on this subject, this chapter aims to explore changes in travel motivation to pilgrimage sites in Japanese society across different historical contexts, mainly during the post-war period. It will also discuss the increasing presence of western visitors in Japanese pilgrimage sites, who are motivated by a variety of factors not limited to institutionalized religion.

Concept(s) of Japanese Pilgrimage

Stoddard explains that 'converting the concept of pilgrimage into an operational definition that will apply equally well to all cultural settings is difficult' (Stoddard, 1997, p. 49). Therefore, it is necessary to first discuss the meaning of pilgrimage in Japanese language to reach a better

*Corresponding author: nprogano@wakayama-u.ac.jp

understanding of the concept. What does 'pilgrimage' mean in Japan? The matter is not as easy as a simple Japanese-to-English translation. As Reader and Swanson (1997) note, multiple Japanese words can be technically translated as 'pilgrimage'. Some of these words overlap and can be used interchangeably, but other terms are mutually exclusive:

- *Junrei* (巡礼): refers to visiting multiple religious sites that are linked together in a circuit. During the Heian period (794–1185), it implied ascetic practices and visiting miraculous sites, but the term is no longer limited to this use nowadays.
- *Junpai* (巡拝/順拝): less common than *junrei*, but with the same implications, emphasizing praying.
- *Henro* (遍路): a circuitous route. Specifically used to refer to the Shikoku Henro pilgrimage in Shikoku island and its pilgrims.
- *Meguri* (巡り): literally meaning to 'go around'; it refers to a pilgrimage to a number of shrines and temples. It is most used when the sites of a pilgrimage route are not dedicated to the same divinity, but where a group of them are related.
- *Moude* (詣): a visit to worship at a shrine or temple. It may be added as a suffix to the names of sacred sites or areas to indicate a special journey of worship.
- *Mairi* (参り): the process and practice of making special visits to a sacred place that is not part of one's everyday life. It is also widely used to refer to the pilgrimage to Ise.
- *Sankei* (参詣): indicates the process of visiting sacred sites and making pilgrimages. It is not used to refer to a specific pilgrimage.
- *Sanguu* (参宮): refers to pilgrimage to the Ise shrine.
- *Sanpai* (参拝): denotes a more ordinary visit to a local shrine, as well as the act of worship itself.
- *Yuugyou* (遊行): refers to pilgrimage or religious itinerancy. It is commonly related to ascetic practices and religious wanderers such as the *hijiri* (聖).

Another related concept is *utsushi* (写し), which refers to copied pilgrimages. This phenomenon is commonly found in Japan and consists of replicating major pilgrimages in smaller localized sites for people who cannot afford to make the journey to the main site. One prominent example is the replication of the Shikoku Henro pilgrimage in Sado Island, Niigata prefecture. Pilgrimage replication is, however, not exclusive to Japan and can be observed in other religions such as Christianity (Reader, 2014). Finally, there are pilgrimage-related terms that are mostly used in the context of mountains. For example, *mineiri* (峰入) refers to specific mountain retreats and is used in relation to Shugendo (修験道), a syncretic Japanese mountain asceticism (Reader and Swanson, 1997). There is a close relationship between mountain asceticism and pilgrimage in Japan, since some of the routes, such as Shikoku Henro or Kumano Kodo, were originally sites for ascetic practices and not popular sites that attracted a large number of worshippers. Interestingly enough, coastal settings play a relatively small role in pilgrimage, despite Japan being an island country. Gorai (1989), however, notes that pilgrimage sites such as Shikoku Henro, Ise and Kumano Kodo contain elements of ancient coastal worship, which are often overlooked nowadays, perhaps because of the strong cultural impact of Chinese ideas on sacred mountains (Koizumi, 2001).

Regarding their structure, pilgrimages can be based on multiple sites that are linked together (Kadota, 2013). These multiple-site routes do not have a single 'centre' but, instead, encompass a determined geographical area where the sites that should be visited are located. Another characteristic is that there is a greater scope in performance of their itineraries in the sense that pilgrims have the choice of visiting the sites at their own convenience and even complete the pilgrimage in a multiple number of journeys over a period of time.

Historical Transformations of Japanese Pilgrimage

Spiritual travel has a long tradition in Japan in the form of pilgrimage to the different sacred sites of the country, having its origins in the ritual travels of mountain ascetics along the Kii peninsula and Shikoku island, as well as the official visits to the Ise Grand Shrine. During the Heian period (794–1185), emperors and aristocrats travelled from the capital city of Kyoto to the mountain ranges of Kumano located in the Kii peninsula. Their travel included ritual purifications and other religious practices, but contained leisure. Politics was also a motivation for the aristocracy, since performing pilgrimages helped to reinforce central authority, obtain symbolic power and maintain relations with local governors and clergy (Moerman, 1997). In the Kamakura period (1185–1333), the new military class joined the aristocracy in the pilgrimage practice. The Muromachi period (1336–1573) witnessed the emergence of mass pilgrimages, with large numbers of peasants undertaking the Kumano pilgrimage as well. During this period, the Ise pilgrimage also began to gain popularity among larger sectors of society (Hoshino, 1981). During the Edo period (1603–1868) the government established travel restrictions that prevented the population from freely travelling, but allowed travel under special conditions such as pilgrimages (Guichard-Anguis, 2009; Funck and Cooper, 2013). This fact, coupled with improved travel conditions and quality of life, led to the popularization of travel (Hoshino, 1981), making mass tourism a combination of pilgrimage and sightseeing. The most visited pilgrimage destination was the Ise Grand Shrine, which eventually replaced Kumano in popularity (Funck and Cooper, 2013; Reader, 2014). While Ise peregrinations became more of a leisure travel, travellers strongly motivated by religion (Funck and Cooper, 2013) visited other pilgrimages such as the Saikoku pilgrimage, the 33 Kannon temples in the west, and the Shikoku Henro. Pilgrimages to Mount Fuji were also popular during the Edo period. Organized by local pilgrimage associations (*kou* 講) located across the country, the pilgrimages were composed of common farmers and townspeople (Earhart, 1989). According to Koizumi (2001), the introduction of western notions of leisure and hiking to Japan during the late 19th century brought an increasing interest in climbing sacred mountains for scientific research and leisure. Thus, pilgrimage locations began to be visited for new secular motivations.

Immediately after the end of World War II, the socio-economic hardships that Japan faced decreased the overall number of pilgrims (Hoshino, 1981; Reader, 2014). During the post-war period, the Saikoku pilgrimage gradually declined in popularity and instead the Shikoku Henro pilgrimage started to attract a growing number of visitors from the 1960s due to changes in its public depiction, and transport and infrastructure development (Reader, 2014). In particular, stakeholders such as mass media, transport companies and tourism offices played a key role in transforming Shikoku Henro from a far-away, dangerous place to a pleasant tourism destination where visitors can take on journeys of self-discovery in a 'traditional' Japanese setting (Mori, 2005; Reader, 2014). Because of its new image as a repository of traditional culture, which is believed to be lost in urban areas, it could effectively become the 'national' pilgrimage (Reader, 2014). In this sense, the post-war 'pilgrimage boom' originated, in part, as a search for identity by the Japanese people (Hoshino, 1981). Leisure hiking to sacred peaks located in pilgrimage circuits also began to increase as socio-economic conditions improved (Koizumi, 2001).

The UNESCO World Heritage Sites (WHS) also have a relationship with tourism development in Japanese pilgrimage sites. In the 2000s, Japan experienced a nationwide 'world heritage boom' as local communities realized the potential of tourism development for economic revitalization through a UNESCO designation (McGuire, 2013; Sawamura, 2016). The national government, through the Agency for Cultural Affairs, supported this movement by conducting public recruitments for enlisting WHS candidates (Sawamura, 2016). Naturally, pilgrimage sites and routes are no exception to the world heritage boom. The first Japanese pilgrimage route designated as a UNESCO World Heritage Site (WHS) was the Sacred Sites and Pilgrimage Routes in the Kii Mountain Range, in 2004, encompassing a large system of pilgrimage routes and temples located in the Kii Peninsula of Honshu island. This designation had an influence on other pilgrimage routes such as Shikoku Henro, whose related stakeholders have started to work towards a UNESCO designation (Mori, 2005; Okamoto, 2015) (Fig. 10.1.). Christian churches located in the Nagasaki prefecture were also designated in 2018. Although the sites were not connected through pilgrimage routes, different courses have been created to organically connect the properties and commodify them for tourism consumption (Matsui, 2013; Okamoto, 2015).

Fig. 10.1. Temple in Koyasan. (Photo: Ricardo Nicolas Progano)

While valuable for tourism development, UNESCO designations can generate conflict with local traditions. McGuire (2013) points out that UNESCO's assumptions about shared, universal human values conflicted with Mount Omine's traditional prohibition of women. Mount Omine, part of the Sacred Sites and Pilgrimage Routes in the Kii Mountain Range, has traditionally banned women from climbing it. However, UNESCO's designation has put pressure on the banning, which in turn has prompted a segment of religious practitioners to question whether UNESCO has the right to overturn a 1300-year-old tradition. They also voiced concerns that if the ban is lifted and women and then day hikers are allowed, the sacredness of Mount Omine will be lost and it will become a profane mountain like Mount Fuji, which has been climbed as a leisure activity since World War II (Earhart, 1989; Okamoto, 2015). The prohibition of women is, however, not an exclusive tradition from Mount Omine and existed in other parts of Japan (Nomoto, 2006). Jimura (2016) notes that, due to the UNESCO designation, religious practices have been hindered due to the instainstallation of signposts. In Shugendo, senior practitioners called *sendatsu* (先達), traditionally act as religious guides and pathfinders along the mountainous trails. However, the new signposts undermine the role of the *sendatsu* because they show the location names and the correct paths to follow.

Finally, secular pilgrimage has also expanded, reaching both profane and sacred sites. In Japan, pilgrimage is featured in popular culture elements such as anime, movies, television series and novels; thus fans 'sacralize' places like Shinto shrines that appear as settings in different series (Okamoto, 2015). Ishii (2010) describes different examples of real-life Shinto shrines featured in anime series which became assiduously visited by anime fans who did a 'sacred pilgrimage' to these locations. The local communities and the shrines themselves responded to these sudden flows of visitors by offering merchandising or organizing events. In some cases, the local community and the fans became closely connected (Okamoto, 2015).

Contemporary Spirituality in Japanese Pilgrimage

During the post-war period, Japanese society experienced a process of both secularization (Reader, 2012) and non-affiliated spirituality (Okamoto, 2015). In the 1980s, the New Age movement reached Japan, where it is known as Spirit World (*Seishin Sekai* – 精神世界). In contrast to its development in the West, its arrival did not involve the social critical elements, becoming more consumption-oriented, but still lacking unified organizations or teachings. The current popularity of feng shui, yoga, astrology and fortune telling is due in part to the influences of New Age ideas (Borup, 2015). The media played a central role in the expansion of the Spirit World movement. For example, the use of the word 'spiritual' instead of the Japanese word *rei* (霊) is because of the presence of spiritual counsellors in television programmes, in particular Hiroyuki Ehara, a self-proclaimed spiritual counsellor (Ito, 2003; Horie, 2009; Kotera, 2011). The relationship between spirituality and media in Japan goes back to the 1910s with its presence in newspapers and radio, but has been deepened by the introduction of spirituality-related content on television during the 1970s. The 1995 Tokyo subway sarin attack, perpetrated by the cult Aum Shinrikyo, produced a reduction of these television contents until the 2000s, when spirituality-related themes were presented in a lighter, non-affirmative entertainment format for life improvement. This was done in order to distance from negative associations to dangerous cults (Horie, 2009; Kotera, 2011). In this context, the word 'spirituality' (スピリチュアリティ) gained popularity over the native *rei* and even *Seishin Sekai* as it avoided these negative connotations. Also, because the word has a foreign origin, it was perceived as more advanced and fashionable (Horie, 2009; Borup, 2015). This shift also changed the social segment interested in these matters, appealing to younger generations, especially women (Borup, 2015).

In this context, contemporary spirituality has also reinterpreted pilgrimage sites and given new meaning to them. The concept of 'powerspot' (パワースポット) first emerged in 1986, with the phenomenon starting to grow exponentially from 2006 (Suga, 2010). The media coverage of different powerspots in television and magazines, as well as the trend of celebrities visiting them, were important factors for its expansion (Okamoto and Kawasaki, 2012). The term generally refers to specific places where some people believe mental or conceptual 'energy' or 'power' gather, which can confer different worldly benefits like good fortune, healing or luck in romance. They are akin to 'power places', a concept that emerged in the West during the 1960s (Stausberg, 2011). The 'powerspot boom', which started in the 2000s, is particularly popular among women, which is the segment targeted by media outlets (Okamoto, 2015). Publications tend to present the power spots with photographs, along with data on access and opening hours. They may also present information on accommodation, local specialties and food, making them similar to tourist guides (Tsukada and Omi, 2011; Okamoto, 2015). Local administrations and tourism organizations noticed the 'powerspot boom' and began to rebrand their own destinations as powerspots (Reader, 2012; Okamoto, 2015). In some cases, conflicts arise because visitors' motivations and behaviour may not be in line with what the site administrators consider suitable (Okamoto and Kawasaki, 2012; Okamoto, 2015).

The concept of *iyashi* (癒し), which refers to physical and mental restoration, has also experienced high notoriety since 1988 (Matsui, 2013) and is associated with different products and services. This '*iyashi* boom' has also influenced tourism promotion and activities in pilgrimage sites. For example, the four prefectures of Shikoku island collaborated with railways operator J.R. Shikoku to promote Shikoku Henro as a place for healing (Matsui, 2013). Since the second half of the 1990s, there has been an increase in mass media presenting Shikoku Henro as a place of *iyashi* in place of traditional Buddhist narratives (Mori, 2005), with this process accelerating since the 2000s (Kadota, 2013). Closely related to *iyashi* is the concept of *yomigaeri* (revival – 蘇り), which acquired a new meaning akin to physical and mental restoration from the toils of daily life. As it can be observed, both concepts are related to notions of wellness.

Another important element in contemporary pilgrimage is self-discovery (*jibunsagashi* – 自分探し). Mori (2005) notes that mass media outlets such as magazines and television portray Shikoku Henro as a place where, through walking the 88 temples of the pilgrimage, visitors can achieve experiences of self-discovery and realization, not strictly following Buddhist interpretations of the site. This concept is defined in very vague terms and is bound to have diverse meanings for different people, thus being attractive to a wider audience than traditional Buddhist narratives. It also shifts the focus from visiting the sacred sites to the journey itself as a process of becoming.

The embracing of concepts such as *iyashi*, *yomigaeri* and powerspot is also related to the fact that they are strongly place-oriented, so they are easily applied in tourism promotion. However, travelling to seek a cure or restoration is an ancient practice which is often revived in contemporary forms (Connell, 2011). In particular, powerspots can be understood as a more positive, non-religious version of traditional worldly benefits, known as *genze riyaku* (現世利益) (Borup, 2015). By performing a pilgrimage to Kumano, nobles not only expected abstract rewards related to the afterlife, but also worldly favours such as healing from illnesses and disabilities, or a rise in aristocratic rank (Koyama, 2000).

At the same time, secularization processes have not been absent in Japanese society. Reader (2012) notes that growth of urbanization and education levels since the end of World War II have gone hand-in-hand with a steady decrease of faith among Japanese, especially among the young. Buddhist temples and Shinto shrines located in rural areas are declining rapidly because of depopulation and secularization. The decrease of new recruits to continue the path of older monks has also endangered the future of temples. In urban areas, the general distrust and lack of interest in religion denotes a similar negative situation for religious institutions. In this context, the flourishing of pilgrimage sites such as Shikoku Henro is not due to religious revival, but to the promotion of their heritage, wellness, sightseeing and cuisine. Also, not all pilgrimages have experienced an increase in pilgrims, as sites such as the Saikoku pilgrimage and the Chichibu pilgrimage in the Kanto region have declined in modern times (Reader, 2012). Due to the decline in popularity of traditional religion, its clerical members have tried to create a more positive and joyous image of it, participating more in social affairs and events (Borup, 2015).

Western Visitors to Japanese Pilgrimage Sites

The widespread interest of western societies in Japanese religions, particularly Buddhism, can be traced back to the end of the 19th century. The World's Parliament of Religions, which took place in Chicago in 1893, was a pivotal event in this regard. Here, American audiences, disenchanted by Christianity and supporters of scientific rationalism, were presented to discourses about 'scientific' Buddhism, which emphasized the existence of common grounds between western science and 'pure' Buddhism, dissociated from its religious practices and beliefs in supernatural phenomena (McMahan, 2004). This scientific Buddhism contained different elements found in Protestantism, such as emphasis on text, personal experience and social activism, while combining it with other elements agreeable to rationalist audiences such as the absence of a personal god, and the presentation of Buddha as a skeptic that encouraged experimentation. This westernized version of Buddhism changed their perspective on this religion, which was previously portrayed as nihilistic and decadent in the West (McMahan, 2004). The discourse about the alleged scientific basis of Buddhism still holds popularity among western audiences (McMahan, 2004; Borup, 2015). Along with secularization, contemporary spirituality plays an important role in the motivations of western travellers, who are nowadays inclined to conceive spirituality in individualized, non-affiliated terms and experiment with beliefs outside of their cultural sphere (Luckmann, 1967; Okamoto, 2015). Contemporary spirituality also expanded alternative religious movements such as the New Age movement, a counter-culture that emerged during the 1970s in the United States (Hanegraaff, 2000; Timothy and Conover, 2006).

As mentioned, the individualization of spirituality led to exploration of religious beliefs outside their cultural sphere. Researchers, for example, point out the interest of Indian Ayurveda medicine and yoga practices in western society, as well as their role in tourism development (Ravichandran and Suresh, 2010; Bowers and Cheer, 2017). Ayahuasca, a hallucinogenic plant mixture utilized during aboriginal ceremonies in South America, has also attracted the attention of western visitors (Fotiou, 2016). Westernized versions of Zen Buddhism, understood as an individualized and non-institutionalized form of engagement with the self, also enjoy a steady popularity among western audiences. In contrast, Buddhism, as mentioned before, is the antithesis of trendy individualized spirituality in Japan (Borup, 2015). Naturally, western visitors also travel to pilgrimage sites motivated by reasons beyond religion or spirituality, such as wellness, hiking and cultural heritage.

In this context, different pilgrimage sites in Japan currently attract a growing flow of western tourists, who are in turn an attractive tourist market for Japanese localities. This is, for example, the case of the Shingon Buddhism headquarters of Kongoubu-ji (金剛峯寺). The site is commonly known by the name of the mountain where it is established, Koyasan, which is located in Wakayama prefecture. The site was founded in 819 AD by the revered monk Kukai (空海), also known as Koubou Daishi (弘法大師). Different reasons contributed to Koyasan's tourism development in contemporary society. While Koyasan is located in a high mountain surrounded by forests and relatively far from main urban hubs, infrastructure constructed during the 20th century, such as roads, trains and cable carts, facilitate access for tourists. Access was also facilitated by the end of the prohibition of women in Koyasan at the end of the 19th century, thus allowing women to not only visit the Buddhist complex but also to live and work in the town. The post-war economic development also increased the number of domestic visitors to Koyasan, who stayed the night utilizing temples' lodges services (*shukubou* – 宿坊). Also, the inscription of Koyasan as a World Heritage Site in 2004 as part of the Sacred Sites and Pilgrimage Routes in the Kii Mountain Range helped to increase its popularity among overseas tourists (Matsunaga, 2014). An increasing number of western tourists, particularly from France, have been visiting Koyasan in recent years. The WHS designation, while not the sole reason (Jimura, 2016), was an important milestone in international tourism development. Westerners are reported to be very interested in its elements of pilgrimage and religion. They are also attracted to Koyasan as it offers an experience not commonly found in everyday life, such as *shukubou* and monastic cuisine (*shoujin ryouri* – 精進料理) (Matsunaga, 2014; Jimura, 2016). However, their visits do not mean that western travellers formally convert into Buddhism. Instead, this should be understood as a blend of playful leisure, intercultural learning and individualized spirituality.

Kumano Kodo, a multi-site pilgrimage route located in the Kii peninsula, is another pilgrimage site popular among westerners. It is structured around three grand shrines, which constitute the travellers' final goals: the Kumano Hongu Taisha, the Kumano Hayatama Taisha and the Kumano Nachi Taisha. Each of these grand shrines is dedicated to one of the three deities of Kumano (*Kumano Gongen* – 熊野権現), which are also related with three different Buddhas. These shrines are linked through the three main routes of Kiji, Iseji and the Omine Okugake; thus forming a circuit of pilgrimage routes along the Kii peninsula that also connects to other important sacred sites, such as Ise and Koyasan. After different tourism and restoration efforts undertaken from the 1970s, the Kumano Kodo pilgrimage routes and main shrines were inscribed in the UNESCO World Heritage Site Sacred Sites and Pilgrimage Routes in the Kii Mountain Range cultural landscape in 2004. The designation provided local communities with the opportunity to strengthen its tourism industry in order to revitalize the regional economy and attract both national and international tourists. In particular, Tanabe city (Wakayama prefecture) has been focused on western visitors by promoting properties related to Kumano Kodo (Fig. 10.2.). As part of its tourism policies, Tanabe city decided to focus on western individual tourists and small groups, because they usually have longer stays and spend more money than domestic tourists. The Bureau made it easier to travel around the area by setting signs, restaurant menus and bus timetables in English.

Fig. 10.2. Kumano Kodo during winter. (Photo: Ricardo Nicolas Progano)

It also conducted workshops with the local community in order to improve their English conversation skills and established its own online reservation system.

Conclusions

As reviewed, the motivations that mobilized the Japanese people to visit pilgrimage sites is varied and has experienced major shifts across history. While motivated by traditional faith, politics and sightseeing in pre-modern times, 'spirituality' for Japanese people has gone through major transformations in the post-war period. Institutional religions are perceived with distrust, and secularization processes accentuate non-religious elements in tourism development such as heritage, wellness and cuisine as attractions for the average visitor. At the same time, individualized spirituality notions, spread by mass media and marketing outlets, influenced the motivations of visitors to pilgrimage sites. Western visitors have also mobilized towards pilgrimage sites in Japan for a variety of motivations. Reinterpretations of Japanese religions, as well as fluid spirituality, guided them outside their own cultural sphere towards remote areas such as of Shikoku Henro and Kumano Kodo. The interest in Japanese religions, both in academic studies and tourism experience, albeit through westernized interpretations, is also another motivational factor. Also, and similar to their Japanese counterparts, traditional leisure activities such as hiking, sightseeing and cultural heritage play an important role. Secular pilgrimage also plays an important role in domestic tourism, due to the influence of popular culture and its spread through the internet. In this sense, motivations for pilgrimage in contemporary Japanese society is a complex issue that cannot be reduced to simple religion–secularization dichotomies. As tourism development grows and social attitudes towards spirituality continue to transform, researchers can certainly find interesting study venues in Japan to further advance our understanding of pilgrimage in Asia.

References

Borup, J. (2015) Easternization of the East? Zen and spirituality as distinct cultural narratives in Japan. *Journal of Global Buddhism* 16, 70–93.

Bowers, H. and Cheer, J.M. (2017) Yoga tourism: commodification and Western embracement of eastern spiritual practice. *Tourism Management Perspectives* 24, 208–216. DOI: 10.1016/j.tmp.2017.07.013.

Connell, J. (2011) *Medical Tourism*, 1st edn. CAB International, Wallingford, UK.

Earhart, H.B. (1989) Mount Fuji and Shugendo. *Japanese Journal of Religious Studies* 16(2-3), 205–226. DOI: 10.18874/jjrs.16.2-3.1989.205-226.

Fotiou, E. (2016) The globalization of ayahuasca shamanism and the erasure of Indigenous shamanism. *Anthropology of Consciousness* 27(2), 151–179. DOI: 10.1111/anoc.12056.

Funck, C. and Cooper, M. (2013) *Japanese Tourism – Spaces, Places and Structures*, 1st edn. Berghahn Books, New York.

Gorai, S. (1989) *Yuugyou and Pilgrimage*, 1st edn. Kadokawa Shoten, Tokyo (in Japanese).
Guichard-Anguis, S. (2009) The culture of travel (tabi no bunka) and Japanese tourism. In: Guichard-Anguis, S. and Moon, O. (eds) *Japanese Tourism and Travel Culture*. Routledge, London and New York, pp. 1–15.
Hanegraaff, W.J. (2000) New age religion and secularization. *Numen* 47(3), 288–312.
Horie, N. (2009) Spirituality and the spiritual in Japan: translation and transformation. *Journal of Alternative Spiritualities and New Age Studies* 5, n.p.
Hoshino, E. (1981) *Pilgrimage: Phenomenology of the Sacred and Profane*, 1st edn. Kodansha, Tokyo (in Japanese).
Ishii, K. (2010) Anime, *miko* and Shinto shrines. In: Ishii, K. (ed.) *Where is Shinto Going?* Perikansha, Tokyo, pp. 270–274.
Ito, M. (2003) *Contemporary Society and Spirituality*, 4th edn. Keisuisha, Tokyo (in Japanese).
Jimura, T. (2016) World heritage site management: a case study of sacred sites and pilgrimage routes in the Kii mountain range, Japan. *Journal of Heritage Tourism* 11(4), 382–394. DOI: 10.1080/1743873X.2016.1146287.
Kadota, T. (2013) *Ethnography of Pilgrimage Tourism - The Consumed Religious Experience*, 2nd edn. Shinwasha, Tokyo (in Japanese).
Koizumi, T. (2001) *The Birth of Hiking - Why Did People Start Climbing Mountains?* 1st edn. Nakamatsu Shinsho, Tokyo (in Japanese).
Kotera, A. (2011) What is powerspot? preliminary research on the study of its social background and reception. *Touyou Eiwa Jogakuin Daigaku Jinbun Shakaigakka Ronshu* 29, 87–110.
Koyama, Y. (2000) *Kumano Kodo*, 1st edn. Iwanami Shoten, Tokyo (in Japanese).
Luckmann, T. (1967) *The Invisible Religion*, 3rd edn. Macmillan Publishing, New York.
Matsui, T. (2013) *Language and Marketing: A Social History of the "Healing Boom" in Japan*, 1st edn. Sekigakusha, Tokyo (in Japanese).
Matsunaga, Y. (2014) *Koyasan*, 2nd edn. Iwanami Shoten, Tokyo (in Japanese).
McGuire, M.P. (2013) What's at stake in designating Japan's sacred mountains as UNESCO World Heritage Sites? Shugendo Practices in the Kii Peninsula. *Japanese Journal of Religious Studies* 40(2), 323–354. DOI: 10.18874/jjrs.40.2.2013.323-354.
McMahan, D.L. (2004) Modernity and the early discourse of scientific Buddhism. *Journal of the American Academy of Religion* 72(4), 897–933. DOI: 10.1093/jaarel/lfh083.
Moerman, D. (1997) The ideology of landscape and the theater of state: Insei pilgrimage to Kumano (1090-1220). *Japanese Journal of Religious Studies* 24(3-4), 347–374.
Mori, M. (2005) *The Modernization of Shikoku Henro: From a 'Modern Pilgrimage' to a 'Healing Journey'*, 1st edn. Sogensha, Tokyo.
Nomoto, K. (2006) *Landscape Study on Gods and Nature*, 6th edn. Kodansha Gakujutsu Bunko, Tokyo (in Japanese).
Okamoto, R. (2015) *Pilgrimage to Sacred Sites: From World Heritage to Anime Setting*, 1st edn. Chuokoron Shinsha, Tokyo (in Japanese).
Okamoto, R. and Kawasaki, N. (2012) Meiji Jingu – Kiyomasa well. In: Okamoto, R., Hoshino, E. and Yamanaka, H. (eds) *Sacred Pilgrimage Tourism*. Koubundo, Japan, pp. 142–145.
Prime Minister of Japan and Cabinet (2012) Tourism nation basic promotion plan. Available at: http://www.mlit.go.jp/common/000208713.pdf (accessed 16 October 2019).
Prime Minister of Japan and Cabinet (2016) Tourism vision for supporting the Japan of tomorrow. Available at: http://www.mlit.go.jp/common/001126598.pdf (accessed 16 October 2019).
Ravichandran, S. and Suresh, S. (2010) Using wellness services to position and promote brand India. *International Journal of Hospitality & Tourism Administration* 11(2), 200–217. DOI: 10.1080/15256481003732873.
Reader, I. (2012) Secularisation, R.I.P.? Nonsense! The 'rush hour away from the gods' and the decline of religion in contemporary Japan. *Journal of Religion in Japan* 1, 7–36. DOI: 10.1163/221183412X628370.
Reader, I. (2014) *Pilgrimage in the Marketplace*, 1st edn. Routledge.
Reader, I. and Swanson, P.L. (1997) Pilgrimage in the Japanese religious tradition. *Japanese Journal of Religious Studies* 24(3-4), 225–270.
Sawamura, A. (2016) World heritage registration and tourism trend (revised edition): analyzing 10 cases in Japan. *The Journal of Economics of Niigata University* 100, 117–128.
Stausberg, M. (2011) *Religion and Tourism: Crossroads, Destinations and Encounters*. Routledge, London and New York.

Stoddard, R. (1997) Defining and classifying pilgrimages. *Geography Faculty Publications* 2, 41–60.

Suga, N. (2010) Shinto shrines as power spots. In: Ishii, K. (ed.) *Where is Shinto going?* Perikansha (in Japanese), Tokyo, pp. 232–252.

Timothy, D.J. and Conover, P.J. (2006) Nature religion, self-spirituality and New Age tourism. In: Olsen, D. and Timothy, D.J. (eds) *Tourism, Religion and Spiritual Journeys*. Routledge, London and New York, pp. 39–55.

Tsukada, H. and Omi, T. (2011) A deluge of 'religious' information on new religions, power spots, funeral services and Buddhist statues. *Nanzan Institute for Religion & Culture Bulletin* 35, 24–47.

UNWTO (United Nations World Tourism Organization) (2011) *Religious Tourism in Asia and the Pacific*. UNWTO, Madrid.

11 New Narratives in Valuistic Travel and Identity: *Goshuin* Culture and Religious Mobility in Contemporary Japan

Shin Yasuda*
Takasaki City University of Economics

Introduction

Religious mobility (such as pilgrimage and other related religious practices) is now strongly influenced by consumer society. With the rapid spread of global capitalism and commodification, religious practices and symbolism have also transformed to fit their respective surrounding environments. As Ian Reader notes, pilgrimages are now embedded in a context of markets, consumer activity, publicity and promotion. They operate not just in the marketplace but also through it (Reader, 2013, p. 8). Moreover, he also recognizes 'pilgrimage as a broader phenomenon encompassing not just journeys and (so-called) sacred places but also the wider context within which those places are located and those journeys are carried out' (Reader, 2013, p. 20), which leads to an 'intrusion of market forces' (Reader, 2013, p. 11). Therefore, 'pilgrimage is located in and operates through the marketplace, and it is through the mechanisms of the marketplace that constructs such as the sacred are materialised and successful pilgrimage sites emerge' (Reader, 2013, p. 195). In this situation, the social transformation of religion should be discussed in relation to religious commodification in contemporary society.

Within the discussion of religious commodification, various researchers consider the transformation of traditional religious mobility by analysing its novelty and continuity (Olsen, 2019). Some literature insists on the decline of religiosity and the insignificance of religion in society (Featherstone, 1992). In accordance with the spread of commodities in the religious field (Eade and Sallnow, 2013), other studies consider this to be a new form of religiosity that fits contemporary consumer society by taking into account marketing and consumer behaviours based on individual motivations, demands and preferences (Olsen, 2013). In the process of emerging religious commodification, 'taking religion as consumer goods has enabled individuals to personalize their piety and religious sensibility' (Kitiarsa, 2010, p. 570) through the 'democratization' of religious authority (Reader, 2013, p. 93). In this context, religious mobility is also based on the degree of individual commitment towards the practice, which can be described as the degree of 'religious capital' defined as '[the] degree of mastery of and attachment to a particular religious culture' (Stark and Finke, 2000, p. 120).

As the theory of religious economy conceptualizes, 'religious capital' is described as an individual commitment to embody 'spiritual

*syasuda@tcue.ac.jp, shinmarcie@gmail.com

knowledge, abilities, skills, tastes, and other dispositions' (Schilderman, 2011a, p. 43), and 'the investment in religious capital is a function of the perceived benefits as a return on investment' (Schilderman, 2011b, p. 183). In this sense, the development of religious commodification has encouraged 'self-promotion and status enhancement' through individual investment (Reader, 2013, p. 116).

However, religious capital reflects the social contexts and common good by negotiating the surrounding environment (Schilderman, 2011a, p. 43). 'Religious capital refers to the available skills and abilities to direct and control ideal goods and interests, whereas its symbolic aspects legitimize (and in a sense 'disguise') the social power source of the status attributes that are connected to these goods and interests' (Schilderman, 2011b, p. 43). In this sense, religious capital 'stipulates its character as a commodity that has a certain value in exchange, and therefore may contribute to some shared or common good' (Schilderman, 2011a, p. 46). By developing social interactions in the field, individual practitioners have been encouraged to change their financial and physical resources in the provided social context and environment.

To create a social context and shared benefits that funnel individual financial and physical resources into religious mobility, religious commodification promotes a cost-benefit approach to individual religious investments and visualizes benefits from the return on those investments (Iannaccone, 1992). As Hans Schilderman indicates, literature on religious capital emphasizes 'the resource-character of religion: it is primarily something to gain from, be it spiritually or socially, but always depending on proper investment as fuelled by rational calculation' (Schilderman, 2011b, p. 44). In this situation, 'the public self-validation of a religion refers to a loyalty toward rules and norms that offer stable benchmarks for the organized assembly of believers' (Schilderman, 2011a, p. 48). Religious capital, therefore, strongly reflects its context of social interaction, recognized as individual 'social investment' (Schilderman, 2011a, p. 48).

In the discussion of religious mobility under religious commodification, researchers strongly emphasize the social aspect of individual investment through resource-character and cost-benefit approaches for the socialization of individual investment. In this case, consumers of religious commodities 'do not simply or merely consume, they also produce' (Stephenson, 2010, p. 190; Reader, 2013, p. 116). Thus, it is not suitable to analyse new forms of religious mobility from the perspective of their motivations, religious attitudes or identity. Rather, it is more expedient to consider the perspective of investment in religious capital to achieve religious mobility. To construct individual religious mobility, individuals invest in religious activities using their available capital, such as time, wealth, expenses, human relationships and knowledge.

In the case of Japan, the spread of religious commodification has equally sparked serious discussion about contemporary religiosity in the country. Like pilgrim passports and certificates for the Camino de Santiago, collecting *goshuin* (shrine or temple seal) stamps has a long tradition in Japanese society. These stamps attest to the authenticity of religious experiences and mobilities. Despite Japanese society's secularism in modern times, religious practice and pilgrimage routes have sustained themselves by successfully transforming their purpose from religious to a form of Japanese local culture. Although there are several discussions about whether collecting *goshuin* stamps is religious, the practice is now accepted in popular Japanese culture and widely practised by the Japanese.

This chapter, therefore, examines the recent development of the religious mobility of the *goshuin* movement in Japanese society from the perspective of the commodification of religious mobility. On the aspect of individual investments for religious capital returns, it focuses on the characteristics of religious mobility in *goshuin* practice and the social significance of individual investments in the *goshuin* movement.

This chapter is an empirical case study based on a qualitative approach to content analysis (Neuendorf, 2002; Krippendorff, 2004). It discusses *goshuin* practitioners' narratives that were collected from intermittent field research as participant observation, which began in 2015 and extends to the present in various Shinto shrines and Buddhist temples in Japan, especially in Kanto and Kansai areas. Related articles in newspapers and magazines, tourism guidebooks and previous literature are also provided.

Goshuin, Religious Mobility and Commodification of Religion in Contemporary Japan

In the case of Japan, religious mobility under the influence of religious commodification creates new forms of social mentality, as well as religiosity. Although the country is secular and acknowledges the minimal importance of religion (Buddhism and Shintoism) in public spaces and social life, since the 2000s the practice of *goshuin* (shrine seal or temple seal) collection has become a massive social movement (Tsuban, 2016; Abe, 2017). Many shrines and temples are now eager to promote new *goshuin*, akin to launching new religious products in the market, and Japanese travel agencies are actively advertising such 'products' to encourage the consumption of religion and culture. In this highly commercialized environment, Japanese people of all generations are collecting *goshuin* to show their devotion. This practice is forming new religious and cultural routes, based on the production, consumption and collection of the individual footprints of these specific mobilities.

The *goshuin* or *shuin* (shrine seal or temple seal) has a long religious tradition in Japanese Shintoism and Buddhism (Fig. 11.1). A seal stamp is placed in a specific pilgrimage book (called *goshuinchō* or *nōkyōchō*) with red stamps called *shuin*. The date, the name of the temple or shrine, and some words of worship are imprinted or handwritten in calligraphy. Each religious site has its own stamps and words. Some shrines and temples have various forms of *goshuin*, according to dates, events and other occasions. The *goshuin* is the proof of the visit to a religious place or reading of Buddhist sutras. The cost of *goshuin* is around 300–500 yen (2.5–4 euros), and some special *goshuin* cost more. The special 40-page book for *goshuin* (called *goshuinchō*) costs from 1500 to 2000 yen (13–18 euros) (Fig. 11.2). It takes more than 20,000 yen (about 170 euros) to fill all the pages with *goshuin* stamps.

Some historians say that the practice of collecting *goshuin* stamps dates back to the Heian era (794–1185 AD). As Shintoism and Buddhism are strongly syncretized in Japanese history, *goshuin* practice has been widely accepted in both religions (Tsuban, 2016; Abe, 2017). However, it was popularized during the Edo period (1603–1868 AD) because of increased mobility and travel throughout the country.[1] In those days, *goshuin* were stamped on scrolls (*kakejiku*) and pilgrims' shirts, as well as in the special book (Reader, 2013, pp. 151-152).

Presently, *goshuin* are stamped in almost all Shinto shrines and Buddhist temples, and there is a special distribution counter within these religious sites.[2] After the prayer or the reading of Buddhist sutra, people go to the counter and ask the staff to write *goshuin* on the pages of the *goshuinchō*. In the case of group visits, a delegate collects the *goshuinchō* and brings them

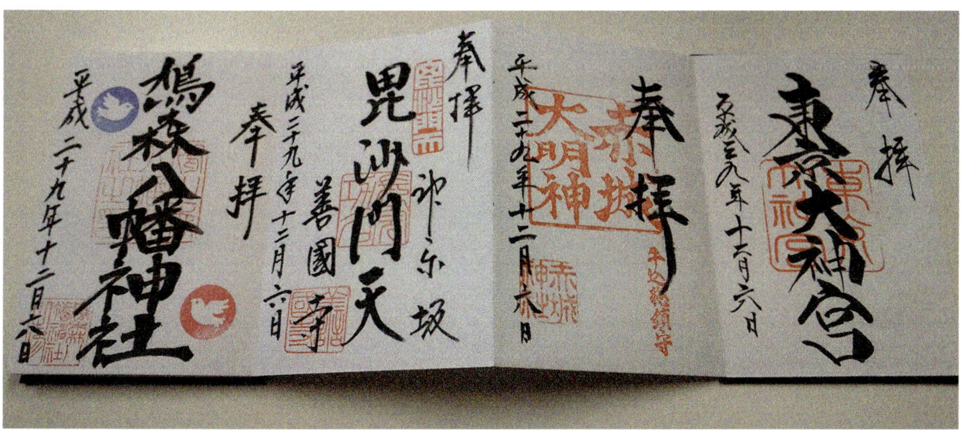

Fig. 11.1. *Goshuin*. (Photo: the author)

Fig. 11.2. *Goshuinchō.* (Photo: the author)

to the counter (Reader, 2005). At religious sites, staff, Buddhist monks and Shinto priests serve the *goshuin* to visitors as part of their religious devotion. In the busy season, shrines and temples increase staff to handle the large number of requests. Alternatively, they prepare *goshuin* beforehand.

The practice of *goshuin* collection does not have a specific form, rule or pattern. However, there are some specific religious routes. For instance, two major Buddhist pilgrimage routes, Shikoku Henro (pilgrimage to 88 sites in Shikoku island in west Japan) and Saigoku Kannon (pilgrimage to 33 Buddhist temples), have strict forms and manners of *goshuin* collection and pilgrimage rituals, which have strictly showed the route for pilgrimage and ways of chanting specific Buddhist sutras (Reader, 2005). On the other hand, other pilgrimage routes do not have such strict rules. Rather, all kinds of visitors are welcome to freely show their own devotion.

In this context, people easily participate at their own expense, i.e. time, money and effort. Some pilgrims are hobbyists, who are motivated by an interest in walking and travelling. Others are deeply committed to the pilgrimage as a sign of their religiosity and strong devotion. These diversities in pilgrimage practice make it difficult to determine the religious foundation of contemporary *goshuin* culture.

The *Goshuin* Movement and Contemporary Japanese Travel Culture

Despite the long tradition of *goshuin* culture in Japanese religions, it is a niche practice in society. It was historically practised by some elderly people, devout individuals and particular local communities. However, the recent expansion of the *goshuin* movement in Japan is somewhat different from the traditional religious and local cultural practices that preceded it. The current popularity is driven by various stakeholders involved in religious consumption.

Since the end of the 2000s, new stakeholders in the tourism media industry have actively promoted *goshuin* culture to popularize tourism to Shinto and Buddhist religious sites. Various guidebooks and travel magazines have published articles promoting *goshuin* collection as a new hobby and cultural activity enriching social life. Some guidebooks introduce the merit of *goshuin* practice to readers who want to increase devotional practices in their daily lives, learn Japanese religious culture, deepen their spiritual experience, and solidify their religious identity (Fig. 11.3). Moreover, mass media actively promote *goshuin* practice as part of a new culture of curiosity about Japanese religions.

Initially, tourism media actively targeted young women in their twenties and thirties who

Fig. 11.3. Some specific guidebooks for *goshuin*. (Photo: the author)

are not familiar with *goshuin* culture. In collaboration with women's magazines and using anime characters, like Hello Kitty and Rilakkuma, these media outlets also published new types of *goshuinchō* to attract this demographic. Some stakeholders such as travel magazines and tour operators collaborated on campaigns with various media outlets to attract a wider range of fans to the culture. As a result, young women in their twenties and thirties became the primary consumers of *goshuin* collection. As the younger generation has actively committed to the movement, other generations have begun to join as a hobby. These generations spread this new type of travel culture without any specific religious motivation. This exemplifies Japan's new tourism culture under religious commodification.

As the number of participants in *goshuin* practice increased across Japanese society, local government and religious sites have enhanced co-operation with the tourism industry and travel media to promote the *goshuin* movement. In accordance with local Shinto shrines and Buddhist temples, some local governments created new religious and cultural trails based on *goshuin* collection. As a result, these new religious and cultural trails continuously attract new leisure visitors.

With the development of the *goshuin* movement in larger society, religious circles in Japanese Shintoism and Buddhism have also sought to attract visitors and to strongly promote *goshuin* in their shrines and temples. Some of them actively advertise their *goshuin* by introducing new *goshuin* and related events. They introduce their *goshuin* in tourism guidebooks and travel magazines to attract fans. Moreover, Shinto shrines and Buddhist temples organize promotion committees to co-operate with the development of *goshuin* culture throughout the country. For instance, to encourage visiting both Shinto shrines and Buddhist temples, Shinbutsu Reijo Kai (Committee for Religious Places in Shintoism and Buddhism in the Kansai region) co-ordinates various pilgrimage routes, publications and special *goshuinchō* (Shinbutsu Reijo Kai (Committee for Religious Places in Shintoism and Buddhism), 2020). Such collaboration between the two religions is common throughout the country.

With regard to the motivations behind *goshuin* collection, religious figures have interpreted the movement from a religious perspective. The development of the *goshuin* movement has encouraged them to develop new modes of communication with their visitors and enhance the religious experience. As some religious figures who are actively committed to the *goshuin* movement frequently mention, serving *goshuin* is a significant expression of their devotion and their mission to share their religious teachings (Tsuban, 2016). In this sense, priests and monks consider serving *goshuin* as a part of their religious training and an enlightening practice for visitors.[3]

Although there is a huge gap in perspective among collectors, religious figures and other stakeholders, such as the travel media industry and other tourism industry, in *goshuin* practice, the development of the *goshuin* movement creates new forms of religiosity within the current context of religious commodification.

Personal Forms of Spirituality in *Goshuin* Mobility

As the *goshuin* movement has become widely accepted in Japanese society, various discussions related to *goshuin* collection practices have emerged among collectors. Despite the absence of any firm rule and manner for the practice in the doctrines of Japanese Shintoism and Buddhism, some people began to insist on 'proper' and 'authentic' *goshuin* practices.

Many participants in this study narrated their own personal commitment to particular religious sites, referring to their feeling of the 'existence of supernatural power' or a 'spiritual atmosphere'. They do not mention their practices in a particular religious context. As *goshuin* practice does not have any firm ritual except some traditional pilgrimage trails, each person freely chooses their own destination according to their preferences and circumstances. The choice of destinations, therefore, varies on the basis of personal motivation, which can include secular *goshuin* practice. Some narratives collected in this study recommend tying specific achievements to the collection of particular *goshuin*. To sustain their motivation, they introduced particular supplications: for example, for business success, wishes for good fortune and good health, etc. As Japanese shrines and temples have been strongly connected to specific supplications and rewards, some participants select their destinations according to the content of their supplications. One narrative even advised people not to have strong motivations or supplications in the *goshuin* practice; rather, it is efficient to participate in the practice casually and to sustain the practice over the long term without strong commitment to formal religious cultures in Shintoism and Buddhism (Edoraku Henshuubu, 2014). Although some participants have strong motivations to visit a particular place within the *goshuin* practice, many people do not have a specific reason or motivation to commit to the movement. This recommended 'weak' motivation is widespread in the *goshuin* movement, and participants clearly express this attitude. As a result, the purposes, behaviours and narratives of *goshuin* practices are diverse. They are not encouraged to make a firm commitment or to conform to formality (Fig. 11.4).

The diversity of individual religious commitment is clearly visible in the collection practices and usage of *goshuinchō*. Despite the absence of firm religious doctrines on the practice, some participants of the *goshuin* movement have conducted serious discussions about the 'proper' manner of *goshuin* practice and its 'authentic' form. Some of my field agents who participate in the movement as practitioners insist that their personal form of *goshuin* practice (such as prayers, and the manner and etiquette of obtaining *goshuin*,name, address and current situation and the expression of respect to religious staff) contributes to the development of *goshuin* culture. Although some guidebooks and articles advocate a 'proper' way of practice, by showing some basic knowledge of *goshuin* practice (Chikyuu no Arukikata Henshuushitsu, 2013), they have invented their own etiquette. One narrative says that participants should talk with gods in shrines and temples, by sharing their name, address and current situation (Chikyuu no Arukikata Henshuushitsu, 2017) . This practice is not evident in Shinto or Buddhist doctrines. Another narrative reports that people should visit entire branch shrines and temples in the same locations to express religious commitment; there is also no religious evidence to support this practice. In addition, these *goshuin* collectors strongly insist on contributing to religious sites by giving *saisen* (a small donation), purchasing religious goods, learning *yuen* (the history and religious significance of the site), purchasing local products, or having lunch near the religious site. They are committed to spending time and money in the local area to express their commitment to *goshuin* culture.

Fig. 11.4. The queue for waiting for *goshuin* in a Shinto shrine, central Tokyo. (Photo: the author)

In the usage of *goshuinchō*, various participant narratives show the diversification of *goshuin* practice in the movement. For instance, some participants insist that people should divide pilgrimage books between Shinto shrines and Buddhist temples, in order to prevent clashes between Shinto *kami* (gods) and Buddhist *hotoke* (buddhas) in the *goshuinchō*. However, others advise that people do not need to divide seal books between religions because Shintoism and Buddhism in Japan have a long tradition of syncretism, and *kami* and *hotoke* peacefully live together in the book. According to them, it is customary in Japanese society to combine *kami* and *hotoke* in one *goshuinchō*. Neither perspective is based on religious doctrine.

Other examples of *goshuinchō* show the expression of individual devotion in *goshuin* practice. Some people own specific *goshuinchō* from particular religious sites, to which they have a strong sense of devotion. By repeatedly collecting *goshuin* from the same religious places, they express their strong commitment and devotion to those places. Continuous commitment constructs a certain sense of religiosity.

Although their individual commitments are diverse, people in the movement have tacit norms to avoid deviancy and strongly condemn deviators. Despite the lack of religious justification for an 'authentic' practice in *goshuin* collection, collectors frequently criticize others who do not adhere to certain manners that are considered 'authentic' ways of showing commitment to a shrine or temple. One narrative says, 'they are pseudo visitors, who do not obey [the] "proper" manner of etiquette in the religious places'. Others generally condemn a lack of respect. One of my field agents showed her strong abhorrence for people who do not respect the manners and etiquette of *goshuin* practice that most of the participants follow. She said, 'they are pseudo practitioners who ought to receive divine punishment from the gods'.

Another serious discussion among participants is the 'free-rider' problem. As the *goshuin* practice became a social movement, reflecting social status as much as individual commitment, some stakeholders started special collection services. These services entail the purchase of completed *goshuinchō* or the hiring of individuals to collect *goshuin* for a fee. Some participants utilize these services to gain special *goshuin* or to flaunt the degree of their commitment. However, this outsourcing service is strongly condemned among other collectors who do not consider it 'proper' and believe this practice denigrates the *goshuin* stamp itself. Although there is evidence that a historical 'trade of *goshuinchō* developed in surrogate pilgrimage (*dairi sampai*) in which commercial firms offer the service of providing people – for a substantial fee – with completed scrolls and pilgrim's book from pilgrimages' (Reader, 2013, p. 157), participants today strongly condemn the historical practice as cheating the gods by not physically going to the shrine to collect the stamps. One of my field agents also told me that it was not proper to obtain *goshuin* without spending one's own time, money and effort to collect it. Thus, obtaining *goshuin* entails exerting personal effort.

Moreover, some participants criticize people who do not contribute to the local economy surrounding religious places, labelling them pseudo-visitors who diminish the value of *goshuin* culture. In this sense, *goshuin* culture is recognized not only as an individual commitment to religious practices but also as a socially significant contribution to the local economy and culture.

Although the narratives show the absence of formal practice and the importance of individual commitment, there are strong, tacit norms that govern how to invest individual resources to show commitment and contribution to 'proper' *goshuin* culture. As an example of religious mobility, *goshuinchō* provide a means to visualize participation and adherence to 'proper' *goshuin* practice, thereby enhancing participants' religious capital and increasing stakeholders' financial capital as well. This situation can be described as the new form of Japanese religiosity, based on individual commitment to *goshuin* culture. Erik and Scott A. Cohen conceptualized this as 'hot authenticity' that can be conducive to different types of relevant personal experiences (Cohen and Cohen, 2012).

Transaction Cost and Cost-effectiveness of Religious Investment under Religious Commodification

As various narratives mentioned, the tacit norms of religious consumption in *goshuin* culture are strongly linked to the socialization of individual commitments to enhance the local economy and culture. This contradictory environment is a result of 'religious investment' under religious commodification.

Previous literature on religious commodification shows that religious consumption can be an investment activity to self-validate and earn religious capital (Iannaccone, 1992; Reader, 2013). In such an environment, religious investment seeks to enhance individual benefit and reduce invested costs. This follows the cost-benefit approach in the theory of religious economy (Stark and Finke, 2000). Green (2015) indicates that the development of religious markets promotes 'adaptation' and 'innovation' in the process of social interaction between producers and consumers. These social interactions function as a new public sphere. In the process of negotiation in the public sphere, new forms of consumer flows of time, money and effort emerge. Social networks in the market optimize the redistribution of customer resources and benefits.

In this situation, stakeholders seek cost-effective religious consumption to change their financial and physical status in the provided social context and to benefit the common good. Negotiating with the environment serves to minimize transaction costs for the practice and create a certain social norm among the stakeholders (Williamson, 1998, p. 27). As some researchers indicate, social norms like religion and culture cultivate trust and morality among individuals, radically diminish transaction costs and enhance communication among the stakeholders (Ensminger, 1997; Williamson, 1998).

In the recent *goshuin* movement in Japan, various narratives for constructing 'proper' forms of *goshuin* practice have also contributed to a certain social context and sense of common good. As the benefit of individual investment is diverse and ambiguously defined, according to individual motivations and purpose, it is expedient strategy for collectors to reduce the cost of religious investments. An enhanced social significance that is based on cost-effectiveness informs a particular way of religious consumption being deemed the 'proper' way. In this aspect, religious identity and community are not recognized from the perspective of religious affiliation; rather, they are based on individual consumption and investment in religious capital. This is in line with the theory of religious economy.

Summary

This chapter examined the recent *goshuin* movement in Japanese society from the perspective of the commodification of religious mobility.

As for the characteristics of religious mobility in the *goshuin* movement, this recently developed movement demonstrates the popularization of religious mobilities and traditional religious practices in secular Japanese society. Each person forms their own way of commitment through the construction of personal forms, manners and patterns of *goshuin* practice. On the other hand, various narratives collected during this study indicate that individual commitment informs the 'proper' way of commitment to the *goshuin* movement.

The development of individual commitment in the *goshuin* movement has strongly encouraged individual investment in self-validation and status enhancement. Various stakeholders have been encouraged by the market forces of 'proper' *goshuin* culture to create a social context and define a common good to reduce the transaction costs of these individual religious investments. As cost-effective strategies enhanced the *goshuin* movement, stakeholders created a certain consensus to develop participants' social impact. This impact is significant for both community and individual identity.

In conclusion, the development of religious mobility in the *goshuin* movement has encouraged new forms of religiosity through the enhancement of cost-effective individual investments for religious capital. These types of micro-level social interactions in the *goshuin* movement have become the basis for the contemporary way in which religiosity is embedded within consumer society in Japan.

Notes

[1] In the Edo period, *goshuin* was also recognized as the document of travel permission, because mobility beyond the feudal region was strictly controlled except for the purpose of pilgrimage (Reader, 2005, p. 22).

[2] Jodo Shinshu (one of the major Japanese Buddhism schools) is not serving *goshuin* because of their religious teaching.

[3] Some researchers point out that another reason for this movement is that religious institutions in the country are now burdened with heavy management costs because of the decline in charity and donations from their followers (Tsuban, 2016). Given this situation, *goshuin* is one of the main income sources for their sustainable management.

References

Abe, M. (2017) Jinja ni okeru goshuin no tsuushi (History of 'Goshuin'). *Jinjahonchou Sougou Kenkyuujo Kiyou* 22, 75–100.

Chikyuu no Arukikata Henshuushitsu (2013) *Goshuin de meguru Edo Toukyou no Koji (Goshuin Pilgrimage for Old Temples in Edo and Tokyo)*. Daiyamondo Sha (in Japanese), Tokyo.

Chikyuu no Arukikata Henshuushitsu(2017) *Goshuin de meguru Kanto no jinja (Goshuin Pilgrimage of Shinto Shrines in Kanto Area)*. Daiyamondo Sha (in Japanese), Tokyo.

Cohen, E. and Cohen, S.A. (2012) Authentication: hot and cool. *Annals of Tourism Research* 39(3), 1295–1314. DOI: 10.1016/j.annals.2012.03.004.

Eade, J. and Sallnow, M.J. (eds) (2013) *Contesting the Sacred: The Anthropology of Pilgrimage*. Wipf and Stock, Eugene, Oregon.
Edoraku Henshuubu (2014) *Toukyou Shuuhen Goshuin meguri Tabi: Otome no Jisha Annai (Goshuin Travel in Tokyo and Suburb: Guidebook of Temples and Shrines for Young Ladies)*. Meitsu Shuppan (in Japanese), Tokyo.
Ensminger, J. (1997) Transaction costs and Islam: explaining conversion in Africa. *Journal of Institutional and Theoretical Economics (JITE) / Zeitschrift für die gesamte Staatswissenschaft* 153(1), 4–29.
Featherstone, M. (1992) Cultural production, consumption, and the development of the cultural sphere. *Theory of Culture* 2, 265–292.
Green, N. (2015) *Terrains of Exchange: Religious Economies of Global Islam*. Oxford University Press, Oxford, UK.
Iannaccone, L.R. (1992) Religious markets and the economics of religion. *Social Compass* 39(1), 123–131. DOI: 10.1177/003776892039001012.
Kitiarsa, P. (2010) Toward a sociology of religious commodification. In: Turner, B.S. (ed.) *The New Blackwell Companion to the Sociology of Religion*. Wiley-Blackwell, Hoboken, New Jersey, pp. 563–579.
Krippendorff, K. (2004) *Content Analysis: An Introduction to its Methodology*, 2nd edn. Sage, Thousand Oaks, California.
Neuendorf, K.A. (2002) *The Content Analysis Guidebook*. Sage, Thousand Oaks, California.
Olsen, D.H. (2013) A scalar comparison of motivations and expectations of experience within the religious tourism market. *International Journal of Religious Tourism and Pilgrimage* 1(1), 1–5.
Olsen, D.H. (2019) Religion, spirituality, and pilgrimage in a globalising world. In: Timothy, D.J. (ed.) *Handbook of Globalisation and Tourism*. Edward Elgar Publishing, Cheltenham, UK, pp. 270–283.
Reader, I. (2005) *Making Pilgrimages: Meaning and Practice in Shikoku*. University of Hawai'i Press, Honolulu, Hawai'i.
Reader, I. (2013) *Pilgrimage in the Marketplace*. Routledge, London.
Schilderman, H. (2011a) Religious capital and public accountability: challenges to empirical theology. In: Francis, L. and Ziebertz, H.G. (eds) *The Public Significance of Religion*. Brill, Reiden, The Netherlands, pp. 41–63.
Schilderman, H. (2011b) Varieties of religious solidarity. In: Francis, L. and Ziebertz, H.G. (eds) *The Public Significance of Religion*. Brill, Reiden, The Netherlands, pp. 169–190.
Shinbutsu Reijo Kai (Committee for Religious Places in Shintoism and Buddhism) (2020). Available at: https://shinbutsureijou.net/ (accessed 26 April 2020).
Stark, R. and Finke, R. (2000) *Acts of Faith: Explaining the Human Side of Religion*. University of California Press, Berkeley, California.
Stephenson, B. (2010) *Performing the Reformation: Public Ritual in the City of Luther*. Oxford University Press, Oxford and New York.
Tsuban, H. (2016) Gendaijin no goshuin meguri koudou to shuukyou ishiki ni tuite: Kongo no kenkyuu ni mukete (Contemporary behaviours of goshuin pilgrimage and religious motivation: an essay for the further research). *Gendai Shuukyou Kenkyuu* 50, 151–161.
Williamson, O.E. (1998) Transaction cost economics: how it works; where it is headed. *De Economist* 146(1), 23–58. DOI: 10.1023/A:1003263908567.

12 Contemporary Perspectives of Hindu Pilgrimage in India: The Experiential Exposition

Rana P.B. Singh[1]* and Pravin S. Rana[2]

[1]*Asian Cultural Landscape Association and Banaras Hindu University, India;* [2]*Faculty of Arts, Banaras Hindu University, Varanasi*

Introduction: Hindu Tradition of Pilgrimage

Experiencing the power of place through acts of pilgrimage is a central feature of Hinduism (cf. Jacobsen, 2013). For Hindus, pilgrimage is a sacramental process that both symbolizes the participation of the pilgrim in the spiritual realm and actively establishes a two-way relationship between the pilgrim and the divine. Many pilgrimage places draw devotees through their reputation for granting some specific spiritual, social or material blessing, usually expressed in terms of purification and the healing of soul, mind and body (Stoddard, 1997). Moreover, Hindu pilgrimage is also a social duty, a rite of passage, and a way of gaining favour, which 'equally involves searching for spiritual experience in special places and learning that these material places lie outside the spiritual, mystical, true reality' (Sopher, 1987). The liminal 'faith-scape' that is so created encompasses sacred places, sacred time, sacred meanings, sacred rituals and associated rituals and performances. The number of Hindu sanctuaries in India is so large and the practice of pilgrimage so ubiquitous and rooted in its cultural traditions that the whole of India can be regarded as a vast sacred space organized into a system of pilgrimage centres and their hinterlands (Bhardwaj, 1973).

The focal points for Hindu pilgrimage travel are called *tīrthas*. Each Hindu pilgrimage is a *tīrthayātrā* (*tirtha* journey) and the geographical manifestation of each *tīrthayātrā* evokes a new kind of landscape that, for the devotee, overlays sacred and symbolic meaning upon a physical and material base. Hindu pilgrims often conceive their sacred journeys as an earthly adventure that combines spiritual seeking and physical tests (Sax, 1991). If touring is an outer journey in geographical space, then pilgrimage is the geographical expression of an inner journey. If touring is something largely oriented to pleasure seeking (and/or the satisfaction of curiosity), then pilgrimage is something that combines spiritual and worldly aspirations in places where the immanent and the transcendent mesh (Singh Rana and Haigh, 2015). Following the path of pilgrimages, 'the flow and intensity of the pilgrims, the followed sacred routes, sacred places and sacred rituals performed therein are imbued with the kinaesthetic and transformative possibilities at the core of religious experience' (Pinkney and Whalen-Bridge, 2018). In Hindu tradition, within the

*Corresponding author: ranapbs@gmail.com

purview of cultural complexity that is rooted in history and still continued and practiced, resultantly 'personal motivations for religious travel are complex, contingent, and cover the spectrum from introversion to extroversion, physical travel to inward journey, and so on' (Pinkney and Whalen-Bridge, 2018, p. 1).

The *Mahābhārata* (*c.*1000 BCE) may be considered as the first Hindu book providing glorious descriptions of sacred places. It clearly indicates that going on pilgrimages (*tīrthayātrās*) is superior to sacrifice (Bhardwaj, 1973). It narrates the story of the five brothers and their wives who, in old age, performed pilgrimage as penance, for peace, having no attachment to worldly things, for purification from bad karma, and perfection to complete the life cycle – all leading to liberation from the transmigration (*moksha*). This notion of pilgrimage regulates the system even today. The *Matsya Purāṇa* (*c.*400 CE) enumerates a large number of sacred places with descriptions of associated schedules, gestures, dreams and auspicious signs and symbols. Similarly, other *purāṇas* also describe sacred places and pilgrimages and associated merits (Singh, 2020, p. 79).

The visit to sacred sites and ritualscapes, the effort to understand and deeply experience, the work to keep the mind and heart open to the sacred manifestation inherent therein, all charge one's being with vision, with insight and with purpose (cf. Jarow, 1986). This is the real pilgrimage – 'an enacting of an internal process in the external world'. Here, issues of pilgrimage cycle, motives and the journey, and experiential reflection, are narrated in the context of historicity and contemporality, and finally the context of relevance and perspective for the future are also suggested.

Pilgrimage Cycle: Initiation to Completion

As in other parts of the oriental world, Hindu pilgrimage is a guiding force that can lead to the unifying of divinity and humanity (i.e. the search for wholeness). Pilgrimage is a *way* to heal the body and the soul by walking and experiencing the spirit of place (*genus loci*). A pilgrim has rightly expressed, 'I enjoyed every moment in the complete peace and solitude. Walking became a meditation and every step was teaching me to be mindful.'

The pilgrimage journey in Hindu tradition represents the physical act of visiting the holy place and, at the same time, a mental attitude (belief) and a moral act with firm determination (vow-taking, *saṅkalpa*). Pilgrimage is also part of a vow-system that refers to the two-way temporal notion of 'binding' (*saṅkalpa lenā*) before the start of the journey, and the 'taken off' (*saṅkalpa choḍānā*) after completion of the journey. The purpose and motives that are responsible for the act of pilgrimage are to be placed into two broad categories: specific mundane motives where deity is the focus of pilgrimage, particularly to have blessings for the solution of a problem the pilgrim is afflicted with; and to earn overall religious merit, generally difficult to define and explain, but which may include holy bathing on a specific astronomical occasion, the *darshana* (auspicious sight of the deity), or visiting holy men (see Diehl, 1956; Bhardwaj, 1973).

Most Hindu pilgrims aspire towards transcendence of the mundane and often express their experiences in purely spiritual terms, perhaps as a dialogue with their deity or greater self. Such deep pilgrimages, whether motivated by a sense of duty, hope or devotion, are rites of passage involving a cycle (*maṇḍala*) of three stages: *initiation* (awakening, from the time one decides to take the journey to the beginning of the journey), *liminality* (the voyage itself and experiences involved), and *re-aggregation* (the homecoming and sharing experiences as well as the upward path of liberation (*moksha*) (Fig. 12.1)). This is a sacramental process that involves the pilgrim in the liminal dimensions of each *faithscape*. It seeks to develop a two-way relationship between the pilgrim and the divine, howsoever conceived, and it offers two means of departure, one back to the mundane world and one to the spiritual realms (Singh, 2011a). In the Hindu tradition of pilgrimage, starting from a believer (person) and progressing to the ultimate, at least four layers exist, but they are interconnected through sacred space, sacred time, sacred function and a sacred psyche system (Fig. 12.2), while in a social context it refers to a march from individual to universal humanity. The act of pilgrimage starts from inner space (*home*) to outer territory (*kshetra*), and later, in reverse manner,

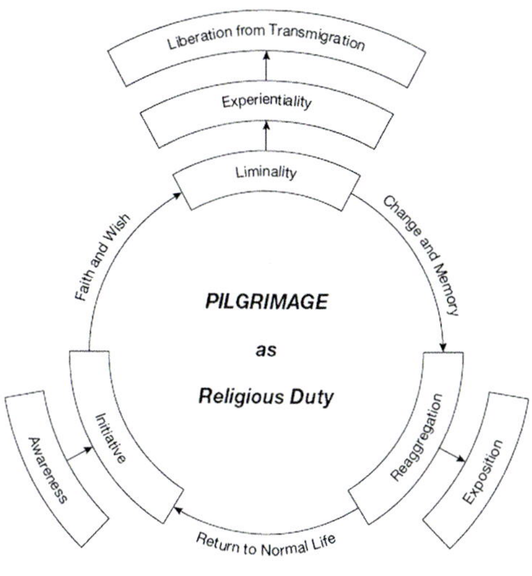

Fig. 12.1. Pilgrimage as religious duty in Hinduism: 'Pilgrimage Maṇḍala'. (From Singh, 2011a)

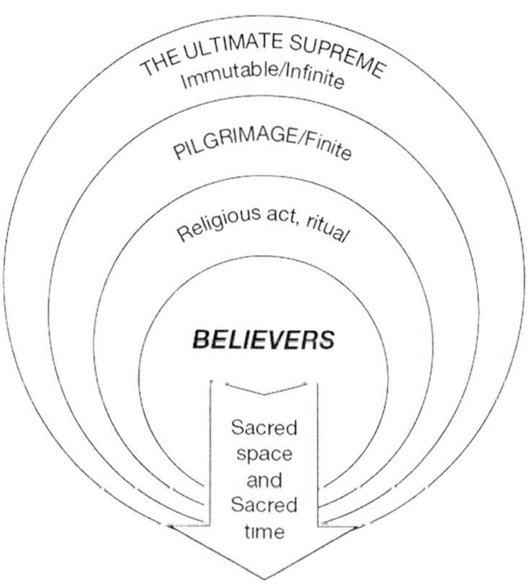

Fig. 12.2. Hindu outlook on pilgrimage. (From Singh, 2011a)

returning to home. Pilgrimage is a *way* to heal the body and the soul by walking and opening the soul to the spirit inherent in Mother Earth.

These two process systems bind sacred space, time, territory and religious functionaries together into a sacrosanct spatial organization. Pilgrims return to their normal life and share experiences with other members of society. This forms a cyclic frame of travel known as 'Pilgrimage *Maṇḍala*', which combines almost all the facets and attributes of pilgrims' behaviour (as individuals and as part of a pilgrims' group): rituals for self-awakening and links with the divine spirit; symbols that mark the signposts for revitalization; motivation through belief systems and cultural backgrounds; and maintenance of identity of oneself and as a pilgrim, and vice versa.

In common Hindu metaphysical belief, the physical and spiritual worlds are identified as two parallel layers of existence, and only by faith and revelation can one perceive their interconnectedness. Contemporary studies have shown that patterns of human movement in pilgrimage may take on cosmological significance, as journeys to shrines are identified with macrocosmic cycles or movement among astral bodies (see Crumrine and Morinis, 1991). That is how the Hindu pilgrimage is narrated as a dialogue of 'imagining, memorizing and understanding' between an *outer* realm of vision and an *inner* realm of self (Singh and Rana, 2020). This process also generates the emergence of a *maṇḍala*, making a round where the point of return is synonymous with the point of departure; they start and return from the same point.

Today, most Hindu sacred places are dominated by hybrid spaces that blend the religious and the mundane in complex, often contradictory forms. Each 'sacredscape' of sacred spaces, religious ritual performances and religious functionaries (cf. Vidyarthi *et al.*, 1979) is embedded within the socio-economic-environmental attributes of the mundane world and so creates the wholeness of a geographical 'faithscape' (cf. Singh, 2013, p. 69).

Faith is central to the desires, vows and act of pilgrimage; nevertheless pilgrimage is also a process whereby people attempt to understand the cosmos around them (Singh, 2006, p. 220). According to Eliade, 1959), 'for those who have a religious experience all nature is capable of revealing itself as cosmic sacrality. The cosmos in its entirety can become a hierophany'. Thus, for the pilgrim, every moment and every step is sacred. The holy places and shrines (or any other built structure) are only symbolic destinations (Singh, 2002, p. 187).

Motives and the Journey

We 'form a sense of our-selves and the sense of our-place at a varying range of scales (*spatiality*), times (*sequentiality*), functions (*activity*), forms of mobility (*pilgrimages*), quests (*sacrality*), and our mental states (*belief systems*). We begin from the local scale, and here we first experience the sacred message through the spirit of place or *genus loci*' (Singh and Rana, 2020). The human quest to find peace and experience sacred space drives faith-building and the desire to travel. According to theological notions, sacred places are consecrated or 'illuminated by faith', 'which because of their religious content, become the inevitable subject of auspicious visits' (Singh, 2011b). Such places are considered sublime (*deva sthāna* in Hindu traditions) and holy, and their holiness is consolidated as a result of continuity of pilgrimage by devotees and their performance of rituals. In this way, such places are made constantly alive through the process of making, remaking and interfacing the sublime qualities and those understood to lead to soul healing.

In addition to spiritual gains, Hindu pilgrimages have always concerned the gaining of social status and the relief of worldly cares (Haigh, 2011). In the ancient texts, very many of the blessings described concern the relief of sins or the fulfilment of wishes for health, wealth, success and so forth (cf. Jacobsen, 2013). This is how these places are known for salvific qualities (*puṇyadāyī*) and soul healing (*ātma-ārogyakar*); obviously, pilgrimage is a process to receive and get that merit (*phala*).

Motives for pilgrimages, as stipulated in the *purāṇas*, may be categorized into four broad groupings: *putreshaṇā* (desire for a son), *vitteshaṇā* (desire for wealth), *mukteshaṇā* (desire for liberation or atonement from sin), and *lokeshaṇā* (desire for worldly gain). But all centres of pilgrimage have the merits of general purification and bliss (Singh, 2020, p. 79). In most cases, the act of pilgrimage is performed as an expression of devotion to a deity and ultimately to gain blissful satisfaction through the fulfilment of a particular desire (Singh, 1997).

Through pilgrimage and deep feelings/faith man can transform the materialistic identity into a cosmic integrity at a place. Such a place becomes holy where 'wholeness' is preserved. The mythological stories merge the divinity's acts and life into a divine environment making a spiritual sphere of *faithscape* more meaningful, and that is more the subject of understanding than narration. And, when more specifically the meaning comes from sacredness, the place converges to *sacredscape*.

Pilgrimage is a process and performance that links inherent, manifested, mythicized and perceived meanings – commonly referred to as cyclic nature – worship and satisfaction to human quest through interaction and reciprocal non-verbal communication with the divine nature. Hindu pilgrimage connotes one of the five cultural-coded moral duties (*pañcha-kriyās*) with a noble aim to live virtuously (*dharma*), celebrate holy days (*utsava*), honour the rite of passage (*saṁskāra*), and ultimately receive merit/fruits of peace (*shānti*) making life gratifying (*dhanya*) with sublime pleasure (*ānanda*) (cf. Davidson and Gitlitz, 2002). In ancient times, when primordial religions predominated in India, the prehistoric geographical features like rock, old trees, caves (the womb of the world), rivers (the life-giving fluid of the world), and mountains (the boundary between earth and sky, and the source of weather) housed spirits who exerted significant influence over human behaviour (Davidson and Gitlitz, 2002). Of course, due to transformation of landscapes and cultures with the passage of time, many other traditions superseded, but they all made their imprint and also maintained the insight and spirit that existed in the past – this is what is called continuity of unity among cultural diversities, which is the essence of Indian culture. The Hindu paradigm of pilgrimage has served as a spirit of awakening and a binding force for 'communion of minds'.

Of course, conventional travel to holy places has been considered as pilgrimage during which pilgrims, with a strong religious or spiritual motivation, undertake long journeys from distant places on foot to arrive at a sacred site. Although the goal of such travel is to *see and be seen by* the deity (and saints) and seek blessings, pilgrims also engage in rituals that are considered necessary by clergymen, priests or religious functionaries for the purpose of travel. That is how the profane attributes support the sacred; they become part of the sacred and help to reciprocally regulate the cycle of sacred-profane.

When the latter dominate in this system, there develops touristic pilgrimage. In this context, the famous proclamation of *a pilgrim being half-tourist* and *the tourist being half-pilgrim*, is reasonably accepted; however, it would be better to use the term 'touristic pilgrimage' (*paryatan-tīrthatan*) and 'pilgrimage tourism' (*tīrthayātrā-paryatan*) in contemporary contexts where one can find qualitative characteristics of 'sightseeing' at sacred and religious places undertaken on holiday (Shinde and Katia, 2014).

In a survey conducted in 1972–73, interviewing 500 pilgrims, more than half had come to Varanasi (one of the holiest cities in India) for special performances, and it would seem that these respondents exemplify the typical pilgrim who travels to a distant place because the object of devotion is located there (Vidyarthi *et al.*, 1979, p. 136). A recent survey (2019) on the same lines was made by the authors, who supported the above observation, indicating that this seat of pilgrimage has still strongly kept its sacred character. Although the journey itself may be meaningful, the primary reason for movement is merely to get from one place to another (i.e. from home to the pilgrimage place) to acquire religious merit (*puṅya*). However, due to its special merit for ancestral rites, another sector of the interviewed pilgrims reported that they undertook a circumambulation (*parikramā*) associated with funeral ceremonies. Pilgrims often have multiple reasons for their religious journey, including fulfilling the vow taken for getting some relief from sufferings (Singh, 1997). Since ancient times, a passion for placement has been basic to Hindu thought. Sacred place as 'storied place' is eulogized in Hindu mythology, or oral epics, with divine connotation – there intersects myth and *terra firma*. Most of the holy and pilgrimage places are such distinct places recoding mythology and beliefs about their association with gods and their activities (*līlās*). Of course, 'valuistic journey is defined as an expression of valuistic ideals, together with demonstration of identity, such journeys help to develop or change personal or social identity' (Liutikas, 2013, p. 43-44). However, in Hindu tradition such journeys are mostly for religious or spiritual gain, and are supported by the profane (secular), and after a complex web of rituals the profane gets transformed into the sacred, and this cycle continues.

Recent surveys of pilgrims in the five pilgrimage cities of north India (see Table 12.1), clearly show predominance of spiritual merit as the main object (recorded by over half of the respondent pilgrims). However, this predominant motive varies according to the special character or personality of the holy place, e.g. Gaya, famous for ancestral rites; the main motive to visit this place is ancestral ritual (recorded in 65%), while Vindhyachal as the regional centre

Table 12.1. Motives of pilgrims in the five holy places of India (%).

	Main purpose/gain (*lābha*)	Varanasi	Prayagraj	Gaya	Ayodhya	Vindhyachal
1.	Spiritual merit (*puṅya*)	58	52	12	53	23
2.	Blessing of holy men (*āshisha*)	10	11	8	21	7
3.	Ancestral ritual (*shrādha*)	14	14	65	9	9
4.	Tonsure (*mundana*)	4	4	5	2	28
5.	Fulfilling vows (*manauti*)	2	3	2	2	20
6.	Liberation (*moksha*)	4	4	3	3	1
7.	Self-peace (*shānti*)	2	4	1	4	1
8.	Relief from suffering (*mukti*)	1	5	1	3	2
9.	Donation/charity (*dāna*)	2	1	2	1	1
10.	Touristic pilgrimage (*yātrā*)	3	2	1	2	8
–	Percentage of respondents	100	100	100	100	100
–	Total respondents (N)	432	380	320	418	290

[a]Personal surveys for different festivities (2016–2019).

of goddess worship, following the local and folk tradition, is popular for tonsure and fulfilling vows (recorded in 48% of respondent pilgrims). Some places, where natural scenery like hills, falls, rapids and caves exists near the holy places, we can see touristic-pilgrimage becoming more common in comparison with other places (Table 12.1). This observation corroborates with the ancient epic of the *Mahābhārata* (13.108.16–18), which states: 'Just as certain limbs of the body are purer than others, so are certain places on the earth more sacred – some on account of their situation, others because of their sparkling waters, and others because of the association or habitation of saintly people.' These cantos also mention the generalized and distinctive rules, ways and codes of conduct to be followed during pilgrimage at various places, as well as the hierarchy and degree of sanctity of various places. One of the respondent pilgrims told us: 'The objective of pilgrimage takes my concern for more meaning when I realize the purpose of life. Life is meant for becoming free from the wheel of *samsara*, which means the continuous cycle of birth and death. It is for making spiritual advancement and to perceive our real identity as human beings'.

Experiential Reflection

In Hinduism, where religious acts, rituals and festivities dominate the life-pattern, peoples' belief systems and philosophy of life on the scale of 'religion-action' (*dharma-karma*) continue by ritualization processes and need to be purveyed in the light of psyche-world and historical tradition. The meanings, symbols, faith and feelings involved all shed light on different issues of realization and revelation. In Hinduism, for those who have a religious experience, all nature is capable of revealing itself as cosmic sacrality; thus, the whole cosmos becomes entirely a manifestation at the end. This further refers to the sense of passing from individualism (a pilgrim) to universal human kindness (cosmic entity).

Pilgrimage gives overall merit for general well-being and relief from misdeeds and sins. In the ancient world, a life was a constant battle to get food, fuel and shelter to sustain existence. Later, as a means to repent from sins performed in a critical situation or even ignorantly, the act of pilgrimage developed by the need to believe in a journey for understanding the spirit of place and miracles. Only recently it has been accepted that man has not developed sacred places; rather he has discovered the sacred places. The pilgrims generally look upon sacred sites and shrines as sources of help – *auxilium* – in the present life and security in the divine realm where the soul goes after death, or where a decision is made for his rebirth. That is how the pilgrimage may be considered as a spiritual insurance policy or blessing.

What matters on a pilgrimage journey is not the moment as such but the *form of notion* – the feeling, faith and idea of beingness in the sacredscape, of being for the place. This beingness has its own 'mode of spatialization' or its own kind of 'placialization' and 'sacralization' – its own way of getting back into place (cf. Casey, 1993). If formation and knowing identity in the place is the primary quest, knowing how to be linked is also requisite. This *how* consists of a variety of ways, rituals and performances in Hindu belief systems.

With respect to walking, there appear two types of pilgrims: (a) *foot-pilgrims*, those following strict rules and walks (see Fig. 12.3), and mostly seeking guidance from priests and clergymen who cite glorifying mythology and religious tales from the texts; and (b) *pilgrimage-tourists*, those who use vehicles and loosely follow the textual tradition or common rules and avoid worship. Foot-pilgrims enjoy sacred journeys as an earthly adventure from one place to another that entails the combined effects of a spiritual quest and physical hardship – by walking, suffering or avoiding temptation. Foot-pilgrims repeat and experience the authenticity (figuratively and even literally) of walking in the footsteps of their ancestors whose descriptions are narrated in mythology.

Most of the foot-pilgrims have expressed feelings similar to the exposition as narrated by Kumar (1992):

> I enjoyed every moment in the complete peace and solitude. Walking became a meditation and every step was teaching me to be mindful. I relaxed to the soul of my own breath as it issued into the deep silence surrounding me. Breathing in I inhaled the warmth of the air, the smell of the wet grasses, the coolness of the water, the

Fig. 12.3. A group of foot-pilgrims taking rest, Pañchakroshi Yatra, Varanasi. (Photo: by Rana P.B. Singh)

> purity of Nature: I was breathing the universe into myself.

Pilgrimage-tourists also derive a similar sense of cosmic integrity, while all the pilgrims enjoy the journey experientially as well. Jay (1991)) notes that 'though intense in their devotion, they clearly were enjoying the recreational aspect of their experience as well'. Thus, pilgrimage operates as a cultural and spiritual tradition on many levels for the individual pilgrims involved. Of course, 'where one goes on pilgrimage, for what purpose and what one does in the sacred place are choices strongly influenced by the individual characteristics of the pilgrim, i.e., caste, class, sex, sect and so on' (Morinis, 1984, p. 281). Metaphysically, the purpose of pilgrimage 'is to come into harmony with the greater unity of all life so that you can become who you are and then serve others according to who you are' (Swan, 1990, p. 221).

The most popular ritual at any temple is offering food to the god and taking it back as *prasāda* (offered food to gods) as blessing. Presenting some general principles of contemporary *prasāda* illustrated by examples from four Hindu pilgrimage sites, while analysing *prasāda*'s usage in well-known Sanskrit scriptures (especially the *purāṇas*), it is argued that it should also to be conceived as a foundational concept for making sense of Hindu religious life in south-Asian terms (Pinkney, 2013). After completion of pilgrimage, and returning home, the *prasāda* will be distributed among friends and neighbours, and by this benefit they will also acquire some merit from the pilgrimage (*phala*).

Kumbha Melā: the Biggest Bathing Pilgrimage

Festivals (*melās*) at sacred sites in India are a vital part of Hindu pilgrimage traditions. Celebrating a mythological event in the life of a deity or an auspicious astrological period, *melās* attract around 450 million pilgrims from all over the

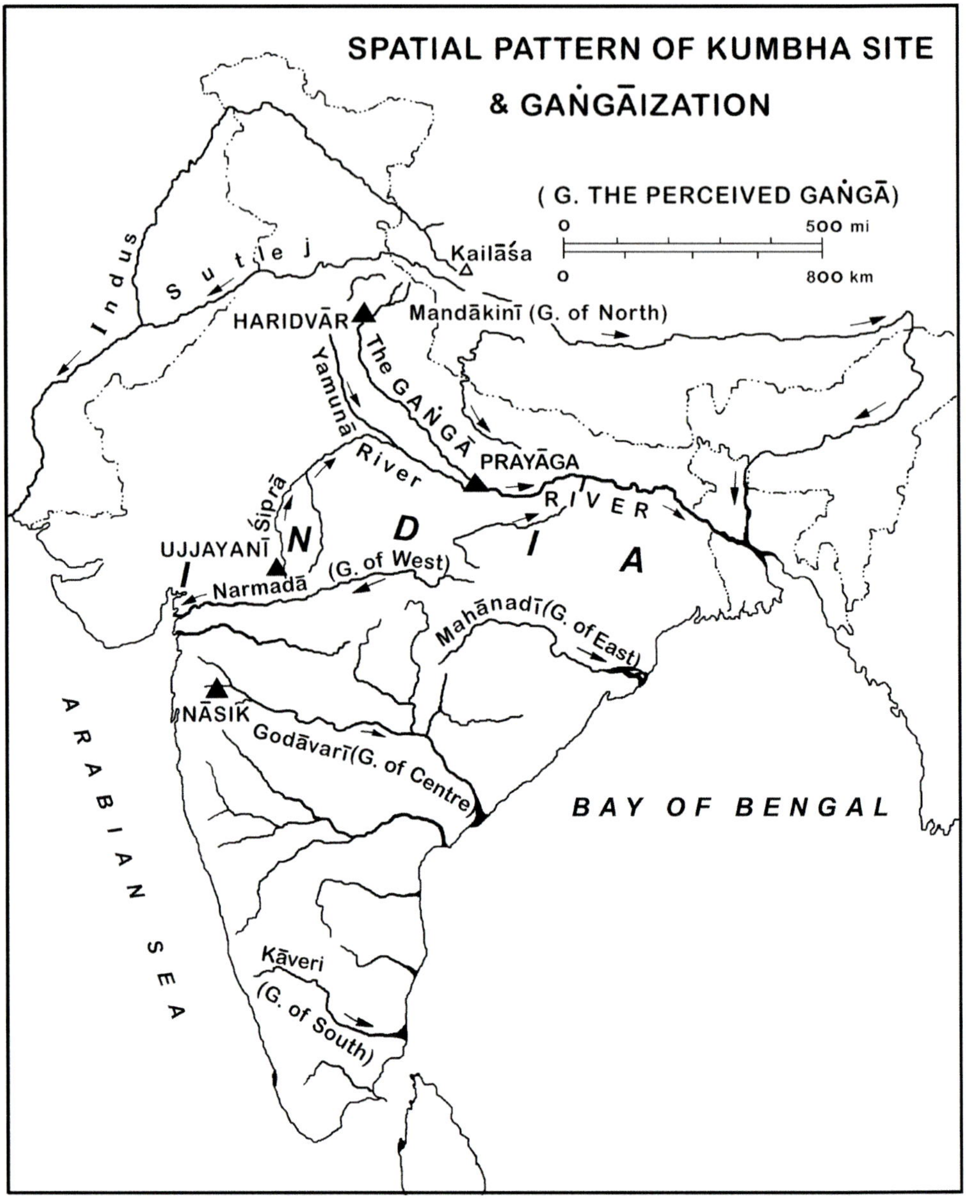

Fig. 12.4. Kumbha Melā sites in India. (From: Singh, 1987, reprinted with permission)

country (cf. Singh Rana and Haigh, 2015). The greatest of these, the Kumbha Melā, is a riverside festival held four times every 12 years, rotating between Prayagraj (Allahabad), located at the confluence of the rivers Ganga, Yamuna and the mythical Sarasvati (cf. Dubey, 2001), Nasik on the Godavari river, Ujjain on the Shipra river, and Haridvar on the Ganga river (Singh, 1987; cf. Fig. 12.4). Bathing in these rivers during the Kumbha Melā is considered an endeavour of great merit, cleansing both body and spirit.

The Kumbha Melā in the past has played the role of propagating new ideas and inventions, and still is a religious stage on which different groups copete for power and identity and operate in multiple directions (Bechler,

2013). Since the end of the 20th century, this pilgrimage festival has been controlled by the government, and has now turned into a global festival attracting people from different religions and even from different parts of the world. The Kumbha Melā is inscribed as representative of the Intangible Cultural Heritage of Humanity in the UNESCO World Heritage List 2017. The event encapsulates the science of astronomy, astrology, spirituality, ritualistic traditions, and social and cultural customs and practices, making it extremely rich in knowledge and continuity of tradition, which are transmitted through ancient religious texts, oral traditions, historical travelogues and texts produced by eminent historians.

The Prayagraj (Allahabad) and Haridvar festivals are routinely attended by millions of pilgrims (13 million visited Allahabad in 1977, some 18 million in 1989, over 68 million in 2001, over 74 million in 2013, and over 105 million in 2019), making the Kumbha Melā the largest religious gathering in the world. It may also be the oldest. The 2019 (15 January–4 March) event was spread over an area of 45 sq. km and an 8 km length of bathing *ghāts* (stairways). Twenty-two million people resided in a temporary tent city divided into 22 sectors, and infrastructural facilities included 122,500 toilets, 20,000 sanitation workers, 20,000 dustbins, 90 parking lots for 500,000 vehicles, 22 pontoon bridges on the two rivers, 500 shuttle buses and 22 hospitals (with 450 beds in each) (cf. Singh and Rana, 2020; also Singh, 2020, p. 85). This is the best-known example of pilgrimage-tourism, or spiritual tourism in Hindu traditions, where both sacred and profane meet.

The idea of sacrosanct energy at the junction of the two heavenly powers is preserved in the tradition of the special sacred bath (Fig. 12.5) and associated rituals. On the special, astrologically auspicious, occasion of Kumbha Melā, when Jupiter enters into Aries or Taurus, and the Sun/Moon into Capricorn, India's greatest fair is held at Prayagraj (Allahabad), at the confluence of the Ganga and Yamuna rivers. This takes place about every 12 years (Singh, 2020, p. 84). A Canadian visitor, who has grasped the meanings and feelings evolved there, expresses his views based on his experiences (in February 1989): 'It went on as it has for centuries, I found myself wondering if I was watching an eternal remains aware of its need for renewal, for purification, for the revival of the spirit. I wondered as I watched again as the streaming water of the Ganga met the streaming masses of humanity in the wonderful dance where outer meets inner and becomes a living symbol of something more' (Bryant, 1994, p. 39).

Fig. 12.5. Sacred bathing by Sadhus in Prayagraj on 14 January 2019. (Photo: by Rana P.B. Singh)

The forthcoming and planned Kumbha Melā at Haridvar (in the state of Uttarakhand, north India), to be held from 14 January to 26 May 2021, is expected to be much bigger in scale, infrastructure and involvement of the devotees and will be considered the biggest such festival on the earth. It is estimated that around 120 million people will attend; thus it will exceed the world record of 105 million at Prayagraj (2019), and will be an extraordinary example of faith-healing (cf. Singh, 2020, pp. 84-85). The focal theme of this greatest-and-grand 'Sacred Water' Melā is 'Green Kumbha – Clean Kumbha'. The Melā City will be spread over an area of 130 sq. km; preparation of the infrastructure and facilities is already in progress under the patronage of the state government. For this project, a grant of INR 22,000 million (US$ 315 million) has been sanctioned.

For many of the groups, like the ascetics, pilgrimage simplifies life, setting one free from the attachment to possessions and money, or family relations. Pilgrimage may be an integral part of life in the goal to achieve mastery over the body, living a life of self-denial, as in the case of Sadhus in the Kumbha Melās. As well as spiritual well-being, perhaps another value of pilgrimage is achieved: therapeutic value for many sufferers may be achieved at sacred places.

Hindu Pilgrims' Experiences: Interfaces among Power, Legitimacy and Reciprocity

Based on experiential, participatory and expositional information, Hindu pilgrims' experience should be theorized in the interlocking network of phase-wise changes in terms of interfaces among power, legitimacy and reciprocity (Fig. 12.6). Here 'power' refers to 'a relationship among cultural actors in which one cultural actor A can get another cultural actor B to do some form of pilgrimage that B would not have otherwise done. The strength and basis for power is maintained by the 'continuity of tradition' through cultural traits (i.e. life-culture) – identified as A in Fig. 12.6. The power of cultural continuity serves as 'coercive force' maintained by the three forms of religious incentives – material, solidary and purposive.

Legitimacy is conceived as a generalized perception or adaptability of the religious actions of the social groups that are desirable, proper or appropriate within some traditionally and culturally constructed systems of norms, values, beliefs and code of conduct. This is represented by the system of tourism-recreation in the purview of Life-World – shown as B in Fig. 12.6. The basis for legitimacy can be individual, organizational and/or institutional.

Life State	*Mental state*	*Condition*	*Ultimate*
A **Continuity of Tradition**: Cultural Traits (Life-Culture)	3 Desire (self-search) *ātmānusaṅdhānātmaka*	4 Difficult *ākrāntya-janya*	7 **Definitive** ***ānubhūti*** **(revelation)**
B **Tourism-Recreation**: Socio-Economic (Life-World)	2 Discretionary *ātmanirṇayātmaka*	5 Dominant *anubhava-janya*	
C **Spiritual Quest**: Belief-systems (*Life-Philosophy*)	1. Dormant *asakriyātmaka*	6 Dependent *āshrita-janya*	

Fig. 12.6. Hindu pilgrims' experiences and interfaces: channelizing the network.

Reciprocity is a transformative channel interlinking the passage of 'succession-sustenance-sustainability' and maintains counter-actions between the pilgrims (sacred) and the functionaries (profane) involved therein, i.e. it influences the other and is influenced *by* the other; this way this process is constantly going on in the frame of spiritual quest under the belief systems (i.e. Life-Philosophy, shown as C in Fig. 12.6). These three together make the platform of 'life state'. The last one influences the dormant condition (*asakriyātmaka*, 1), which, passing through the stage of discretionary (*ātmanirṇayātmaka*, 2), reaches to the stage of desire (self-search; *ātmānusaṅdhānātmaka*, 3). These three mental states will further form three successive conditions: difficult (*ākrāntya-janya*, 4), dominant (*anubhava-janya*, 5) and dependent (*āshrita-janya*, 6). Thus, ultimately, it reaches to the resultant 'definitive' (revelation, *anubhūti*, 7). This way, in the six phases of passing the mental state and condition, pilgrims ultimately receive the final form of faith healing, i.e. from initiation of 'realization' (*anubhava*) to the final state of 'revelation' (*anubhūti*).

Concluding Remarks

In the cosmos, spiritual forces delineating special places as particularly important and sacred were as natural as they were inevitable. These special places were also 'the places of ritual and ceremony, the ones in which the sacred was enacted in daily life; and in the act, the essential mystery and divinity of the universe reaffirmed' (Swan, 1992). Pilgrimage as one of the highest forms of ritualization gives 'meaning' – be it collective or generic, psychic or biological, mental and spiritual. A phenomenal particularization in this context is proposed by Heidegger (1956, p. 78-79), that sacred sites are having 'being-in-place'; to put it other way, that the spirit of place (*genus loci*) is reflected in its manifested spirituality and narrated vividly in mythology, so frequently cited and strongly believed by devout Hindus.

The Hindu pilgrimage sites are regulated, revived and sustained by *darshan* (auspicious glimpse) and *pujā* (prayer and worshipping), which form a complex web of rituals. During the medieval period, many digests and treatises were written that elaborate the still-practised rituals, describing the glories and merits, rules of performances and items used. Rituals play an important role in every sphere of humankind. In Hindu religion rituals they start from birth (*janma*) and are completed upon death (*mrityu*) – though they may also extend after death (post-*shrādha*). The Indian ritualistic cities of Varanasi, Prayagraj, Gaya and Ayodhya represent an aesthetic and unique type of cultural landscape that includes historical-religious traditional performances, mythology and faiths, custom, folklore, festivities, pilgrimage routes and multi-cultural religious sites that are assembled in the formation of ritual landscapes (Singh and Rana, 2020). In performing sacred journeys, visiting shrines and performing rituals, pilgrims 'proceed inwards from multiplicity to unity, just as in contemplation', and when they return to their home the process becomes reversed, thus developing a cyclical process between worldly activities and spiritual feeling.

Basic Issues of Hindu Religious Tourism: Potential and Prospects

1. Hindu pilgrimage places (*tīrtha-sthāna*) are the repository of the ancient cultural tradition and will be continued as the most used destination for religious activities and social harmony, thus serving as Intangible Cultural Heritage.
2. These sacred places possess the vitality of interconnecting Humanity and Nature; and are therefore to be used as the force for environmental awakening and related planning measures, including development of Intangible Cultural Heritage.
3. Psychology needs to better understand the value of pilgrimage to human life as it may be one of the most important ways that we can discover our meaning and deeper quest, find health consciousness and be inspired, as well as build reverence for nature. Reassessing the glorious past and understanding religious traditions and their present viability 'can bring back the necessary balance for us to develop and evolve into mature beings. This would not only benefit Indian society

but also the world at large' (Srivastava *et al.*, 2013, p. 286).

4. Under the national mission (Government of India) of PRASAD (Pilgrimage Rejuvenation and Spiritual Augmentation Drive), pilgrimage places (*tīrtha-sthāna*) may be developed as spiritual foci for religious tourism, within the purview of four ways: philosophical, organizational, heritage-awakening and managerial.

5. Together with holy-heritage centres, pilgrimage centres will serve as a catalyst in inclusive Sustainable Development Goal 11.4, which emphasizes efforts to protect and safeguard the world's cultural and natural heritage.

References

Bechler, S. (2013) Kumbha Mela: millions of pilgrims on the move for immortality and identity. In: Hüsken, U. and Michaels, A. (eds) *South Asian Festivals on the Move*. Harrassowitz Verlag, Wiesbaden, Germany, pp. 297–316.

Bhardwaj, S.M. (1973) *Hindu Places of Pilgrimage in India*. University of California Press, Berkeley and Los Angeles, California.

Bryant, M.D. (1994) River of grace: the Kumbha Mela as sacred place. *Environments* 22(2), 35–40.

Casey, E.S. (1993) *Getting Back into Place. Toward a Renewed Understanding of the Place-World*. Indiana University Press, Bloomington, Indiana.

Crumrine, N. Ross. and Morinis, Alan. (eds) (1991) *The Pilgrimage in Latin America*. Greenwood Press, Westport, Connecticut.

Davidson, L.K. and Gitlitz, D.M. (2002) *Pilgrimage: From the Ganges to Graceland: An Encyclopaedia*. 2-vols. ABC-CLIO, Santa Barbara, California.

Diehl, C.G. (1956) *Instrument and Purpose. Studies on Rites and Rituals in South India*. C.W.K. Gleerup, Lund.

Dubey, D.P. (2001) *Prayaga: The Site of the Kumbha Mela*. Aryan International, New Delhi.

Eliade, M. (1959) *The Sacred and the Profane*. Harper & Row, New York.

Haigh, M.J. (2011) Interpreting the Sarasvati Tirthayatra of Shri Balarāma. *Itihas Darpan, Research Journal of Akhil Bhartiya Itihas Sankalan Yojana* 16(2), 179–193.

Heidegger, M. (1956) An ontological consideration of place. In: Heidegger, Martin. (ed.) *The Question of Being*. Twayne Publ., New York.

Jacobsen, K.A. (2013) *Pilgrimage in the Hindu Tradition: Salvific Space*. Routledge, London and New York.

Jarow, R. (1986) *In Search of the Sacred: A Pilgrimage to Holy Places*. Quest Books, Wheaton, Illinois.

Jay, E.J. (1991) Panch Kosi Yatra: a sub-regional pilgrimage of southern Chhattisgarh, India. In: Jha, Makhan. (ed.) *Social Anthropology of Pilgrimage*. Inter-India, New Delhi, pp. 99–117.

Kumar, S. (1992) *No Destination: An Autobiography*. A Resurgence Book/Green Books, Cambridge, UK.

Liutikas, D. (2013) Experiences of pilgrimage in Lithuania: expressions of values and identity at new destinations. *Journal of Tourism Consumption and Practice* 5(2), 43–60.

Morinis, A. (1984) *Pilgrimage in the Hindu Tradition*. Oxford University Press, Delhi.

Pinkney, A.M. (2013) Prasada, the gracious gift, in contemporary and classical South Asia. *Journal of the American Academy of Religion* 81(3), 734–756. DOI: 10.1093/jaarel/lft022.

Pinkney, A. and Whalen-Bridge, J. (eds) (2018) *Religious Journeys in India: Pilgrims, Tourists, and Travelers*. SUNY Press, Albany, New York.

Sax, W.S. (1991) *Mountain Goddesses. Gender and Politics in a Himalayan Pilgrimage*. Oxford University Press, New York.

Shinde, K.A. and Katia, R. (2014) A cross-cultural comparison of weekend trips in religious tourism: insights from two cultures, two countries (India and Italy). *International Journal of Religious Tourism and Pilgrimage* 2(1), 17–34.

Singh Rana, P.B. and Haigh, M.J. (2015) Hindu pilgrimages: the contemporary scene. In: Brunn, StanleyD. (ed.) *The Changing World Religion Map, vol. 2: Sacred Places, Identities, Practices and Politics*. Springer Nature, New York and London, pp. 783–801.

Singh, R.P.B. (1987) Towards myth, cosmos, space and mandala in India. *National Geographical Journal of India* 33(3), 305–326.

Singh, R.P.B. (1997) Sacred space and pilgrimage in Hindu Society: the case of Varanasi. In: Stoddard, R. and Morinis, A. (eds) *Sacred Places, Sacred Spaces: The Geography of Pilgrimages*. Geoscience & Man, vol. 34. Dept. of Geography and Anthropology, Louisiana State University Press, Baton Rouge, Louisiana, pp. 191–207.

Singh, R.P.B. (2002) *Towards the Pilgrimage Archetype: The Pañchakroshi Yatra of Banaras. Pilgrimage & Cosmology Series: 3*. Indica Books, Varanasi.

Singh, R.P.B. (2006) Pilgrimage in Hinduism: historical context and modern perspectives. In: Timothy, D. and Olsen, D. (eds) *Tourism, Religion, and Spiritual Journeys*. Routledge, New York and London, pp. 220–236.

Singh, R.P.B. (2011a) Holy places and pilgrimages in India: emerging trends & bibliography. In: Singh, Rana P.B. (ed.) *Holy Places and Pilgrimages: Essays on India*, Planet Earth & Cultural Understanding Series, Pub. 8. Shubhi Publications, New Delhi, pp. 7–56.

Singh, R.P.B. (2011b) Sacredscapes & sacred places and sense of geography: some reflections. In: Singh, Rana P.B. (ed.) *Sacredscapes and Pilgrimage Systems*. Shubhi Publications, New Delhi, pp. 5–46.

Singh, R.P.B. (2013) *Hindu Tradition of Pilgrimage: Sacred Space and System*. Dev Publishers & Distributors, New Delhi.

Singh, R.P.B. (2020) Sacrality and waterfront sacred places in India: myths and making of place. In: Ray, C. (ed.) *Sacred Waters: A Cross-cultural Compendium of Hallowed Springs and Holy Wells*. Routledge, London & New York, pp. 77–90.

Singh, R.P.B. and Rana, P.S. (2020) Faith and place: Hindu sacred landscapes of India. In: Edensor, T., Kalandides, A. and Kothari, U. (eds) *The Routledge Handbook of Place*. Routledge, London and New York, pp. 95–108.

Sopher, D.E. (1987) The message of place in Hindu pilgrimage. In: Singh, R.L. and Singh, Rana P.B. (eds) *Trends in the Geography of Pilgrimages. Homage to David E. Sopher*. National Geographical Society of India, Varanasi, pp. 1–17.

Srivastava, C., Dhingra, V., Srivastava, A. and Bhardwaj, A. (2013) Morality and moral development: traditional Hindu concepts. *Indian Journal of Psychiatry* 55(6), 283–287. DOI: 10.4103/0019-5545.105552.

Stoddard, R.H. (1997) Defining and classifying pilgrimages. In: Stoddard, R. and Morinis, A. (eds) *Sacred Places, Sacred Spaces: The Geography of Pilgrimages*. Louisiana State University (Geoscience and Man, vol. 34), Baton Rouse, Louisiana, pp. 41–60.

Swan, J.A. (1990) *Sacred Places. How the Living Earth Seeks our Friendship*. Bear & Co. Pub., Santa Fe.

Swan, J.A. (1992) *Nature as Teacher and Healer. How to Reawaken Your Connection with Nature*. Villard-Random House Books, New York.

Vidyarthi, L.P., Jha, M. and Saraswati, B.N. (1979) *The Sacred Complex of Kashi. A Microcosm of Indian Civilisation*. Concept Publishing Co., New Delhi.

13 Modern Full-sensory Experiences and Pilgrimage Fulfilment in Malta's Ancient Temples

Dane Munro*
Institute for Tourism, Travel and Culture, University of Malta, Malta

Abstract

Within valuistic journeys, the rush of the experience has turned into a long-term trend (Liutikas, 2012), also called experiential tourism. This umbrella-term covers much ground, and pilgrimage and faith-based travel are certainly among them. In valuistic journeys, the quality of the experience is not only a great expectation but has practically become a demand.

The basis of this demand is the phenomenon of fulfilment, which runs as a *fil rouge* throughout pilgrimage (Munro, 2017). Literature confirms the need for destinations and sites to be endowed with originality, character, spirit of place and authenticity (Zuntz, 1971; Rountree, 2002, 2003; Tate, 2008; Reader, 2015), although the value of the latter topic, authenticity, is currently still debated among scholars. Kunaeva (2012) suggests that in the realm of pilgrimages there is a certain visible trend of the pilgrim as tourism consumer who wants a (nearly guaranteed) authentic experience but not at the cost of safety and loss of too much comfort. Timothy (2011, p. 107–108) remarks that under the constructivist umbrella there are many stakeholders who are engaged in creating religious sites and experiences, all having a different idea of authenticity and how it must be interpreted; for instance, existential authenticity is based on how it is experienced in relation to one's self-definition.

Fulfilment is expected in faith-based travel among those who visit the archipelago of Malta with the purpose of a spiritually enriching visit to the cultural heritage in Malta left by its Neolithic inhabitants roughly 7600 and 4500 years ago. These megalithic edifices are called 'temples', for two reasons. The first one is that there is probably no better word for it (Malone *et al.*, 2005), and secondly, a temple is a place, a cult centre, where regularly rituals or ceremonies were held (Whitehouse, 2004a, b, 2014). At present, the Maltese temples are used by a variety of spiritual travellers to Malta who are using these ancient temples as temples in the modern context. They find these temples extremely inspiring, providing them with a full sensory experience and spiritual fulfilment.

Introduction

This chapter is one of the research results of continuous fieldwork regarding faith-based, spiritual and religious tourism and pilgrimage, which I

*dane.munro@um.edu.mt

have been conducting for the last 15 years. I have come across hundreds of people whom I have interviewed, and I have also observed many people engaged in the act of pilgrimage. Occasionally, on invitation, I participated in various forms of pilgrimage, allowing me to form generalist and specialist researched opinions. Historical backgrounds have always been importantessential starting points for my research, as it is essential to know where something came from, to be able to better understand the conceptual and contextual frameworks.

Throughout history, up to the present day, millions of people of many religions, faiths and convictions have undertaken pilgrimages, making it a universal or near-universal appearance of valuistic travel. Pilgrimage, a journey with a benefit or a value, involves, inevitably, sites, distance and reverence (and economic benefits for those people operating along the route or at the sites). Pilgrimage is something that has stayed with us, as it is intrinsically linked to answer questions regarding perennial human needs, such as matters of health, safety, food, shelter, fertility, safe birth and the future (Rountree, 2003), or care of the soul and the hereafter.

In the here and now, according to Davie (2010), religion is still alive and prospering in the post-modern and post-secular world, albeit in modified and new forms, often removed from historic churches. Pilgrimage is the oldest, most crisis-resistant, enduring and resilient form of travel, no matter what kind of social developments are occurring. Timothy and Olsen assert that faith-inspired travel is 'one of the most significant types of tourism in the world today by volume and prevalence' (2006, p. 276), to which the research of Morinis (1992, p. 1) can answer that contemporary pilgrimages draw the largest regular and peaceful crowds on earth. Benefits, certainly in the context of valuistic journeys, can be measured through the register and intensity of a person's fulfilment brought about by a pilgrimage.

Malta has experienced a continuation of religion over millennia, as it has been part of the Neolithic society, the Bronze Age world, the Classical civilization, the Judeo-Christian culture, the Arabic realm and, after all that, till the present day, the Roman Catholic sphere. It is assumed, for the sake of argument of faith-based tourism, that all these cultures were familiar with pilgrimage and tourism in various degrees of intensity; some are still of influence to Malta as a destination for faith-based travel. There is a variety of sacred sites in Malta, both rural and urban. Of those, there are at least 590 Christian sites, ranging from caverns to cathedrals. There are numerous sites for secular pilgrimage related to, e.g., World War II or Dark Tourism, while other sites are, at the same time, places of interest for cultural tourism, such as the aforementioned Neolithic temples. The latter are important for the adherents of the new religions movements (NRMs) too.

The increase in global interest for pilgrimage in the last decade, especially faith-based travel of NRMs, coincides with the tourism sector's habit to compartmentalize and subdivide specific forms of tourism into niche markets and special interest tourism (SIT), in order to differentiate and to distinguish it from the ordinary mass tourism product (McKercher and Chan, 2005).

At present, the concept of a pilgrimage, that of a journey with a benefit, still holds strong, whether it applies to the ancient form, i.e. sacred or religious, or to a modern, secular form of travel or a stay at a particular site. The idea of a benefit has become dominant in the last couple of years. Service providers and travel agents are fencing with words such as 'meaningful', 'memorable experiences', 'connected experiences' and 'magical travel' (see for instance Airbnb and Uber websites), placing alternative meanings on valuistic travel, while many travellers are simply demanding and expecting it. I can envisage tech-savvy entrepreneurs coming up with a kind of 'rent-an-ancient-temple' app for pilgrims in the near-future.

New Religions Movements (NRMs)

Since NRMs are a relatively recent addition in academic circles, a brief introduction is presented here. NRMs are a very diverse and somewhat secretive segment of the religious landscape and at present still at its fringe in Malta. Worldwide, however, NRMs have become recognized religions in many countries and its adherents seek to fulfil needs which the historic Churches cannot fulfil.

Many NRMs are based on historical values of religions past or on symbolical or perceived values of the Neolithic or druidic past, such as seeking relationship with one's ancestors, predicting the future, fertility and the celebration of the cycles of life and death (Hutton, 1999). In eastern Europe, for instance, some groups go back totheir claimed or ideologically perceived cultural roots before Christianity arrived in their countries (Aitamurto and Simpson, 2013). Some NRMs are, what Hutton (1999) regards as a new approach, not based on past or existing religions, while others are what Rountree (2002, 2003) labels as 'designer' religions, based on modern needs. Adler (2006 [1979]), regarded as the first American academic working on modern paganism, has demonstrated that the new religious movements in the USA consist of a wide array of freshly minted religions, explaining their values, character, foundation and development (1999). Harvey (2009) describes the NRMs as 'like all lived religions, paganism is messy, and terms that might refer to discrete phenomena in systematic thinking are commonly entangled or braided in reality', which reflects the reality within NRMs, since there is no overall structure like in monotheistic religions.

Reader (2015) argues that when new religions develop, they also look for places to go to for their pilgrimages or spiritual travels. For faith-based tourism to Malta, this widens the scope of places of interest enormously, as the whole scope of the Maltese cultural heritage can be included. Adler (2006 [1979]) also realized that Wicca had been built on a wrong interpretation of archaeological and historical facts, but that it was a normal occurrence of all new religions to go in all directions before they consolidated. In the case of Malta, the Neolithic sites are real, but the religions which are brought to these sites by NRM adherents are not grounded in historic reality; the replication takes place on the ritual level, as NRMs, like the Mother Goddess movements, accept that the Maltese temples form part of the origin of their faith, and that the Mother Goddess, or Nature, is the focal point of the pilgrimage (Christ, 2004). A unifying factor for the NRM groups may be the potency of the mystery which is empowered by the lingering vestige of the divine, which attracts pilgrimages of this kind.

NRMs range from monotheistic views, including many Christian elements, to polytheistic understandings. Hutton (1999) agrees with Graham Harvey that the religion of a modern (western English-speaking) pagan can be attractive to both a 'romantic rationalist' as also to 'full-blown theists of a wholly traditional kind', with many opinions in between. For faith-based tourism to Malta, this widens the scope of places of interest enormously, as the whole scope of the Maltese cultural heritage can be included. For over a decade, Heritage Malta, the manager of the Neolithic sites in Malta, has introduced a specialized policy, in which the conditions are set for using the Neolithic sites as places of worship outside regular tourism visiting hours. This has helped in making these places accessible and is also a means of generating extra income for maintaining these sites.

Diversity

NRM adherents seek to fulfil needs which the historic churches cannot; thus some groups go back to their claimed or ideologically perceived cultural roots before Christianity arrived in their countries (Aitamurto and Simpson, 2013). Many are based on historical values of religions past or on symbolical or perceived values of the Neolithic or druidic past, such as seeking relationship with one's ancestors, predicting the future, fertility and the celebration of the cycles of life and death (Hutton, 1999). In the case of Malta, the Neolithic sites are real, but the religion is not grounded in historic reality; the replication takes place on the ritual level, as the NRMs accept that the Maltese temples form part of the origin of their faith, and that the Mother Goddess, or Nature, is the focal point of the pilgrimage. A unifying factor for NRM groups may be the potency of the mystery which is empowered by the lingering vestige of the divine, which attracts pilgrimages of this kind.

Several organizations exist, and the most relevant for this thesis are, among others, the neo-pagans, neo-druids, adherents of pagan witchcraft, goddess spirituality, and such like. Wicca also includes traditional, hereditary or hedge witchcraft, which suggests that many of their beliefs and rituals are fundamentally similar (Hutton, 1999). By now there are many forms of NRM, maybe in number only outdone by Protestant denominations.

Harvey (1997, p. vii) summarizes the NRM religious outlook as 'a religion at home on Earth, an ecological spirituality, a somatic philosophy of life'. This covers the majority of modern western nature religions, such as Wicca (the modern witchcraft created mainly by Gerald Gardner), druidry, contemporary western Shamanism, goddess spirituality, eastern European variations (going back to the religions of the time before the Teutonic Order forced Christianity upon them) and heathenism (based on Germanic and Scandinavian traditions). NRM adherents usually do not follow one single path but are eclectics who follow a customized path consisting of elements of various traditions, but sharing a number of ethical principles.

However, there are very common features in NRM, and Rountree (2010a) distinguishes the following main elements:

- reverence for nature, its life force and its cycles of life, death and regeneration;
- the concept of goddess and god as dual gender system, complementary expressions of the divine reality;
- personal freedom and autonomy, encapsulated in the Wiccan Rede: 'Do what you will, but harm none';
- everything is connected, all is matter and energy.

Fulfilment

Fulfilment is a big, overall *leitmotif* and many studies on pilgrimage, throughout all ages and religions, have revealed the human yearning for fulfilment (Rappaport, 1971; Csikszentmihalyi, 1990; Morinis, 1992; Bender, 1998; Notermans, 2008; Bhattacharjee and Mogilner, 2013; Turner and Turner, 1978; Rountree, 2002; Timothy, 2002, 2011; Timothy and Olsen, 2006).

Davidson and Gitlitz (2002) explain that 'pilgrims are persons in motion, passing through strange lands, seeking fulfilment'. In particular, Morinis (1992) says that pilgrimage is a paradigmatic and paradoxical human quest, both outward and inward, a movement towards ideals known but not achieved at home. Can we therefore define fulfilment? Concerning valuistic journeys and pilgrimage, it may better and more accurate to describe fulfilment than making attempts to catch it in a definition. Pilgrimage is an image for the search of fulfilment of all people inhabiting an imperfect world. The latter issue, the imperfection of our lives, leads to a longing for fulfilment, and after we have obtained some sort of fulfilment, we want to enjoy it as long as possible.

In this context, important matters are questions such as belief, hope, confirmation of belief, prayer for oneself or for others, reflection, enlightenment, purification, personal salvation, embodiment, transcendence, comfort of the soul and the tangibility of religious sentiment. Then there are psychological matters, for example fulfilling a wish, or a quest to be closer to God or one with God, seeking solutions for unresolved matters, happy marriage, ambitions, self-definition, happiness, initiation, reconciliation, inner harmony, transition, enchantment, authenticity, (self-)acceptance, self-actualization, healing from spiritual injury, restoration, reconstruction of one's life, a better fortune in life or empowerment. Others are about well-being, e.g. healing, health, fertility and safe birth. Further issues could be visitors' satisfaction, searching for something close to a miracle, looking for a better or purer road to somewhere, searching for the truth, or simply enjoying the moment.

Fulfilment in valuistic journeys is also related to identity (Liutikas, 2012) and social identity theory. Identity theory focuses on social structural arrangements and the link between persons, while social identity theory focuses on characteristics of situations in which the identity may be activated. In short, everybody has several identities and certain circumstances can trigger such an identity (Schwartz *et al.*, 2012). When does one take on the identity as a pilgrim – the moment the idea is born?, during the booking stage?, or is it prompted at the first step at the Camino? And how long after the journey is over does one remain with the pilgrim's identity? The research of Diener *et al.* (2009) proves without doubt that good health has the greatest and most long-lasting positive effect on happiness and fulfilment. Of course, one of the goals of pilgrimage has always been the regaining of good health.

In analogy to Veenhoven's (2003) research into happiness, fulfilment is a conscious state of mind and in the context of faith-based travel it implies motivation to set out on a trip to achieve something. It is argued in this chapter that when visitors are properly motivated and their mindset is positively charged towards the goal of fulfilment, this drive can be incentivized.

Experiential Travel

Experiential travel is the new buzzword in tourism, a term advanced by what tourism lawmakers and hoteliers probably regard as the most disruptive innovation in tourism, Airbnb. Within pilgrimage, one can make a differentiation between fulfilment in prescribed religious and lived religious settings. In the prescribed religious settings, fulfilment follows a structure and a regimental traditional pattern, e.g. during the Hajj, fulfilment is drawn from following a prescribed itinerary and prescribed activities in Medina and Mecca, concentrating on Abraham and the Prophet Mohammed. Likewise, during the Good Friday processions, people think about or relive what Jesus Christ did or what was done to him, during the Stations of the Cross, along the Via Dolorosa and at Golgotha in Jerusalem. In Malta, the shipwreck of St Paul is also celebrated annually, as this event is the first identity marker of the Maltese (Sant Cassia, 1993, p. 358-9). This fulfilment follows the rules and is often part of group dynamics, including those of identity and social identity theory. Such experiences during a prescribed or structured pilgrimage can be very intensive, while the connection with tourism is at present unavoidable due to travel and hospitality arrangements.

On the other hand, there are the unstructured faith-based, religious or spiritual experiences and fulfilment within the realm of the lived religion, away from the control of the prescribed religions. It is within such a framework of lived religious experience, free of demands and control, that the attraction of Malta as a destination for pilgrimage or, in the wider sense, of faith-based tourism becomes apparent.

Authenticity

In practical terms, the intrinsic qualities of the destination are important triggers for fulfilment or disappointment. Factors such as authenticity, continuation of the past or traditions, site management, spirit of place, a synergy consisting of many variables – may have the ability to bring their own kind of incentive to kindle fulfilment to the visitors. But quite often it remains a visitors' satisfaction. However, to reach a step higher, that of fulfilment, or leading to a life-changing situation, a catalyst must be involved. When there is the right atmosphere at a site, when all the synergies are delivering to maximum, fulfilment is in easy reach. Authenticity is, as previously mentioned, a topic of debate. In the case of faith-based tourism, authenticity probably means how fulfilling the experience is within a site, and not if it is still in its original state.

How the right atmosphere can be created depends on the quality of the site and its authenticity. Much depends on the occasion, the ritual or how the senses are stimulated. This kind of stimulation can come through art, decorations and architecture as visual attractions, but also through light, sound and smell. There remains the question of energies and vestige being felt by faith-based respondents in the various categories, including Roman Catholics and Protestants (Shackley, 2001; Mazumdar and Mazumdar, 2004; Notermans, 2008; Ambrosio, 2010; Scott, 2010; Tay and Diener, 2011). Feeling energy or being energized are common remarks, the feeling of a vortex or the sensation of a whirl are regularly mentioned sensations. A high sense of fulfilment was registered when people were alone or within their exclusive group at a chapel, temple or a natural site, engaging in a ritual with their peers without being disturbed or overtly observed by others. Religious fulfilment outside the prescribed religion touched faith-based visitors at moments when they expected it the least. Others experienced closure on difficult emotional issues and health/healing related matters, including the death of a loved one.

Pilgrims' Accounts

Pilgrims' accounts are full of descriptions of when people reach such a significant moment in their lives when arriving at the shrine, the enjoyment of a moment, the experience of healing, the emotions during the travel and the sense of fulfilment after all was done. In Freller (1995) there is mention of a travelogue account of 1745 by a Spanish traveller and author, Bartholom Pancorbo de Ayala, where he describes the atmosphere of St Paul's Grotto in Malta bringing 'about a sweet movement to the heart'. The account of Harris (1999) regarding Lourdes is full of such accounts, whereby Our Lady of Lourdes graced the pilgrims with healing, acceptance, fulfilment and moments of joy and unity.

From Rountree (2010b) two accounts are taken, related to NRM fulfilment in literature. One adherent relates that when she is at the temples of Mnajdra and Ħaġar Qim: a feeling of deep calm comes over her (Fig. 13.1). It may be that the beautiful landscape surrounding the temple and the adjacent seascape have a positive effect on her fulfilment. In short, the feeling of calm described by this particular person is a feeling of connection with the world and its past, which is, however, still touching us today (Rountree, 2010b, p. 197-9).

A person professionally connected to the temples, during an interview, confirms witnessing visitors being healed from physical and psychological stress, as if the spirit of the temple builders was trying to tell them something vital. Observations were made, as a temple custodian, of many rituals, including hand-holding (marriage), baby-naming ceremonies (equivalent to baptism, originally a pagan ceremony) and coming-of-age ceremonies – all rituals in which

Fig. 13.1. Inside Ħaġar Qim megalithic temple. (Photo: D. Liutikas)

the Mother Goddess is asked for blessing, dancing, praying, singing etc.

In the following paragraphs, there are some NRM pilgrims' accounts, obtained during my fieldwork, through consentient interviewing and observing.

A Canadian follower of the Mother Goddess movement is absolutely convinced that the Great Goddess exists and is alive, deep inside everybody. She describes the Goddess as Earth and Nature, a breathing, living, nurturing, life-giving power. With all the troubles going on in the world – climate change, economic disaster, violence against women and children, continuation of slavery, and violence caused by armed men in the name of a male god – she thinks that the point is rather proving itself: 'We really need to go back, or forward, to a world of peace and tolerance. The Great Goddess gives and teaches people everything they need to know for a balanced and respectful life. The temples of Malta are revered and sacred spaces, designed for the purpose of the Great Goddess, a centre from which love and spirituality spread out.'

Such statements indicate that the belief in a Great Goddess has been connected with social commentary, with disapproval of destruction of nature, greed and cruelty, which are also themes in Christianity.

For the NRMs, however, there is little doubt that the Virgin Mary is indeed a continuation of a divine female concept. An NRM pilgrim says: 'For us Wicca witches, there is a direct connection from the Great Mother to Mary as a giver of life and healing. We revere them both in our own way but rest assured that we have great respect for Mary, much more than the Evangelicals of my youth in Germany.'

Similarly, a group of NRM women were observed in the church of Mosta, when, during the feast of the Ascension of Our Lady on 15 August, the Marian statue was exhibited in the church, ready to be carried in procession. They touched the statue and gave hand-kisses to it, but they also stroked their long hair along the statue, as if their hair functioned as a *brandeum* (any object having touched something sacred and can be taken home), and they could get closer to the Mother Goddess. These intense multi-sensory experiences are also found in a study carried out by Wells (2011) on medieval devotion, in which she concludes that 'the senses became evermore influential on the fabric of the church building itself, changing as a direct result of the pilgrimage experience', and that 'sensation was the means by which belief was to be experienced'. As Brown (1981) also has remarked, it seems that a multi-sensory experience happens when believers feel that the sacred flows from certain objects of veneration into their very soul, with or without a *brandeum*. A sacred site which can arouse such intense multi-sensory experiences will often be visited by others, in the hope also of experiencing such effects.

The Spirit of Place is sometimes also referred to as a sacredspace. Jackson and Henrie (1983) have defined it as 'that portion of the earth's surface which is recognised by individuals or groups as worthy of devotion, loyalty or esteem'. Religions have been instrumental in the mystico-religious division of spaces into sacred and profane (p. 94). A site, whether created by people or by nature in the past, has acquired a spirit of place, which is maintained by its users. The sample of this was taken during a private visit to Ħaġar Qim and Mnajdra of an NRM group of seven middle-aged women from the UK and the USA, all academics. As a loosely organized Mother Goddess group, they travel the world to visit Mother Goddess sites. They come well prepared, read about the temples and meet a number of times before departure. The group members have an open mind and want to listen to the temple itself and not to bring any preconceptions, philosophies or ideologies to the temple, as they want to open up to its wisdom. In their view, the temple is the holder of knowledge and a place of history. A group member remarks, 'I just need to "be" at a site with no preconceived ideas or thoughts and become aware of past histories, beliefs and circumstances, with no judgements attached.'

Mnajdra is probably the most atmospheric of all the temples on Malta (Fig. 13.2). The surrounding landscape is a gentle slope, with low hills on one side and the sea on the other. It is one of the few places in Malta where there are no modern buildings (besides the modern tent covering the temples) and where traffic cannot be heard. A group reacts:

> The landscape was breathtaking, the view over the sea, the colours of the landscape and the variation of the vegetation. For me there was

Fig. 13.2. Mnajdra megalithic temple. (Photo: D. Liutikas)

> definitely a 'pull' from both sites to the little island of Filfla. The landscape prepared us for the visit, and when we were ready for the temple, and she for us, she asked us in.

For faith-based tourism, spaces are of the essence. The material circumstances of the temple features, its decorations and architecture, and the landscape it is in, give a sense of belonging and identity. An NRM group member says:

> It is a fantastic privilege to be in these temples. Especially now it is getting darker, the rounded, feminine forms of the temple speak louder than words. There is great mystery in the shape and form of the temple, the vestige of the past, everything must have had a function. I feel at ease here. I try to image what it could have been in the past, performing rituals at those altars and secret chambers. I am totally taken in by the temple and the Goddess.

There were places inside the temple where group members felt that the energy was strongest (Fig. 13.3). A temple is regarded as a living being, and the importance of the sensory experience cannot be underestimated for the sense of fulfilment. For ceremony and ritual, it is considered very important to have sound or toning (Davidson and Gitlitz, 2002; Mazumdar and Mazumdar, 2004; Skeates, 2010; Hamilakis, 2014). Visitors weigh in:

> We felt that there was a vortex at work at Mnajdra, sucking us in. We literally asked the temple to communicate itself through movement, sound, image or body to us, becoming a portal of communication. Temple responded to us, and we were filled with energy, moving powerfully in a counter-clockwise direction. Wandering undisturbed through this site, one of the oldest in the world, was extremely uplifting. We sang, meditated and venerated the Goddess. We also performed probably the oldest way of praying: we stood barefoot in a circle, holding the space, giving space. We thought it was beautiful and uplifting, and extremely energizing. The energy of the temple came through the stone through our feet into our hearts and minds.

Fig. 13.3. Inside Mnajdra megalithic temple. (Photo: D. Liutikas)

Davie (2010), in view of historical European churches, discusses the possible existence of a common factor, an experiential 'feel-good' factor, whatever that may be. The important matter is that the adherents or visitors simply 'feel' something, either the sacred or what Davie describes as a Durkheimian insight, namely that religion or the sacred arouses a sentiment reaching beyond the ordinary filling of an emotional void, without which people would remain dissatisfied (Fig. 13.4). She proposes a hypothesis: late modern Europeans prefer a site where the 'experience of the sacred' is the focal point rather than a 'purely cerebral' occasion, which, in itself, is a shift from obligation to consumption (p. 146), which fits in with the distinction this paper makes between the experiences of the prescribed religion and the lived religion. This points again to the social identity theory, when being at a place of significance triggers off an identity. There is always quite a special atmosphere at places where worshippers have been congregating for many years. Faith-based tourism is, like normal tourism, an intangible experience. A visitor confirms that:

'The privacy in the temples was a great gift; alone in the Mnajdra temples is deeply enriching and greatly fulfilling.'

The following sample represents a group of women observed at the Mnajdra temple. They were all women who left Israel and now live in the USA, formerly belonging to the Christian, Jewish or Muslim religions. They had adopted the Mother Goddess spirituality and had shed their former religions, for reasons of patriarchy, repression of their sexuality and procreativity, and the general feeling that women were treated as second-class beings and property. They stood in the temple in a circle, holding hands and asked the Mother Goddess for the healing of their wounds, their religious traumas, their suffering because of having been a woman in a woman-unfriendly region and maltreatment at the hands of violent men. Then the drums started and they danced first in a circle then in individual rhythms and pace, but still connected

Fig. 13.4. Visitors at Mnajdra megalithic temple. (Photo: D. Liutikas)

to each other. Then they started to dance individually, while invoking the Goddess. They cried, laughed, shouted, in a collective elated state, casting the bad thoughts of the past out and letting in the sweet thoughts of the present. It was regarded as a healing process, closure of a bad period in their lives, through the mysteries of the temple and a warm welcome back home in the arms of the Goddess. After the ritual ended, everyone seemed exhausted but elated, with a sense of new-won freedom, a very precious thing. An NRM academic can relate to the idea of mystery and healing:

> Generally, it is important that their ideals of mystery and religion can root in something old and authentic, in order to undo the uprooted feeling many people have in modern society. For many ex-monotheists within the NRM, the issue is to discontinue the excessive suffering and guilt elements, which they experienced as suffocating, negative and unnecessary for fulfilment or simply to lead a normal life without excess baggage.

'Andy' (fictitious name), a young woman from the USA living in the UK, was introduced to the Mother Goddess about six years ago. She is careful in her comments not to offend the Mother Goddess and says that the Goddess had only one identity but offers many ways to connect with her. Andy travels all over the world to visit sites and to experience the energy and presence of the Mother Goddess. At temples she tries to feel the energy and lets herself be directed by energy flows, to which she moves. She calls this working, and it makes her express herself, and this gives her fulfilment. Sometimes she would just stand still on one spot, rigid and unmovable; in other spots she feels a different flow and makes movements with her arms or whole body. Andy lets herself be directed by the energy and impulses she picks up. She does this until a certain point is reached, when it is enough. This kind of fulfilment is not predictable since it is the Goddess who decides what is enough. Spiritual fulfilment for Andy is when spirit, mind and body are in harmonious alignment with each other and when the environment without is in harmony and mirrors the environment within. It comes about in moments of grace characterized by higher awareness, peace and a sense of bliss and gratitude. A 'successful site' is that which induces a sense of peace, stillness, beauty and harmony within – a beautiful place in nature, a structure or building, which is in harmony with the geometric patterns in nature, a space that is considered sacred and has attracted devotion and prayer. All sites in Malta have the potential to be successful sites, Andy says, but it depends on a number of factors. She achieves the best results and the highest form of spiritual fulfilment when she is allowed to be on her own in a temple. The weather is no issue, she does her working even in the pouring rain. However, noise is a distracting factor and ruins everything. She has learned not to be bothered by remarks of other visitors, they are just words of people missing out on great experiences, she concludes.

Temples and churches belong to the category of conflicted spaces when masses, ceremonies, rituals etc. are held at the same time as tourists are allowed to enter the space, as observed by Cohen (1992) and Smith (1992). Olsen (2006) remarks that the touristification of religious sites consists of an overlap between the religious and the touristic space; he calls this the duality of space. This disparity in the purpose of people visiting a site is not only theoretical (Nyaupane *et al.*, 2015); its practical application helps to improve the total experience at a site, by understanding the different needs, and to avoid conflicts about the use of a shared space, to which Shackley (2001) agrees. Collins-Kreiner (2010) also states that this duality, or 'third space' concept acknowledges, both in implicit and explicit terms, that pilgrims and tourists are interdependent entities and that the social construction of a site can be sacred and secular at the same time. A practical example of duality of space is provided by an occurrence at the Neolithic temple of Ġgantija, at Gozo. On a fine spring day, a group of 12 middle-aged women, in wide simple dresses, sit under some palm trees in front of the west temple of Ġgantija, during regular tourism opening hours. They sit in a circle and pass a small object around. The one who has the object in her hand says something and then passes the object on anti-clockwise. They are passing the so-called 'talking stick', an ancient custom, also used by American Indians. Tourists stay at a distance, understanding it is a ritual, and do not want to disturb. The security guard has no such reservations and lurks suspiciously close by. Afterwards, they get up and go to stand in front of the west temple, their arms in front of them, palms up in prayer. Then they form a chain, holding hands with stretched arms, alternately one facing the temple and the other with their back to the temple. It is the starting position of a dance. With a flowing movement they bend their knees and stretch their legs, while moving their arms slowly up and down. The idea is of a rolling and moving wave, passing in front of the temple back and forth. This is where the laughter and cajoling starts from the other visitors. The women continue with their dance, they are unflappable. After their dance, they enter the temple and start praying, standing in a circle. This is where the security guard starts to panic, as the circle prevents other paying visitors to enter the temple. To his credit, and that of the other visitors, the guard and visitors wait till the coven is ready with their short prayer and leave the temple. I approach them and introduce myself as a researcher. They inform me that they are a Wicca coven from Middlesex. They had not been able to make an appointment with the management to book an hour in the early mornings and went in as normal visitors. Although they were hurt by the laughter, they still enjoyed their rituals and drew fulfilment from it.

Perspective

Instead of a conclusion, I present a perspective on further developments within the faith-based tourism sector. Being discovered by NRMs is a good development in the field of faith-based tourism for destinations, as this carries benefits too.

Seemingly, a point has been reached where the Weberian gloomy prediction of the disenchantment and demystification of the world, where rationality rules and religion is dead (Weber, 1978 [1956]), is being refuted. Re-enchantment and magical moments have found their way into faith-based tourism, and for destinations very much acquainted with religion, matters such as enchantment, enigmatic authenticity and spiritual fulfilment find themselves in a new field of interest. The NRMs are becoming more relevant than ever could have been imagined, as people seek supernatural help where the natural world and the old religions fail to accommodate them. In the process of empowerment and improving life's qualities, many of the old, new and imagined spiritual qualities are being employed (Starhawk, 2011; Notermans, 2008; Hermkens *et al.*, 2009). This ties in with the research of Csikszentmihalyi (1990), Diener *et al.* (2009) and Scott (2010) regarding happiness and people's efforts to strive for meaning, purpose and fulfilment. It also ties in with the observations of Davie (2010) and Reader (2007) that many people do not secularize but simply unchurch, not losing faith but seeking fewer restricting manners of continuation and achieving fulfilment in faith.

In addition to this, there are more arguments why destinations should embrace faith-based tourism. For receiving destinations, it is also a form of tourism which respects cultural, religious and natural heritage and appreciates a cared-for environment, since much worship takes place at sites which fall within the remit of cultural heritage. Faith-based tourism is also searching for authenticity, a real experience to search for and share, creating a fulfilment which has already been found by countless other people before. Economically, engaging in faith-based tourism has many advantages. It is one of the forms of tourism that often prefer the low season, when there is less crowding. The restricted number of people also has another advantage, namely that the spirit of place is not pushed out with the crowds. One of the biggest differences between the traditional Christian pilgrim groups and the NRM groups is that the latter have seemingly more spending power than the former, resulting in booking higher-class hotels and more expensive dinners and wines – a case of quality travel versus quantity. In pilgrimage, as a valuistic journey with benefits, the benefits go both ways, not only for the pilgrim but also for the service providers.

References

Adler, M. (2006 [1979]) *Drawing Down the Moon. Witches, Druids, Goddess-Worshippers and Other Pagans in America*. Penguin Books, London.

Aitamurto, K. and Simpson, S. (2013) *Modern Pagan and Native Faith Movements in Central and Eastern Europe*. Acumen, Durham, UK.

Ambrosio, V. (2010) Development models in Shrine cities and pilgrimage routes. In: *Proceedings of World Congress of Pilgrimages and Shrines*. Santiago de Compostella.

Bender, B. (1998) *Stonehenge: Making Space*. Berg, Oxford, UK.

Bhattacharjee, A. and Mogilner, C. (2013) *Happiness from ordinary and extraordinary experiences. Journal of Consumer Research 41*. University of Chicago Press, Chicago, Illinois.

Brown, P.R.L. (1981) *The Cult of the Saints, its Rise and Function in Christianity*. The University of Chicago Press, Chicago and London.

Christ, C.P. (2004) *Rebirth of the Goddess: Feminist Meaning in Feminist Spirituality*. Routledge, New York.

Cohen, E. (1992) Towards a sociology of international tourism. *Social Research* 39(1), 164–182.

Collins-Kreiner, N. (2010) The geography of pilgrimage and tourism: transformations and implications for applied geography. *Applied Geography* 30(1), 153–164. DOI: 10.1016/j.apgeog.2009.02.001.

Csikszentmihalyi, M. (1990) *Flow, the Psychology of Optimal Experience*. HarperCollins e-book.

Davidson, L. and Gitlitz, D. (2002) *Pilgrimage: From the Ganges to Graceland: An Encyclopedia*. 2 Volumes. ABC-Clio, Santa Barbara, California

Davie, G. (2010) *The Sociology of Religion*. Saga, London.

Diener, E., Wirtz, D., Tov, W., Kim-Prieto, C., Choi, D. *et al.* (2009) *New Measures of Well-being: Flourishing and Positive and Negative Feelings*. Springer e-book.

Freller, T. (1995) *St Paul's Grotto and its Visitors: Pilgrims, Knights, Scholars and Sceptics*. Valletta Publishing, Malta.

Hamilakis, Y. (2014) *Archaeology and the Senses: Human Experience, Memory, and Affect*. Cambridge University Press, Cambridge, UK.

Harris, R. (1999) *Lourdes: Body and Spirit in the Secular Age*. Penguin, London.

Harvey, G. (1997) *Listening People, Speaking Earth: Contemporary Paganism*. C. Hurst Co. Publishers, London.

Harvey, G. (2009) Animist paganism. In: Pizza, M. and Lewis, J. (eds) *Handbook of Contemporary Paganism*. Brill, Leiden and Boston, pp. 393–411.

Hermkens, A.K., Jansen, W. and Notermans, C. (2009) *Moved by Mary*. Ashgate, Farnham, UK.

Hutton, R. (1999) *The Triumph of the Moon. A History of Modern Pagan Witchcraft*. Oxford University Press, New York.

Jackson, R.H. and Henrie, R. (1983) Perception of sacred space. *Journal of Cultural Geography* 3(2), 94–107. DOI: 10.1080/08873638309478598.

Kunaeva, M. (2012) *Sustainable Tourism Management along the Camino de Santiago Pilgrimage Routes. Unpublished Master's Thesis in Tourism*. Haaga-Helia University of Applied Sciences, Finland.

Liutikas, D. (2012) Experiences of valuistic journeys: motivation and behaviour. In: Sharpley, R. and Stone, P.R. (eds) *Contemporary Tourist Experiences: Concepts and Consequences*. Routledge, Abingdon, UK, pp. 38–56.

Malone, C., Bonanno, A., Gouder, T., Stoddart, S. and Trump, D. (2005) The death cults of prehistoric Malta. *Scientific American* 15(1), 14–23 DOI: 10.1038/scientificamerican0105-14sp.

Mazumdar, S. and Mazumdar, S. (2004) Religion and place attachment: a study of sacred places. *Journal of Environmental Psychology* 24(3), 385–397. DOI: 10.1016/j.jenvp.2004.08.005.

McKercher, B. and Chan, A. (2005) How special is special interest tourism? *Journal of Travel Research* 44(1), 21–31. DOI: 10.1177/0047287505276588.

Morinis, A. (1992) *Sacred Journeys*. Greenwood, New York.

Munro, D. (2017) Historical perspectives of shifting motives for faith-based travel. In: Munro, D., Carspecken, L. and McIntosh, I. (eds) *Journal for Religious Tourism and Pilgrimage 5.*Institute of Technology, Dublin, pp. 17–23.

Notermans, C. (2008) Loss and healing: a Marian pilgrimage in secular Dutch society. *Ethnology* 46(3), 217–233.

Nyaupane, G.P., Timothy, D.J. and Poudel, S. (2015) Understanding tourists in religious destinations: a social distance perspective. *Tourism Management* 48, 343–353. DOI: 10.1016/j.tourman.2014.12.009.

Olsen, D.H. (2006) Management issues for religious heritage attractions. In: Timothy, D.J. and Olsen, D.H. (eds) *Tourism, Religion and Spiritual Journeys*. Routledge, Abingdon, UK, pp. 104–118.

Rappaport, R.A. (1971) The sacred in human evolution. *Annual Review of Ecology and Systematics* 2(1), 23–44. DOI: 10.1146/annurev.es.02.110171.000323.

Reader, I. (2007) Pilgrimage growth in the modern world: meanings and implications. *Religion* 37(3), 210–229. DOI: 10.1016/j.religion.2007.06.009.

Reader, I. (2015) *Pilgrimage: A Very Short Introduction*. Oxford University Press, Oxford.

Rountree, K. (2002) Goddess pilgrims as tourists: inscribing the body through sacred travel. *Sociology of Religion* 63(4), 475–496. DOI: 10.2307/3712303.

Rountree, K. (2003) The case of the missing Goddess: plurality, power, and prejudice in reconstructions of Malta's Neolithic past. *Journal of Feminist Studies in Religion (* 19(2), 25–43.

Rountree, K. (2010a) *Crafting Contemporary Pagan Identities in a Catholic Society*. Ashgate, Farnham, UK.

Rountree, K. (2010b) Tourist attractions, cultural icons, sites of sacred encounter: Engagements with Malta's Neolithic Temples. In: Scott, J. and Selwyn, T. (eds) *Thinking through Tourism*. Berg, Oxford, pp. 183–208.

Sant Cassia, P. (1993) The discovery of Malta: nature, culture and ethnicity in 19th century painting. *Journal of Mediterranean Studies (* 3(2), 354–377.

Schwartz, S.J., Luyckx, K. and Vignoles, V.L. (eds) (2012) *Handbook of Identity Theory and Research*. Springer.

Scott, R.A. (2010) *Miracle Cures: Saints, Pilgrimages and the Healing Powers of Belief*. University of California Press, Berkeley, California.

Shackley, M. (ed.) (2001) *Managing Sacred Sites*. Thomson, London.

Skeates, R. (2010) *An Archaeology of the Senses. Prehistoric Malta*. Oxford University Press, Oxford.

Smith, V. (1992) The quest in guest. *Annals of Tourism Research* 19, 1–17.

Starhawk (2011) *The Empowerment Manual: A Guide for Collaborative Groups*. New Society Publishers, Gabriola Island, BC, Canada.

Tay, L. and Diener, E. (2011) Needs and subjective well-being around the world. *Journal of Personality and Social Psychology* 101(2), 354–365. DOI: 10.1037/a0023779.

Timothy, D.J. (2002) Sacred journeys: religious heritage and tourism. *Tourism Recreation Research* 27(2), 3–6. DOI: 10.1080/02508281.2002.11081214.

Timothy, D.J. (2011) *Cultural Heritage and Tourism: An Introduction*. Channel View Publications, Clevedon, UK.

Timothy, D.J. and Olsen, D.H. (2006) *Tourism, Religion and Spiritual Journeys*. Routledge, Abingdon, UK.

Turner, V. and Turner, E. (1978) *Image and Pilgrimage in Christian Culture: Anthropological Perspectives*. Columbia University Press, New York.

Veenhoven, R. (2003) Happiness. *The Psychologist* 16(3), 128–129.

Weber, M. (1978 [1956]) *Economy and Society*. University of California Press, Berkeley and Los Angeles, California.

Wells, E.J. (2011) Saints' cults and their evolution in space and time. *Peregrinations: Journal of Medieval Art & Architecture*3(2), 122–146.

Whitehouse, H. (2004a) *Modes of Religiosity. A Cognitive Theory of Religious Transmission*. AltaMira Press, Lanham.

Whitehouse, H. (2004b) Towards a comparative anthropology of religion. In: Whitehouse, H. and Laidlaw, J. (eds) *Ritual and Memory*. AltaMira Press, Lanham, Maryland.

Whitehouse, H. (2014) To understand religion, think football: sacred beliefs likely arose out of prehistoric bonding and rituals. *Nautilus*.

Zuntz, G. (1971) *Persephone: Three Essays on Religion and Thought in Magna Graecia*. Clarendon Press, Oxford.

14 New Forms of Pilgrimage: World Youth Day Rio 2013

Luciana Thais Villa Gonzalez*
Departamento de Administração e Turismo (UFRRJ), Rio de Janeiro, Brasil

Introduction

The relationship between pilgrimage and tourism has already been widely explored and discussed in the literature; however, in contemporary times, new forms of pilgrimage and pilgrims emerge. The world of pilgrimages is no longer exclusively driven by religious motives. Increasingly, people make pilgrimages to visit battlefields where important historical events took place, tombs of artists and famous personages, and they even participate in virtual pilgrimages (Kaelber, 2005; Povedak, 2008; Dos, 2008; Margry, 2008, 2009; Hyde and Harman, 2011; Lloyd, 2015). Today's pilgrim can take on many faces and be motivated by factors other than those directly related to religion or spirituality. However, in this work we sought to exemplify the theoretical notes with data on pilgrims at a religious event who hardly resemble the typical pilgrim, often seen as someone already in adulthood and who has achieved stability in life (Carneiro, 2003).

World Youth Day (WYD) – the one held in Rio 2013 being the object of this study – is an international Catholic Church event held in different countries around the world every two to three years. The event is aimed at a young audience (up to 30 years) and the final goal is the Pope's meeting – representative of the sacred for Catholics – with youth from around the world. The gathering is made up of a diversity of modern and youthful activities, such as moments of evangelization, entertainment and leisure, against a background of Catholic faith. Shows, and sports, artistic, tourist and religious events, were held over six days in the city of Rio de Janeiro, from 23 to 28 July 2013. As the audience of the festival is young people, the organizers of WYD Rio 2013 used a series of strategies to allow the arrival of these youths from around the globe, such as: homestay and public space accommodation arrangements in Rio, negotiation of 'budget menus' in food-service establishments in the city, and provision of public transportation passes, amongst other incentives, to make the trip more affordable (Gonzalez *et al.*, 2017).

Conforming to the National Confederation of Bishops of Brazil (CNBB, 2013), WYD Rio 2013 had 427,000 people registered, of which 298,900 were Brazilians (70%) and 129,000 were from another 175 countries (30%). The image and profile of these young people for the organization and implementation of the event was shaped by the Church according to the interlocutor with whom it needed to interact to ensure the success of the event. The result of these efforts was visible at the farewell mass celebrated by Pope Francis before 3.7 million people gathered in Praia de Copacabana – icon of Brazilian

*ltvgonzalez@yahoo.com.br

Fig. 14.1. Final Mass at Copacabana Beach, WYD 2013, Rio de Janeiro. (From https://creativecommons.org/licenses/by/3.0/br/deed.en)

tourism (see Fig. 14.1) – making WYD Rio 2013 the event with the largest attendance ever in a Brazilian city (Ministerio do Turismo, 2013).

Thus, events such as WYD Rio 2013 bring to the fore the new relationships between tourism and the sacred, especially when we consider that in this case religion takes over tourism, not the other way round, and how this fact changes pilgrimages and pilgrims by moving them away from traditional models of religious/spiritual travel. In order to illustrate this point, in the next section we present the theoretical frameworks on the subject and how these reflect the changes in the pilgrims' behaviour and worldview, especially in the western world.

Pilgrimage and Tourism, Tourist and Pilgrim – Theoretical Considerations

It is commonplace in academic discussions of pilgrimage to assert that this phenomenon has occurred since ancient times, in the most diverse cultures (Coleman and Elsner, 1995; Bajc, 2012). Other authors, such as Carneiro (2003), refer to studies that identify pilgrimages in the Neolithic period, long before the rise of the major world religions. Another very common point in this debate, as claimed by Bajc (2012), delves into the transformative potential of pilgrimages, both in their individual and collective dimensions; however, the author points out that researchers have had difficulties in fully understanding this process. In Brazil, thorough studies have been conducted focusing on the meanings and objectives reported and assumed by pilgrims and institutions about this phenomenon, especially anthropological studies such as the works by Carneiro (2003) and Steil (2002, 2003).

The works of Victor Turner and Edith Turner, in this sense, were the dominant model in pilgrimage studies for an extended period (Steil, 2002). Turner (1973) and Turner and Turner (1978) identify different motivations of pilgrims: pursuit of penance/sacrifice, gratitude, fulfilment of a promise, to visit the place where a miracle happened or a spiritual leader lived, and also to live a transcendental experience different from ordinary daily life, often achieved by physical and spiritual suffering during the pilgrimage journey and the search for one's own identity or 'even with the aim of achieving a step towards holiness and wholeness in oneself, both of body and soul' (Turner, 1973, p. 32).

Much current research supports these motivations, while some add that the search for adventure, contemplation of the landscape or visitation to the tomb of some historical personage or idol, as well as places where important events occurred,[1] are also common reasons for going on pilgrimage (Margry, 2008, 2009; Liutikas, 2015; Shinde, 2015). Carneiro (2003) states that the motivations of pilgrims are rarely unique; on the contrary, they are multiple, and often of first, second and third order.

In addition to the motivations of pilgrims, Turner (1973) and Turner and Turner (1978) proposed a pilgrimage model that includes three main aspects: pilgrimage sites, usually located on the fringes of political and economic centres; the state of liminality; and the spiritual dimension of pilgrimage – all three understood as integral to the ritual process (Turner, 1973) toward *communitas*.

In the Turnerian model, the displacement of the individual, outside his geographical centre, represented by the journey to the core of pilgrimage and the displacements within it, would be equivalent to the inner journey that the individual undertakes in search of the sacred. This inner search, manifested in the act of pilgrimage, would be a way of escaping routine, an experience in which everyday rules and conditions can be suspended. Thus, the anti-structure would be established, opposing the structure. In this way, the geographical location of pilgrimage sites would always be linked to peripheral regions, away from economic, political and social hubs, a 'center out there' (Turner, 1973), where pilgrims would find the sacred even moving to peripheral pilgrimage centres.

This condition would be liminal, taking people from one initial inner state to another at the end of the journey, enabling social relationships to develop between people of different, often contrasting, social backgrounds. The pilgrims, thus united around the ritual of pilgrimage, experience *communitas*, a state in which social rules and constraints are suspended, giving the impression that all individuals are equal. Turner and Turner (1978) classify the experience of *communitas* into three types: spontaneous, in which feelings of equality and solidarity[2] are stronger and more present and is the only form of the phenomenon that occurs in the anti-structure; normative, when the individuals, organized in groups, attempt to regulate the experience to maintain its characteristics, creating structures; and ideological *communitas*, which would be a utopian situation within which spontaneous *communitas* resides, multiplying and becoming a structure, where universal ideals of equality, peace and friendship would be the norm among individuals.

In pilgrimage studies, the Turnerian model has been hegemonic for many years; however, from the 1990s onwards, research questioning it began to gain traction. One of the earliest and best-known works in this line was published by Michael J. Sallnow in 1981, entitled 'Communitas reconsidered: the sociology of an Andean pilgrimage' – a fierce criticism of Turner's model where the author states that he saw nothing in his study even remotely resembling the Turnerian descriptions, especially regarding the phenomenon of *communitas*. Sallnow (1981) reports that he only observed situations of competitiveness and conflict among all the social agents involved in a pilgrimage in the central Andean region. Ten years later, another work in the same vein, by Eade (1992), in Fatima (Portugal), also received a lot of attention. In 1991, Eade and Sallnow published a collection of articles aimed at discussing various evidences that the Turner model could not be considered universal and that the phenomenon of *communitas* was pure delusion. Thus, the expression 'sacred arena' becomes jargon in the area. These critiques were shared by other authors such as Cohen (1992) and Morinis (1992) and seem to have become predominant.

Another common criticism of Turner's model for understanding pilgrimages as a social phenomenon is the idea that the spaces where they occur are on the fringes of the dominant social structures. Through the study of three popular Thai pilgrimages, Cohen (1992) notes that these places lie precisely within the political and economic power centres, demonstrating that the understanding of religion as a separate sphere does not apply in non-western settings. Shortly before, in the field of Anthropology of Religions, studies that criticized western concepts of religion like those of Asad (1993) became known.

In this sense, from the publication of *Contesting the Sacred: The Anthropology of Christian Pilgrimages* by Eade and Sallnow (1991), Turner's model was called into question, since most published research on the subject from then on presented data challenging the concepts of anti-structure, *communitas* and liminality.

Discussions of the differences between pilgrims and tourists in this context also become significant. Within a framework of contestation, the emphasis on the different motivations of these social actors has gained traction, with some arguing that tourism, as a modern and consumption activity, is a distortion of pilgrimages, stealing their authenticity (Boorstin, 1971), while others claimed that tourism is a modern take on pilgrimage in which 'staged authenticity' plays a fundamental role (MacCannell, 1976). Cohen (1988) argues that, in MacCannell's view, although the tourist always pursues authenticity and believes he attains it, he is often led to inauthentic experiences by the tourist industry, thus creating a *false touristic consciousness*. On this discussion, Cohen (1988, p. 394) states that:

> 'Authenticity' thus takes up a given or 'objective' quality attributable by moderns to the world 'out there.' The only apparent difference between the tourist and the social analyst is that the latter is more circumspect than the former. He is therefore assumed to be able to penetrate beyond appearances, and discover the deception of 'staged authenticity' (MacCannell, 1973) perpetrated by the tourees, or the tourist establishment. (p. 394)

In the authenticity debate, the issue of the differences between tourists and pilgrims was nurtured for a considerable period (Morinis, 1992; Badone and Roseman, 2004; Lloyd, 2015; Shinde, 2015). Although pilgrimage is an ancient social practice, Carneiro (2003) argues that many people use pilgrimage as a pretext to travel, which today would probably be classified as tourism, even if it is not possible to discern a single motive for the realization of displacements such as pilgrimage.

Cohen (1992) also draws parallels between tourism and pilgrimage, believing that these two movements have as their goal the search for references by the individual in societies in which discourses and ways of acting and thinking in the world are multiple and diverse, leading to a certain strangeness and disorientation. According to Steil and Carneiro (2008), tourism and pilgrimage could be presented as explanatory categories and understanding of reality that condense structures of meanings that are constantly updated and re-evaluated in social practice.

Urry, 1996, in turn, argues that tourism is a phenomenon of simultaneous production and consumption in which the language and messages used to promote a tourist destination are part of such a product, of the image the tourist forms in relation to it, thus being an 'anticipation of the experience'. For Carneiro and Freire-Medeiros (2004, p. 107), this view of pilgrimage makes it a product that is a mixture of associated items that seem to work with an ideology and as a set of symbolic goods 'made' by 'producing agents' and by media language. In this sense, pilgrimage and tourism must be understood as realms that intertwine, relate and dialogue, even if there are tensions and conflicts, since not every pilgrimage can be commodified by becoming a product (Carneiro and Freire-Medeiros, 2004).

From these concepts, it becomes clear that a traveller on a pilgrimage can take on various identities such as traditional pilgrim, religious tourist, pilgrim tourist or tourist pilgrim (Steil and Carneiro, 2008). Pilgrimage and tourism share interfaces and similarities in which tourism often emerges as a mediator of the sacred, establishing relationships between tradition and modernity, religion and tourism, sacred and profane (Carneiro and Freire-Medeiros, 2004, p. 108).

Steil (2003) furthers the discussion on pilgrims' identity, stating that these have something of the heroic since their journey through foreign, unknown and inhospitable lands in search of the other and their own sanctification is understood as painful and full of difficulties, leading to a transformation at the end of the pilgrimage (Steil, 2003).

Moreover, Steil (2003) also points out that, currently, the fragmentation of the religious field and the concomitant emergence of diverse beliefs and religions gave rise to a pilgrimage movement as the search for oneself, the 'true self', the sacred, outside the sphere of religious institutions. From this perspective, the outer journey corresponds to the inner journey the pilgrim must undergo to live the 'experience of the sacred'. The author believes that this tendency is strong, because it is linked to the proliferation of religions of the self, built around detachment, self-improvement and full development of the human and divine potential of each individual.

Carneiro (2003) believes that, semantically, the pilgrim is the one who leaves home to go on a journey that will take him to an unknown place. Also, the quest for personal contact with the sacred is one of the main motivations of this type of traveller. Dias (2002, p. 22) warns that the pilgrim is the one for whom the journey to the sanctuary occurs strictly because of his spirituality and that the term is associated, in Brazil, with the individual experience of those who undertake this journey, contrary to a Romero, who usually travels in a group, strengthening community ties. He goes on to say that the pilgrim does not perceive himself as a tourist, even when he uses all the facilities and amenities that this sector offers, since his behaviour is different.

There is also an extension of these concepts by authors who consider the religious tourist as the one who has an outer gaze, emphasizing that they seek the spectacle, the observation of religious events, not participation and immersion in these encounters (Dias, 2002; Steil, 2003; Carneiro, 2003). It is noted that the word 'pilgrim' is more used in administrative-political scope by its managers than by the practitioners themselves (Steil, 2003). For Steil (2003), there is some resistance from religious authorities to accept the term religious tourism for fear of losing control over such manifestations of faith. On the other hand, increasingly, managers and politicians from religious destinations seek to associate religious events with tourism, in order to leverage economic benefits from such occasions.

Dias (2002) raises important considerations about the nature of religious tourism and its relationship with leisure. For the author, the pilgrim who travels exclusively for religious reasons and fulfils an intense programme, participating in cults, celebrations, prayers, among others, cannot be classified as someone in leisure activity, since the demonstration of faith and religious attitude of these participants is of a very strong and significant nature, unlike resting.

However, many pilgrimages are made in leisure time by individuals who choose to participate in these events of their own free will, since no religion forces people to go on a pilgrimage. Likewise, on these occasions, these individuals take advantage to know tourist attractions. The author acknowledges that this situation stems from the complexity and differentiation of leisure activities in contemporary societies, dissociating them from a solely recreational and idle meaning.

As for Badone and Roseman (2004), the pilgrim–tourist dichotomy seems to have been overcome. The authors note that some postmodern approaches have led to profound changes in the understanding of this issue, leading to a different understanding of the concept of 'travel', 'place' and 'theory'. Thus, the concepts of tourist, host and native take on a different perspective, blurring the boundaries, definitions and clear separations.

Collins-Kreiner (2010) suggests that the polarization between pilgrimage and tourism goes through a process of 'dedifferentiation', since current theories clearly show this distinction makes no sense. This same author highlights in her article an important change in the analytical concepts currently employed for the understanding of human displacements: the opinion that not only people, images, ideas, concepts and theories are mobile, but even the places themselves become a new paradigm as 'places are like ships' and people do not settle in a certain location during their travels but move from place to place constantly.

The analysis of this approach therefore focuses on the immobility/mobility of human flows, on the relationship between mobility and the infrastructure that supports this phenomenon, and on the interactions between physical, informational, virtual, and imaginative mobility. Cohen and Cohen (2015) also point to this new theoretical approach, stating that it seeks to understand the relationships established between people and places, and performances and performativity. Other works addressing the concept of 'virtual pilgrimage' follow the same line (Kaelber, 2005; Lloyd, 2015). As for the clash between the analyses inspired by the Turners' work and those following the contentious perspective advocated by Eade and Sallnow (1991), Coleman (2002) argues that reducing the analysis of pilgrimages to *communitas* or sacred arena is to impoverish both the framework proposed by the respective theoretical lines and the very result of their research. *Communitas* and sacred arena should be the starting points for pilgrimage studies and not their ultimate goals (Coleman and Elsner, 1995). In order to highlight these innovative ways of pilgrimage, the following section presents WYD Rio 2013, a modern and religious pilgrimage, which seeks to incorporate into the traditional the current yearnings of youth.

WYD Rio 2013 Pilgrims: New Forms of Pilgrimage

The Catholic Church considers WYD as a modern pilgrimage; however, the structure of the event is very different from that of a traditional pilgrimage. Contrary to most pilgrimages studied by Turner (1973) and Turner and Turner (1978), the WYD event does not take place in the periphery of political and economic centres. In this respect, Norman and Johnson (2011) point out that WYD can be understood as a twist in Eliade's and Turner's paradigms, since host locations are mobile, not fixed in space and time. In this way, just as there is a displacement of the participant to the host city, there is a displacement of the chosen locations, and especially of the representation of the sacred, since the Pope goes to meet the young people, and not the other way around. Profane and sacred, pilgrim and Pope, move to meet. The centre is not 'out there' (Turner, 1973), fixed and motionless. In this way, the list of cities that have already hosted the WYD clearly illustrates the concept of 'mobile' pilgrimages in space and time, since the journeys are held every two to three years in different cities around the world (Table 14.1).

In this sense, it is perceived that the concept of places understood as 'ships' moving from time to time (Collins-Kreiner, 2010) is valid when analysing WYDs. Articulating all its symbolic capital on pilgrimages, such as Israel, Rome, Santiago, Guadalupe, among others, the Catholic Church chooses 'places', cities to where this symbolism is transferred, preferentially adding to the tourist attractiveness of the chosen

Table 14.1. Host countries of World Youth Day.

Year	City/Country	Attendance	Pope
1984	Vatican City/Vatican	300.000	John Paul II
1984/1985	Vatican City/Vatican	300.000	John Paul II
1987	Buenos Aires/Argentina	1.000.000	John Paul II
1989	Santiago de Compostela/Spain	400.000	John Paul II
1991	Czestochowa/Poland	1.600.000	John Paul II
1993	Denver/USA	500.000	John Paul II
1995	Manila/Philippines	5.000.000	John Paul II
1997	Paris/France	1.200.000	John Paul II
2000	Rome/Italy	2.000.000	John Paul II
2002	Toronto/Canada	800.000	John Paul II
2005	Cologne/Germany	1.200.000	Benedict XVI
2008	Sydney/Australia	400.000	Benedict XVI
2011	Madrid/Spain	1.500.000	Benedict XVI
2013	Rio de Janeiro/Brazil	3.700.000	Francis
2016	Kraków/Poland	2.500,000	Francis
2019	Panama City/Panama	700.000	Francis

Source: adapted from Vatican (n.d.)

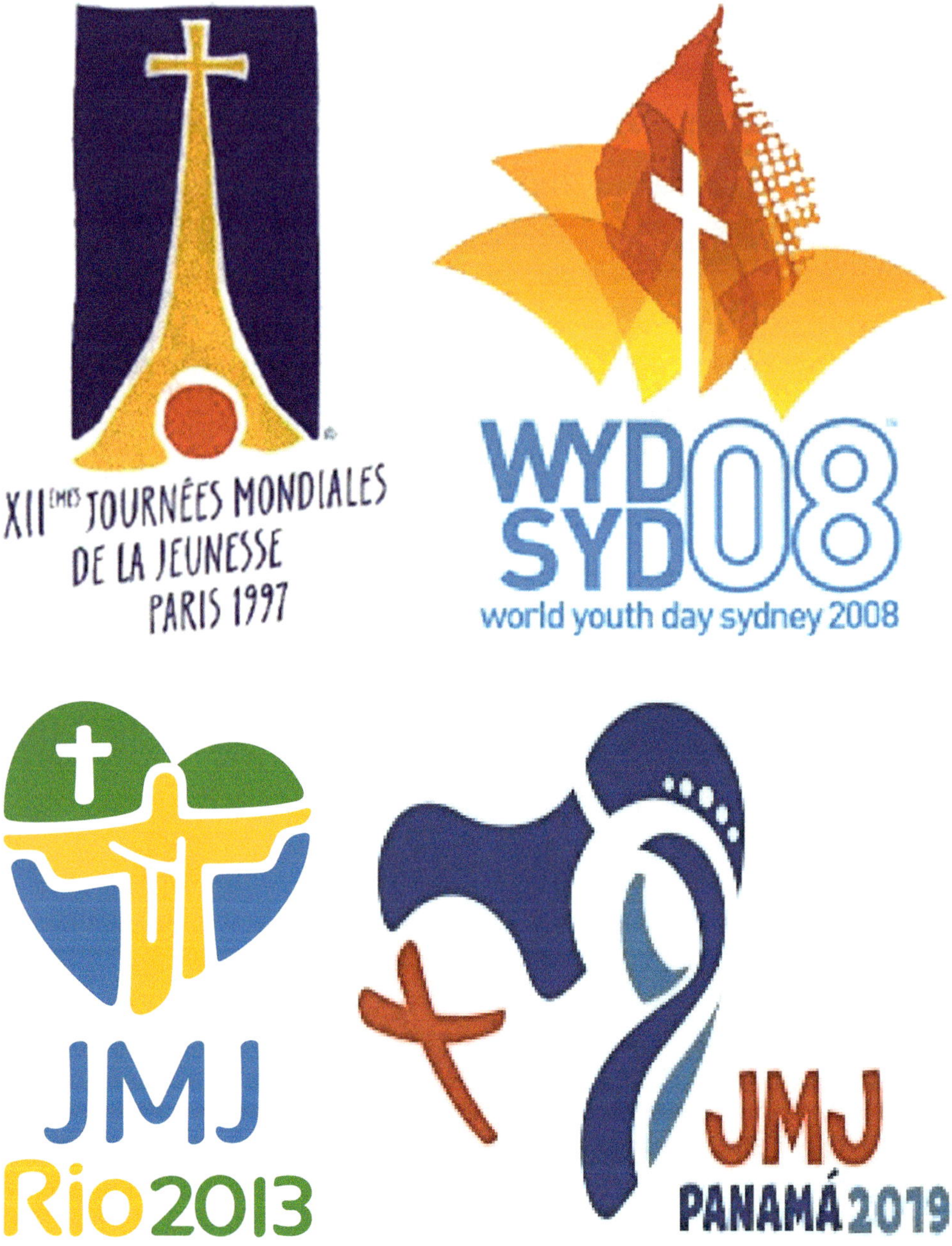

Fig. 14.2. WYD logos of Paris, Rio de Janeiro, Sydney, Panama City in Botucatu, Sao Paulo. (Source: the author)

locations. Since the Paris edition (1997), WYD host cities have strong tourist appeal and the organization of the event has used well-known tourist attractions in promotional campaigns and logos, as shown in Fig. 14.2.

From the images above, it is clear there is association between tourist symbols – profane, desired by young people from all over the world – and the idea of a religious pilgrimage, whose final goal is reaching the sacred, something that at first sight could be unattractive to young people. In the logo of WYD Rio de Janeiro, the largest religious and tourist symbol of the city is associated with the event and the colours of Brazil, a country commonly seen as joyful and hospitable. Similarly, in Sydney (2008) the greatest tourist and artistic symbol of the city, the opera house, was associated with the religious event promoted by the Catholic Church, transmuting the profane into the sacred. This is also the case when we look at the WYD Paris 1998 logo, in which the Eiffel Tower, the world's most visited tourist monument, becomes a cross – the universal symbol of Christianity – re-signifying it in an attempt to overlap the religious and the historical, artistic and cultural spheres.

This strategy makes WYD more attractive to the young audience, since it associates elements of modern culture and contemporary consumption with the experience of religion, considered by many young people to be traditional, even 'old- fashioned'. Thus, business and pleasure combine, the chance to travel to places considered interesting and synonymous with 'well-lived life', while an underlying message says that, despite being modern and young, they are 'differentiated' for seeking a life based on religious values. The same analysis can be applied to the WYD Rio 2013 publicity campaign in Brazil and abroad (Fig. 14.3).

From the host city perspective of WYD participants – those who came to Rio de Janeiro and not the host families and volunteers – the designation of 'pilgrim' was well accepted and the 'tourist' dimension of this individual was seen as natural. Together with the local community, the Church articulated the dimensions 'tourist' and 'pilgrim', with the latter overlapping the former. According to field data of Gonzalez (2016), WYD promotional strategy focused on the term 'pilgrim', with tourism often being classified as 'frivolous', consistent with observations of Norman and Johnson (2011). In a video produced by the WYD team in 2012, focusing on the local population, several young people described in short sentences the 'WYD experience', of which the following were selected:

> I was a testimony to people who used to see me only as a tourist....
>
> I squeeze myself as much as I can to give pilgrims space in the subway car.

Fig. 14.3. Advertising the WYD on Central Station building. (Photo: the author)

> I travel miles and rest in sleeping bags for my faith.
>
> I am a young pilgrim on this journey! Would you take me in?

Quantitative data collected in the survey of Gonzalez (2016) shows that most of the interviewed pilgrims visited or intended to visit tourist attractions. These rates were 93.1% among Brazilian participants and 93.5% for foreigners. Some reports show that participants felt no clash between the identities of pilgrim and tourist as these quotes from Gonzalez (2016) reveal:

> Yesterday, I called my company and postponed my flight, with the journey, the catechesis, masses, early mornings and no [pause]. I want to visit the Christ the Redeemer, the Sugar Loaf, the shrine of the Virgin of the...Brazil [...] Tomorrow, I go to a game in the Maracanã.... (Carlos, pilgrim, 19 years old, Ecuador).
>
> I think that they [the pilgrims staying at the house of the interviewee] enjoyed it a lot, they also enjoyed the key moments of the journey, those who stayed at my home, as the Mass with the [Pope], the Argentines, right? Who had mass with the [Pope], then, I do not know, some peaks you know? Not the others, the others went

everywhere, right [the non-Argentines]? They also wanted to do a lot, kind of tourism, right? (Camila, hostess and volunteer, 25 years old).

The Church thus mobilizes the traditional imagery of the pilgrim as suffering and refusing the convenience of tourism, to win over local community to help in the planning and execution of the event, since Carioca families were invited to host for free pilgrims in their homes. Opening one's home to the 'brother' who represents Christ is more plausible than supporting the trip of someone with 'frivolous tourist interests'. Few people and situations during the fieldwork of Gonzalez (2016) suggested that the dichotomy 'pilgrim/tourist' was present. Church, volunteers, host families, media and public power, in their vast majority, referred to the visitors as 'pilgrims' and not as 'tourists'. The tourism and pilgrimage spheres seemed to be inter-related, since the word 'pilgrims' was used at the same time as their great impact on the 'tourism industry' was mentioned. The newspapers[3] informed the measures Rio's attractions implemented to receive the pilgrims and the municipal administration emphasized the benefits that the event would bring to the city, always mentioning the tourist-economic benefits.

As to the public power, the Church 'played' with the positive impact the WYD could have on the tourism industry, either in terms of revenue or promotion abroad of the image of the city (6400 accredited journalists from 57 countries) (Gonzalez, 2016). This was highly desirable in a time when the city's revitalization project on a global scale was being undertaken because of other mega-events, such as the 2014 FIFA World Cup, the 2016 Olympics and all new works ventures that arose in Rio de Janeiro as a result (Sant´Anna, 2012. The Church states that there was no public investment in the event, only that related to the reception of a head of state such as Pope Francis. However, even if there was no direct investment of the public power in WYD, the mobilization of all the public infrastructure of transport and mobility, security and tourist attractions required specific planning by the latter to make the event possible (Gonzalez, 2016).

These data suggest that the Catholic Church managed to articulate the tourist and religious spheres, using these social facts according to the actor with whom it sought to establish a 'partnership'. Therefore, the different facets were highlighted – traditional pilgrim or more modern and linked to tourism – depending on the situation and purpose, articulating images and symbolism, as we can observe in Fig. 14.4. Tourism is used to attract young pilgrims to a religious event, and the stereotype of a traditional pilgrim is used – introverted and averse to partying – to convince the local population to welcome them with open arms in their homes. The positive economic impacts on the tourism production chain are also remembered whenever public and private sector support is necessary. Thus, we highlight the way the fluidity between profane and sacred, modern and traditional, new and old, young and old, is articulated and shaped so that experiences of evangelization in the contemporary world are possible and successful.

Fig. 14.4. Pilgrims walking to the final holy mass at Copacabana. (Photo: the author)

Concluding Remarks

To promote the discussion about the theoretical path of pilgrimages and how this phenomenon is transformed from the demands of contemporaneity, and the ways in which each individual, in this case the pilgrim, materializes generating spaces and times, and thus new forms of pilgrimage, was the main goal of this chapter.

As such, we sought to show that modern western societies have become increasingly diverse, multiple and complex, making it increasingly difficult to define ideal types in which individuals and collectivities must fit. Therefore, the ancient practice of pilgrimage transmutes itself, responding to the yearnings of the modern subject without losing its space. While some researchers blame tourism for 'stealing the authenticity' of pilgrimage centres, distorting their religious/spiritual purposes, religion also uses tourism and the profane to take these spaces and act in the society as a whole, as the case of WYD Rio 2013 irrefutably illustrates. Pilgrimages today are not limited to religious travel, as tourist travel is not confined to frivolity and consumerism.

Research on pilgrimages and tourism seeks to show how tourism takes over the sacred and fails to look at the other side of the phenomenon, which, as shown here, is also effective. Few references were found on this last perspective and, in this sense, this work seeks to narrow this gap. Moreover, further studies and in-depth analysis of virtual pilgrimages are required as they bring to light the new ways in which individuals experience their realities, currently marked by technological changes. The abandonment of fixed and inflexible types of pilgrims and tourists is essential to the understanding of current tourists' needs and to know how these needs can be met and fulfilled without affecting the other actors in the process. Such an understanding would not only be beneficial to the planning and management of tourist destinations but would also enhance economic impacts and strengthen unique cultural characteristics of pilgrimage sites.

Notes

[1] This chapter is a partial result of a PhD thesis about World Youth Day Rio 2013. All data and theoretical discussion come from this work. For more information check the references section below. Examples of such pilgrimages are those to the tomb of Jim Morrison, the lead vocalist of the American band The Doors, in Paris (Dos, 2008) and to the Gallipolibattlefield in Turkey (Hyde and Harman, 2011), where thousands of Australian and New Zealand soldiers died during World War II.

[2] The Durkheimian sense of solidarity is not used here, which, contrary to Turner's proposal, reflects the social structures.

[3] Just for illustration purposes, we indicate some URLs where these issues are evident, such as a news report in *O Globo* on 29 July 2013 entitled 'Cristo bate recorde histórico com 20 mil visitas em um só dia na JMJ' (Christ breaks historic record with 20,000 visits in 1 day in WYD), 'JMJ bate recorde sobre atração de turistas' (WJD sets record in attracting tourists) from the BAND group on 27 July 2015, and 'Fluxo de turistas do Rio é o maior da história do país' (Flow of tourists in Rio is the largest in the history of the country) from *O Globo* on 27 July 2013. Available at: http://oglobo.globo.com/rio/fluxo-de-turistas-no-rio-o-maior-da-historia-do-pais-9214528 (accessed 1 August 2020).

References

Asad, T. (1993) The construction of religion as an anthropological category. In: Asad, T. (ed.) *Genealogies of Religion: Discipline and Reasons of Power in Christianity and Islam*. Johns Hopkins University Press, Baltimore, Maryland, pp. 27–54.

Badone, E. and Roseman, S.R. (2004) *Intersecting Journeys: The Anthropology of Pilgrimage and Tourism*. University of Illinois Press, Champaign, Iiilinois.

Bajc, V. (2012) Pilgrimage. In: Ritzer, G. (ed.) *Encyclopedia of Globalization*. Willey-Blackwell, Oxford.

Boorstin, D.J. (1971) *The Image: A Guide to Pseudo-events in America*. Vintage, New York.

Carneiro, S.M.C.S. (2003) *Rumo a Santiago de Compostela* (Doctoral thesis). Universidade Federal do Rio de Janeiro. Available at: http://www.ppgsa.ifcs.ufrj.br/teses-e-dissertacoes/rumo-a-santiago-de-compostela-os-sentidos-de-uma-peregrinacao-moderna/

Carneiro, S.C. and Freire-Medeiros, B. (2004) Antropologia, religiao e turismo: multiplas interfaces. *Religiao & Sociedade* 24(2), 100–125.

Cohen, E. (1988) Authenticity and commoditization in tourism. *Annals of Tourism Research* 15(3), 371–386. DOI: 10.1016/0160-7383(88)90028-X.

Cohen, E. (1992) Pilgrimage centers. *Annals of Tourism Research* 19(1), 33–50. DOI: 10.1016/0160-7383(92)90105-X.

Cohen, E. and Cohen, S.A. (2015) Beyond eurocentrism in tourism: a paradigm shift to mobilities. *Tourism Recreation Research* 40(2), 157–168. DOI: 10.1080/02508281.2015.1039331.

Coleman, S. (2002) Do you believe in pilgrimage? Communitas, contestation and beyond. *Anthropological Theory* 2(3), 355–368.

Coleman, S. and Elsner, J. (1995) *Pilgrimage, Past and Present: Sacred Travel and Sacred Space in the World Religions*. British Museum Press, London.

Collins-Kreiner, N. (2010) Researching pilgrimage. *Annals of Tourism Research* 37(2), 440–456. DOI: 10.1016/j.annals.2009.10.016.

Conferencia Nacional dos Bispos do Brasil (CNBB) (2013) *Balanco final da JMJ Rio2013: publico recorde de 3,7 milhoes de pessoas em Copacabana*. Available at: http://www.cnbb.org.br/index.php?option=com_content&view=article&id=12522:balanco-final-da-jmj-rio2013-publico-recorde-de-37-milhoes-de-pessoas-em-copacabana&catid=251&itemid=126

Dias, R. (2002) O turismo religioso como segmento do mercado turistico. In: Dias, R. (ed.) *In Turismo Religioso*. Alnea, Campinas, pp. 7–37.

Dos, E. (2008) Rock and roll pilgrims: reflections on ritual, religiosity, and race at Graceland. In: Margry, P.J. (ed.) *Shrines and Pilgrimage in the Modern World*. Amsterdam University Press, Amsterdam, pp. 123–142.

Eade, J. (1992) Pilgrimage and tourism at Lourdes, France. *Annals of Tourism Research* 19(1), 18–32. DOI: 10.1016/0160-7383(92)90104-W.

Eade, J. and Sallnow, M.J. (1991) *Contesting the Sacred: The Anthropology of Christian Pilgrimage*. Routledge, New York, London.

Gonzalez, L.T.V. (2016) *Estar no mundo, sem ser do mundo: alguns casos sobre a Jornada Mundial da Juventude Rio 2013* (Doctoral thesis). Universidade do Estado do Rio de Janeiro.

Gonzalez, L.T.V. Mariz, C.L. and Manz, C.L. (2017) Jornada Mundial da Juventude Rio 2013: ressignificando espaços da cidade e identidades religiosas. *Religião & Sociedade* 37(2), 14–37. DOI: 10.1590/0100-85872017v37n2cap01.

Hyde, K.F. and Harman, S. (2011) Motives for a secular pilgrimage to the Gallipoli battlefields. *Tourism Management* 32(6), 1343–1351. DOI: 10.1016/j.tourman.2011.01.008.

Kaelber, L. (2005) Changing paradigms of religious travel: from pilgrimage to the postmodern virtual tour. *Paper presented at the Annual Meeting of the American Sociological Association*, Montreal, Quebec, Canada.

Liutikas, D. (2015) In search of miracles: pilgrimage to the miraculous places. *Tourism Review* 70(3), 197–213. DOI: 10.1108/TR-08-2013-0046.

Lloyd, S. (2015) Sacrilisation of secular pilgrimages as archetypal transformational journeys: advancing theory through emic and etic interpretations. *International Journal of Tourism Anthropology* 4(1), 25–45. DOI: 10.1504/IJTA.2015.067643.

MacCannell, D. (1973) Staged authenticity: arrangements of social space in tourist settings. *American Journal of Sociology* 79(3), 589–603.

MacCannell, D. (1976) *Tourist: A New Theory of the Leisure Class*. Schocken Books, New York.

Margry, P.J. (2008) *Shrines and Pilgrimage in the Modern World: New Itineraries into the Sacred*. Amsterdam University Press, Amsterdam.

Margry, P.J. (2009) Marian interventions in the wars of ideology: the elastic politics of the Roman Catholic church on modern apparitions. *History and Anthropology* 20(3), 243–263. DOI: 10.1080/02757200903112628.

Ministerio do Turismo (2013) Pesquisa revela que turistas da JMJ querem voltar ao Brasil. Available at: http://www.turismo.gov.br/ultimas-noticias/3554-pesquisa-revela-que-turistas-da-jmj-querem-voltar-ao-brasil.html (accessed 1 August 2020).

Morinis, A. (1992) *Sacred Journeys: The Anthropology of Pilgrimage*. Greenwood Press, Westport, Connecticut.

Norman, A. and Johnson, M. (2011) World youth day: the creation of a modern pilgrimage event for evangelical intent. *Journal of Contemporary Religion* 26(3), 371–385. DOI: 10.1080/13537903.2011.616034.

Povedak, I. (2008) The Apostle of Love: the cult of Jimmy Zambo in post-socialist Hungary. In: Margry, P.J. (ed.) *Shrines and Pilgrimage in the Modern World: New Itineraries into the Sacred*. Amsterdam University Press, Amsterdam, pp. 123–142.

Sallnow, M.J. (1981) Communitas reconsidered: the sociology of Andean pilgrimage. *Man* 16(2), 163–182. DOI: 10.2307/2801393.

Sant'Anna, M.J.G. (2012) Sant'Anna, M.J.G. and Santos, A.M.S.Penalva. (eds) *Transformações Territoriais no Rio de Janeiro do Século XXI*. Gramma, Rio de Janeiro.

Shinde, K.A. (2015) Religious tourism and religious tolerance: insights from pilgrimage sites in India. *Tourism Review* 70(3), 179–196. DOI: 10.1108/TR-10-2013-0056.

Steil, C.A. (2002) Uma antropologia da peregrina e do turismo religioso. *Imagina¡rio* 8(8).

Steil, C.A. (2003) Peregrinacao, romaria e turismo. In: Abumanssur, E.D. (ed.) *Turismo Religioso: Ensaios antropologicos sobre religiao e turismo*. Papirus, Campinas.

Steil, C.A. and Carneiro, S.C. (2008) Peregrinao, turismo e nova era: rotas de Santiago de Compostela no Brasil. *Religião & Sociedade* 28(1), 105–124. DOI: 10.1590/S0100-85872008000100006.

Turner, V. (1973) The center out there: pilgrim's goal. *History of Religions* 12(3), 191–230. DOI: 10.1086/462677.

Turner, V.W. and Turner, E.L.B. (1978) *Image and Pilgrimage in Christian Culture: Anthropological Perspectives*. Blackwell, Oxford, UK.

Urry, J. (1996) *O olhar do turista: lazer e viagens nas sociedades contemporâneas*, 2. Studio Nobel/SESC, São Paulo.

Vatican (n.d.) *Holy See Press Office*: World Youth Day. Available at: http://www.vatican.va/news_services/press/documentazione/documents/giornate-mondiali/giornata-mondiale-gioventu_elenco_en.html#I Giornata Mondiale della GioventÃ¹

15 Pilgrimage Traditions to Sacred Sites of Modern Central Asia

Elyor Karimov[1]* and Darius Liutikas[2]
[1]Hofstra University (USA); [2]Lithuanian Social Research Centre (Lithuania)

Introduction

This chapter is about Central Asian holy sites and pilgrimage traditions. Islam has been a defining aspect of life in Central Asia. Sacred places and especially the saint shrines have played a key role in the everyday spiritual life of Muslims for much of the region's history over the last 12 centuries (Abramson and Karimov, 2007). Although there are not only Islamic but also Christian Orthodox and Buddhist holy places in Central Asia (Mendikulova and Nadezhuk, 2018), we will examine the holy places formed by the Islamic tradition.

Researchers agree that religiosity in the post-Soviet Central Asian countries is two-sided. On one hand, it appears that the Central Asian Republics have experienced an Islamic revival and the end of official atheism. On the other hand, this revival has been more symbolic than substantive (Yemelianova, 2018). Recent studies (Ro'i and Wainer, 2018) suggest that Uzbekistan's inhabitants perceive their ethnic and Islamic affiliation as givens that play an inherent role in their collective identity. Though Uzbeks identify themselves as Muslims and believers, they do not mean a strictly orthodox Islam, but rather customs and traditions, almost folklore or appurtenances (Ro'i and Wainer, 2018).

In this chapter, we will review the main places of pilgrimage in Central Asia, as well as their visitors. The situation of pilgrimage during the Soviet era also affects the present. We will end the chapter by describing the significance of traditions in pilgrimage sites. Maintaining national traditions and preserving communal customs fell more heavily on women than on men (Khalid, 2014). However, some traditions are exclusively cherished by women. The ritual *Mushkel' Kushod*, intended for women only, perfectly illustrates the intertwining of Islamic and ancient local traditions.

Pilgrimage Sites in Central Asia

In Central Asia the pilgrimage to sacred sites is called *ziyarat*, and as in Islam in general, it refers to pious visitation, pilgrimage to a holy place, tomb or shrine. Wheeler (2006) notices that various structures at Islamic sanctuaries and shrines are associated with burials and tombs. In the Central Asian context, a pilgrimage site can consist of anything from a natural site like a water source to an extensive complex including the mausoleum of a saint (*mazar*), tomb (*qabr*), mosque (*masjid*), religious school (*madrassa*), cemetery (*qabriston*), and additional buildings

*Corresponding author: ekarimov@yahoo.com

for administration, artisanry, and even tourist facilities (Abramson and Karimov, 2007).

Travel guides dedicated to Central Asia mention the most important pilgrimage and religious tourism sites in the region consisting of Kazakhstan, Uzbekistan, Turkmenistan, Kyrgyzstan and Tajikistan. Lonely Planet (2014) offers the ability to visit 19 mosques, 4 minarets and 32 mausoleums. The Bradt Travel Guide – Uzbekistan (2016) suggests visiting 27 mausoleums and tombs, and 26 mosques in Uzbekistan (Ibbotson and Lovell-Hoare, 2016).

While the actual form of the site can vary (e.g. a water source, tree, stone, mosque, mausoleum, the saint's tomb), most pilgrimage sites consist of at least a tree, a water source and a tomb of a saint, all of which are considered sacred because of their affiliation with the site and miracle stories (*karomatlar*) about the saint who is buried there. The telling and retelling of miracle stories by shrine caretakers and other devotees who essentially re-enact the miracles themselves intensifies the experience for audiences in search of a saint's goodwill, grace or blessing (*baraka*). Sacred places are commonly understood to serve the spiritual needs of all kinds of people, including the sick and childless, who visit shrines to ask for the help of a saint.

We can divide sacred sites in Central Asia into a number distinct categories. For example, Central Asia is famous not only for Islamic sacred sites but also pilgrimage places relating to other religions. At least three shrines of biblical prophets can be found in the territory of Central Asia. Uzbekistan, one of the Central Asian republics, appeared on the BBC news as the 'Land of a thousand shrines' (Uzbekistan: Land of a thousand shrines, 2018): the pilgrims are not just Muslims, because this tomb is believed to be the last resting place of the biblical prophet Daniel, or Daniyar, as Uzbeks call him. However, this tomb in Samarkand is one of numerous places claimed as the final resting place of Daniel. Some say the remains were brought here by early Christians. Others say that Amir Timur (Tamerlane), who founded the Timurid Empire (1370–1506), ordered them to be buried here (Shterenshis, 2002).

The tomb for the Prophet Daniel is 18 metres (59 feet) long. Legend says that Daniel's body grows half an inch per year. The phenomenon of such long tombs of prophets is widespread throughout Asia and related to Islamic conceptions of tomb symbolism (Wheeler, 2006). According to the local legend, on the day Daniel was entombed here, a natural source of water sprung up at that spot, and it is believed by locals that its water has the power to heal (Fig. 15.1).

Sulaiman-Too (also known as Sulayman Rock, or Sulayman Throne, or Tahti-i-Sulaiman) has been venerated as a sacred place since ancient times and is one of the major places of Muslim and pre-Muslim pilgrimage located at Osh (Kyrgyzstan). Named in honour of the prophet Sulaiman (who may or may not be the same person as King Solomon of the Hebrew Bible), the Sulaiman Mountain has been visited by religious pilgrims since well before the advent of Islam (Fig. 15.2).

At Sulaiman's summit there is a mausoleum where, according to legend, the prophet Sulaiman (Solomon in the Bible) prayed, leaving behind an impression of his knees in the rock. The founder of the Great Moghul dynasty, Zahiraddin Muhammad Babur (1483–1530), upon seeing the impression, constructed over it a *hudjra* (small chamber) with a *mihrab* (a niche in the wall that specifies the *qibla*, that is the direction of the Kaaba in Mecca). The Sulaiman-Too Sacred Mountain was added to UNESCO's World Heritage List in 2009.

One of the most sacred sites in Bukhara (Uzbekistan) is the Chashma-Ayub mausoleum. It is located about 30 km north-east of Bukhara. Its name means the 'Spring of Ayyub', referring to the biblical prophet Job (Ayyub to Muslims) who is considered a prophet by Christians, Jews and Muslims alike. According to legend, this is where Job struck the ground during a drought to make a spring appear. Many pilgrims come here to get a drink of the water since their belief is that the water has special healing powers. Chashma-Ayub mausoleum was added to the UNESCO World Heritage Tentative List on 18 January 2008.

Some pilgrimage sites are significant for the entire Muslim world. The tomb of Imam al-Bukhari (810–870), who was an Islamic scholar born in Bukhara (the capital of the Bukhara Region (*Viloyat*) of what is now Uzbekistan), is in Hartang village, 25 km from Samarkand. He authored the hadith collection known as *Sahih al-Bukhari*, regarded by Sunni Muslims as one of the most authentic (*sahih*) hadith collections and

Fig. 15.1. The Grave of the prophet Daniyar. (Photo: Alexandronikos. Public domain, https://commons.wikimedia.org/wiki/) [The_grave_of_prophet_Daniyar.JPG])

as the most authentic and authoritative book after the Koran. Today, Imam al-Bukhari's tomb lies within the Imam al-Bukhari Complex near the town of Samarkand, Uzbekistan. The mausoleum complex consists of Imam al-Bukhari's tomb, a mosque, a madrassah and a library (Tomb of Imam al-Bukhari, 2020).

Another Muslim world-renowned pilgrimage site is the tomb of Imam al-Termezi or Tirmidhi (824–892), who was an Islamic scholar and collector of hadith from Termez (in present-day Uzbekistan). He wrote *al-Jami`as-Sahih* (known as *Jami`at-Tirmidhi* or *Sunan at-Tirmidhi*), one of 'the six books' (*Kutub al-Sittah* – the six canonical hadith collections in Sunni Islam). At-Tirmidhi is buried on the outskirts of Sherobod, 60 km north of Termez in Uzbekistan, and his mausoleum is the site of pilgrimage with significance for the entire Muslim world.

Some of the most popular pilgrimage sites of Central Asia are mausoleums of saints allied with Sufi brotherhoods. The main Sufi *tariqats* (brotherhoods) associated with Central Asia – Yasaviya, Kubraviya and Naqshbandiya – are well known throughout the Muslim world. The sacred graves of the founding fathers of these brotherhoods are known today as the most well-known and popular places of religious pilgrimage. These destinations draw out not only Muslim pilgrims but also tourists of various religions from all corners of the world each year.

Outside of the town of Otrar in southern Kazakhstan (in a region adjacent to Uzbekistan) is the grave of the saint Arslan Baba (Arslan Bab or Arystan Bab), one of the spiritual teachers (*pir*) of Ahmad Yasawi, after whom another of the Central Asia's Sufi brotherhood was named (De Weese, 1996). The mausoleum dates from the 14th century and is constructed over Arystan Baba's 12th-century grave. According to historical data, Amir Timur (Tamerlane) ordered Ahmad Yasawi's tomb to be rebuilt. All attempts were unsuccessful. One day Timur had a dream that if he wanted to have success, he

Fig. 15.2. Sulayman Mountain in Osh, Kyrgyzstan. (Photo: A. Savin. Public domain. https://lt.m.wikipedia.org/wiki/Vaizdas [Osh_03-2016_img06_Sulayman_Mountain.jpg])

should first build a mausoleum over the grave of Arslan Baba (Fig. 15.3).

The Mausoleum of Khawaja Ahmad Yasawi is in the city of Turkestan (Kazakhstan). Ahmad Yasawi was a 12th-century Sufi and an early mystic, known as the founder of the Yasawiyya Sufi order, and his tomb is a world-famous pilgrimage site. Mendikulova and Nadezhuk (2018) state that the cult-memorial complex of Ahmad Yasawi has a very high spiritual and ideological significance. UNESCO recognized the mausoleum as a World Heritage Site in 2003.

Najmiddin Kubra was a 13th-century Khwarezmian Sufi from Khwarezm and the founder of the Kubrawiyya Sufi Order. The Najmiddin Kubra Mausoleum was erected in the first half of the 14th century, and this place became the holiest spot in Kunya-Urgench (Turkmenistan). It was declared a UNESCO World Heritage Site in 2005. Pilgrims from all over the world believe his tomb to have healing properties.

Bahauddin Naqshband was a 14th-century founder of what would become one of the largest Sufi orders, the Naqshbandiyya. The Memorial Complex of Bahauddin Naqshband is one of the most important Muslim shrines. It is located 12 km from Bukhara (Uzbekistan). The Mausoleum of Bahauddin Naqshband is considered as the Central Asian place for so-called small *Hajj*. Pilgrims from different countries come here to ask for the fulfilment of wishes and healing. It is believed that visiting the mausoleum of Bahauddin Nagshband three times is equal to one visit to the shrines of Mecca and Medina (Navruz-zoda and Navruz-zoda, 2018).

At this point, it would be good to recall the pilgrimage sites visited primarily by people of a particular occupation. As it is known, different Sufi brotherhoods were legitimately developed and played an enormous role in the socio-political and spiritual-religious life of Central Asian cities over many centuries. Sufi tariqats reflected and protected the interests of the average strata of society, and it was the ideology of the middle class and first of all the urban environment – small merchants and craftsmen.

Craftsmen were divided into professional shops (guilds). Each shop or guild had its own patron. The persona of any number of Sufi-Sheikhs was the basis of many artisan corporations (Gordlevsky, 1960, p. 289-306). Many of the representatives of Naqshbandiya brotherhood had *nisba*, where were reflected their social origin; for example, the above-mentioned Bahā al-Din Naqshband – the founder of

Fig. 15.3. Arslan Baba Mausoleum, Kazakhstan. (Photo: E. Karimov)

Naqshbandiya. Some have said that the translation means 'related to the image-maker'. Some also consider it to mean 'patternmaker' rather than 'image maker'.

Here is another example: Khwādja Amir Kulāl (or Mir Kulāl) is Naqshbandiyya, a saint who made significant contributions to Naqshbandi Silsila (spiritual genealogy). Amir Kulāl is believed to have been a teacher of Bahā al-Din Naqshband. *Kulāl* means potter. Thus, Amir Kulāl was the holy patron of all potters. Arif Diggarāni was another well-known representative of Naqshbandiya tariqat. *Dig* means *bowl* or *cauldron*. The well-known representative of Naqshbandiya Sufi Order, Arif Diggarāni, was from the settlement Diggaron of the Bukhara region. The inhabitants of this settlement specialized in manufacturing large kettles.

There are pilgrimage sites known only within a particular region, and new sites, which have not yet acquired renown. Central Asia has earned a reputation as 'the land of a thousand shrines' for a reason.

Pilgrims' Motivation

While sacred sites draw pilgrims for multiple reasons and at any time, there are generally several types of visits. Everyday-visits are characteristic of people who live near a shrine and regularly attend mosque. Visitors leave *sadaqa*, usually a small sum of money, in a standard set of places close to the site, such as under stones and trees or by tombstones. Typically, the donation is small, but may vary in accordance with the sacred site's popularity.

Thursdays, Fridays and Sundays are considered to be special days for visiting shrines and tend to bring larger numbers of visitors who come from further afield. On these days, people usually make more elaborate preparations, often with the goal of requesting special favours from God. Typical requests include children (if a woman is having difficulties getting pregnant), a son (if a couple has only female children, since boys are generally more highly valued than girls), a cure for an illness (to supplement other efforts or when other efforts have failed), and economic or social success (Abramson and Karimov, 2007).

Historically, *ishans* as the saint's descendants, would oversee maintenance of the *mazar* (mausoleum of a saint), although *imam-hatibs* of affiliated mosques, fortune-tellers and seers (*folbin*), spiritual healers (*tabib*) and others are more typically on hand and serve as proprietors of the shrine. In many present cases, one will find local or national government-appointed imams or *mutawalis* (shrine administrators), who may not have any blood ties to the saint.

Depending on the shrine's remoteness from population centres, visitors may come alone, as a family or in groups of as many as 20–30 people, in which case a bus is chartered for the trip. Preparations are made in advance for making animal sacrifices, e.g. slaughtering of chickens or rams. Visitors also give a donation. According to tradition, only the designated proprietor of a sacred site may take the money (Abramson and Karimov, 2007).

Another category consists of sacred sites that are famous and may be deemed to have a special power that supplements visits to other lesser-known sites. Usually, these places are more easily accessible by road and can receive large numbers of visitors. On traditional pilgrimage days, one can find visitors from all social strata, including high-level government officials, wealthy businessmen, and famous artists and performers.

Income from donations can be considerable. It is worth noting that the income varies according to the season, where it is lowest during the winter and highest during the summer, when most people are on vacation, and the autumn, when people earn more money from harvests (Abramson and Karimov, 2007).

According to Louw (2007), usually, people go to *ziyarat* on the occasion of the socially recognized rites of passage such as birth, circumcision, marriage and death. They also travel on the occasion of more individual situations characterized by transition and uncertainty, such as sending a son off for military service, or commencement of a new job or a new course of education (Louw, 2007). So pilgrimage becomes important in crisis, or when there is 'tension between their hopes and visions and their limited means of realizing these hopes and visions' (Louw, 2007). Pilgrims bring to the sacred places everyday concerns and problems (e.g. how to pay off debt or how to resolve a dispute at work).

There are special occasions for pilgrimage, such as before making the *Hajj* or *Umra* or in connection with the harvest. There are other reasons, too, for people to make pilgrimages to sacred sites. As a rule, people choose the most popular sites and make visits on the traditional days of the week. These visits are typically treated as holidays and involve commemorative rituals and feasts for up to ten acquaintances at the sites themselves.

Pilgrimage has also become popular for just-married couples, accompanied by their wedding party, to express respect for local traditions or to visit the most respected and popular local sacred sites. At the sites, they follow the same ritual practices that others do when making a pilgrimage – circumambulating the site and touching it (Abramson and Karimov, 2007). Some pilgrim places are related to sacred trees, where people tie strips of cloth as a votive offering (Khalid, 2014). Some pilgrims still try to touch tombs believing that they have magical powers. People know about the *shirk* (idolatry) but, nevertheless, they still want to touch the stones of the tombs.

Visiting the sacred places in Central Asia makes a strong impact on the self-concept of pilgrims since they actually go on journeys for this purpose. Pilgrimage, in this case, helps construct or affirm personal and social identity (Liutikas, 2012). Travelling to the sacred places helps people to discern between the relative values of the different traditions and beliefs. Drawing attention to the history of Central Asia from the past to the present, it is clear that the purpose of such visits is to find some cultural values and different form of identity as religious or national.

Pilgrimage in the 20th Century

During the 20th century, the intellectual and philosophical part of Islam in Central Asia was powerfully damaged. It happened because of the political and ideological circumstances. In other words, since Central Asia was part of the Soviet Union, atheistic ideology became the official policy. The Soviet regime created an official institution – Spiritual Administration of the Muslims of Central Asia and Kazakhstan (SADUM) – in 1943, to administer and control Islamic religious activity in Central Asia. The harsh administrative policy of the USSR applied to all faiths. From year to year, the number of functioning Muslim, Orthodox, Jewish, Protestant and Buddhist places of worship declined. For example, a document from 27 September 1937 discusses the review of petitions for 30 mosque closures, the decision to accept the petition, and the release of the buildings for cultural and educational uses (Document of Central State Archive of Uzbekistan TsGA, f. 86).

And in the case of Islam, the USSR religious policy coincided with the extreme diversity of Islam in Central Asia at the beginning of the 20th century, which began to change, initially due to the activity of an Islamic scholar named Shamidamulla and his group of students, which later came to co-operate with a Bolshevik regime. Officially Islam, increasingly represented by Shami-domulla's teachings, became antagonistic toward the local Hanafi and Sufi traditions. Shaykh Eshan Babakhan (1859–1957) and his son Ziyauddin Babakhanov (1957–1982), the future Mufti of the SADUM, was closely aligned with the Shamidomulla group.

SADUM had a representative in each Central Asian republic and each was headed by a *qazi*. Every qazi was appointed by and subordinate to the mufti in Tashkent, and within each republic all religious personnel (such as imams and muezzins) were subordinate to their respective qaziyats. The Muftiyats carried out organizational and administrative functions and were an official link between Muslims and the public authorities. For example, all religious questions concerning Muslims in Kirghizia were dealt with by the Spiritual Administration of the Muslims of Central Asia and Kazakhstan (SADUM) located in the capital of Uzbekistan, Tashkent. The residence of the spiritual head of the Kirghiz Muslims – the Qaziy – was subordinate to the central administration and was located on the territory of Kirghizia (present-day Kirgizstan), in the city of Osh (Karimov and Abramson, 2009, p. 17-18).

Under Babakhanov's influence, SADUM opposed the traditional Central Asian Sufi form of Islam as being inconsistent with Soviet ideology; advocated reforming local Islam in the direction of fundamentalism, based only on the Qur'an and Hadith; and trained students to study only Arabic philology (neglecting Persian and Turkic sources), the Qur'an and Hadith (but not authoritative books of the Hanafi school). This postulate took root during the 20th century and was adopted by SADUM's religious curriculum, lasting until today.

SADUM, as well as leading religious figures and establishments, was tightly controlled by state apparatus and collaborated with the Soviet government structures through the Commission [or Council] for the Affairs of Religious Cults attached to the Soviet Council of Ministers.

Through an examination of archive documents of the USSR, we can find a number of different documents concerning policies on religion in Soviet Central Asia between the the 1920s and 1960s. Here is one of them from TsGA.[1] A document from 5 July 1946 is a letter from the representative of the above-mentioned Religious Affairs Council of the Uzbek SSR Ibadov to the representative of the Council on Religious Affairs of the Kyrgyz SSR Ahmatov (Document of Central State Archive of Uzbekistan TsGA, f. 2456). In his letter, Ibadov refers to Mufti Babakhanov's complaint to SADUM's representative in Osh who exhibits excessive independence in his affairs. According to Babakhanov's denunciations, Shafoat-khodji Holik Nazar ughli (SADUM's representative in Osh) reopened a number of mazars, including the sacred site 'Tahti-i-Sulaiman', without official permission.

As mentioned above, for centuries, Tahti-i-Sulaiman ('Sulaiman's Throne') in Osh city (Southern Kyrgyzstan) was known as a place of pilgrimage and worship for Muslims, not only from the Ferghana Valley but also from all over Central Asia, Iran and Kashgar. Local Muslims consider this place sacred and many of them believe that three pilgrimages to 'Tahti-i-Sulaiman' are equivalent to making the *Hajj* to Mecca.

In 1957, SADUM had made a decision to officially close the sacred sites throughout Central Asia. Accordingly, SADUM issued a series of fatwas with de-legitimization of 'Pilgrimage and the Cult of Saints', as well as *Mawlyud* (*Maulud al-Nabi*) and other traditional rituals.

There were a number of fatwas of SADUM dating from 1950, 1952, 1957, 1959 and 1960–1966 in which the widespread traditions such as *marsiya-honlar*, *khudoyi*, *qalim*, the tradition of commemoration of the 3rd, 7th, 20th (*yigirma*) and 40th (*qirq*) day of a person's death, as well as the traditions of celebration of *nikah-toy, beshik-toy, sunnat-toy* etc. were announced as being non-Sharia and inconsistent with Islam. Those fatwas were also referred to *Ziyarat*, i.e. pilgrimage traditions as non-Sharia.

Incidentally, the situation remains the same presently. Official Islamic sites, for instance those in Uzbekistan, strictly criticize the practice of such traditions. All such traditions are classified as *'bid'at wa khurofat'*, in other words, as *bid'a* (within the religious practice is considered a sin) and *khurafat* (superstition and ignorance).

With regard to the *Hajj* as official Muslim pilgrimage to Mecca, which is one of the Pillars of Islam, *Hajj* had been supported by SADUM as the only legal form of pilgrimage, but despite SADUM's co-operation with Soviet authorities, the first pilgrimages of Muslims to Mecca and Medina were made only at the end of World War II and in the immediate post-war years. Whatever indulgences in the religious sphere were made during the war by the Soviet regime, however, were short-lived. Curiously, the existing ban on the *Hajj* to Arabia at the beginning of the Soviet power, and then, after its restoration, the scant number of pilgrims, led to a growth in popularity of holy sites located in Muslim areas of the USSR.

A common narrative associated with the destructors of shrines in Soviet times was entrenched in people's stories. According to those stories, various tragic events happened to the people who had tried to demolish sacred places. These people had met with accidents, illness or death (Louw, 2007; Liutikas, 2015).

Dadabaev (2016), after an evaluation of the Soviet past, concludes that experiences under the Soviet government affected the construction of a new religious identity. This new post-Soviet religious identity has different understandings of religiosity from those embraced in either pre-Soviet or Soviet times. Religious identity is constituted by elements that redefine and reimagine the meaning and function of religion in the lives of people (Dadabaev, 2016). On the one hand, the Soviet anti-religious policy, including the closure of mosques and churches, has been evaluated as a negative aspect. These actions rejected historical traditions, rituals and other elements of self-identification. On the other hand, some elements of these anti-religious campaigns were evaluated positively in the post-Soviet period, including those related to the unveiling of women, the introduction of conventional education and the integration of women into societal life (Dadabaev, 2016).

Female Rituals of Sacred Sites in Modern Central Asia

There are different Muslim traditions regarding the opportunities for women to participate in religious rites. In GCC countries (UAE, Oman, Kuwait etc.), as well as Turkey or Maghreb countries, there is information concerning the women's praying section in front of the mosque entrance. However, women in Central Asia have not been encouraged to pray at the mosques. For centuries, the Central Asian clergy tried to argue that the Prophet Muhammad preferred women to pray at home rather than at a mosque, and they referred to a hadith in which Prophet Muhammad supposedly said, 'The best mosques for women are the inner parts of their houses' (Hughes, 2013).

Traditionally, women in Central Asia, India, Pakistan and some other countries have not been allowed to go to mosques even though there is no legal restriction. This could explain partly one of the reasons for the popularity of pilgrimage to sacred sites among women. Unlike men, women in Central Asia were not able to go to the mosques for prayers, so the shrines have become the main place of worship for them. Those worship places are sites where women could partially fulfil their spiritual needs.

The people continued to save all traditions even during the Soviet time since that was part of their identity values. Conservation of religious practice was allowed in the territory of Central Asia in 20th century. Some pilgrimage traditions were born in Central Asia a long time ago. In spite of the difficulties, people continued to practise them, even if they were illegal. Thereby, these traditions continue to exist. One such tradition is the ritual of the Deliverance of the Person from Problems (difficulties) and is practised by women during the period of pilgrimage to sacred places. This ritual is called 'Mushkel' Kushod' which means 'Deliverances

Fig. 15.4. Mushkel' Kushod ceremony. Namangan Region, Uzbekistan. (Photo: E. Karimov)

from Difficulties' (Fig. 15.4). It relates to the mystic saint Bibi Mushkel' Kushod (the *Mistress Destructor of Difficulties*).

Actually, we can see yet another striking difference between the Central Asian tradition and those of the other parts of the Muslim world. For example, 'Mushkel' Kushod' is a very commonly practised tradition in many countries which is associated with Ali (Ali ibn Abi Talib), the cousin and son-in-law of Prophet Muhammad. Mushkel' Kushod is considered to be a Shi'a tradition and is not allowed from the point of view of Sunni orthodoxy. When it comes to Central Asia, Mushkel' Kushod symbolizes a woman – Bibi Mushkel' Kushod – who is considered to be of the Sunni tradition and not the Shi'a. In Central Asian mythology, Mushkel' Kushod is a female saint who is called upon during bad times. It is likely to be linked with an ancient symbol of female sainthood, similar to that of *Bibi Seshanba* or *Mother Tuesday*. Needless to say, in Central Asia, both Bibi Mushkel' Kushod and Bibi Seshanba are purely female rituals.

In accordance with local tradition, to find the way from a predicament, a woman should pray to saint Bibi Mushkel' Kushod. The woman asks for the assistance from *otin-oyi* (in Bukhara region these women are named *oyi-mumo* and *Bibi khalifa* in Tajik). This term means unofficial female imams or, in other words, women who serve as local religious leaders in their communities. The women invite *otin-oyi* and the members of the family or neighbours or friends for Mushkel' Kushod ritual. The number of invited people varies from 10 to 15, but can reach 30, or even more. Invited women take seats on traditional Uzbek *kurpachas* (some type of mattress). There are small, round breads (*non*), dishes with water, *piyola* (cups) with flour, and plates with sugar and dried black raisins (*kishmish*) on the tablecloth. There are also cotton wool and candles on tablecloths. The number of candles can be from 7 to 40. The candles are often handmade.

There is a widespread and popular Central Asian tradition of the lighting of candles during pilgrimage to sacred places. These are special candles call *pilik*. This is a handmade candle, made of cotton wool, winding around a small wooden stick – the cotton wool part of the candle soaked in cotton oil (a very popular kind of oil in Central Asia).

Mushkel' Kushod is usually performed on sacred sites during pilgrimage. An *otin-oyi* or any other woman begins to pray (read *salawat*) during the ceremony. After completion of the prayer, all women together pronounce 'Omin' ('Amin' is a Muslim declaration of affirmation; in Arabic, Amin simply means 'so be it'; to Muslims, it is a reasonable end to any supplication) or Muslim takbir *Allahu Akbar* (Allahu Akbar is an Islamic exclamation called the takbir in Arabic, which means 'Allah is the greatest') and take a handful of *kishmish* (raisins). Women remove wooden sticks from *kishmish*, one by one, and wrap *kishmish* into the cotton wool. They do 40 nubbins or flakes of the cotton wool and put them near. This process also can be accompanied by prayer. Then each woman takes the *kishmish's* tails and wraps each one into the cotton wool. When this procedure is finished, the reading of the Koran stops. Afterwards, a meal is eaten. Often, it can be a salad, traditional soup (*shurva*) and fresh fruits. Tea is given as a traditional drink.

Then the woman cleans the table, leaving only the ingredients required for the ritual, and lights the candles. In this moment the *otin-oyi* starts to read the Koran or tells the legend about saint Bibi Mushkel' Kushod. At the end of the ceremony, a blessing in hope that all wishes will fulfil is pronounced. The women get up. Each of them carries away a handful of handmade flakes, *kishmish*, some sugar and water in a jar or plastic bottle. When they go back home, the women throw the *kishmish*'s sticks in *aryq* (irrigation ditch), simultaneously make a wish and ask help from Bibi Mushkel' Kushod.

Sultanova (2011) points out that seven items are crucial to performing the ritual: water or tea, bread, sweets (honey, sugar, bonbons, or all of these), flour, oil (liquid), a mirror and fire. The ritual assigns specific significance to each item, the symbolism of which originates from Zoroastrian rites; for example, fire is a symbol of faith, flour is a symbol of purity, a mirror is a symbol of contained and open space, honey is the symbol of intended sweet bliss, bread is a symbol of fertility, water of the flow of time and oil is a substance of energy (Sultanova, 2011, p. 169). Special food consisting of rice soup, pilau rice, corn soup, sherbet, pastry balls and little breads is prepared for this ritual.

The techniques of carrying out the Mushkel' Kushod ritual could differ or vary depending on region (Fathi, 2006). Some of Mushkel' Kushod rite participants consider that separate ingredients (for example *kishmish* and sweet water) should be eaten reliving the wish. Some of the organizers of similar ceremonies believe that there is no need to eat in such ritual. Also, the moment varies, when women participating in the rite address *Mushkel' Kushod* with their own requests and wishes. Also, the Koran prayers can be changed. This depends on the person's knowledge of the Koran. Some elements remain unchanged in the ritual: telling the legend about saint Bibi Mushkel' Kushod, the women requesting help, and the food ingredients necessary for the rite.

Conclusion

Pilgrimage in Central Asia has a long historical and cultural tradition. During, for example, the 18th and 19th centuries, one of the most common forms of 'historical' writing was a particular sub-genre of collective hagiography, combining a 'city history' with a catalogue of the town's holy places that served, in effect, as a guide for pilgrims (De Weese, 2000, p. 245).

After many decades of life under Soviet socialism, almost throughout the entire 20th century, Islamic rituals, such as pilgrimages to sacred sites, are gaining popularity in present-day Central Asia. While many of the ritual practices observed today have centuries-old roots, they also have new, contemporary meanings for the Muslims in the region. There are many motivations for pilgrims to undertake journeys to shrines, which include not only important life experiences, but also personal values, a personal and social identity growth, lifestyle, and a social and cultural influence. Pilgrims are now struggling to make sense of social, economic and political changes affecting their lives.

As far as we have learned in this chapter, the people continued to save pilgrimage traditions even during the Soviet period (1917–1991).

Through many difficulties, people continued to practise them legally, or illegally. In this way, these traditions continue to exist. Finally, the conservation of religious practice of pilgrimage has allowed saving not only Islam in the territory of Central Asia in 20th century but also the revival of cultural identity in the present day. Today, the pilgrimage in Central Asia is experiencing astounding growth. Religious and non-religious people make significant efforts to fulfil their spiritual needs, and travel across the sacred sites in search of personal and social identity.

As some Islamic practices have been more openly tolerated and embraced since the independence of Central Asian republics, the countries of Kazakhstan, Kyrgyzstan, Tajikistan, Turkmenistan and Uzbekistan, more people in Central Asia have expressed a new public appreciation for sacred sites and pilgrims from abroad have sought to build ties to kindred spirits, especially given the fact that some shrines located in Central Asia apply to both Muslims and Christians and to representatives of other confessions as well.

Such tradition rituals as Mushkel' Kushod have therapeutic functions – women believe that their wishes could come true and problems could disappear. Another important observation is that such rituals make women's everyday environment a sacred and ritualistic space. Traditional food acquires the power of symbolic transformation. The combined elements of prayer and recreation, Islamic and pre-Islamic symbols, independent and clergy-assisted practices, and the variety of ritual practices that people engage in when making pilgrimages illustrate the varied levels of knowledge and expectations that pilgrims have in Central Asia. At the same time they demonstrate understanding that Islam plays an important role in their national identity and lives and that sacred sites are integral to their cultural values.

Note

[1] The acronym TsGA (*Tsentral'niy gosudarstvenniy arkhiv*) refers to the collections in the Central State Archive of Uzbekistan concerning policies on religion in Soviet Central Asia during the period of the 1920s–1960s. Each archival entry in this chapter provides basic information of historical documents and contains the following symbols:

- fond identification number (f.);
- quantity of inventory lists or *opisi* in the collection (op.);
- number of archival units or *edinitsy khraneniia* (ed. khr.);
- total pages of documents or *listy* (l.).

References

Abramson, D. and Karimov, E. (2007) Sacred sites, profane ideologies: religious pilgrimage and the Uzbek state. In: Sahadeo, J. and Zanca, R. (eds) *Everyday Life in Central Asia Past and Present*. Indiana University Press, pp. 319–338.

Dadabaev, T. (2016) *Identity and Memory in Post-Soviet Central Asia: Uzbekistan's Soviet Past*. Routledge.

De Weese, D. (1996) The Mashâ'ikh-i Turk and the Khojagân: rethinking the links between the Yasavi and Naqshbandi Sufi traditions. *Journal of Islamic Studies* 7(2), 180–207. DOI: 10.1093/jis/7.2.180.

De Weese, D. (2000) Sacred history for a central Asian town: saints, shrines, and legends of origin in histories of Sayrām, 18th–19th centuries. *Revue des mondes musulmans et de la Méditerranée* 8, 245–295.

Document of Central State Archive of Uzbekistan TsGA (f. 86) op. #10, ed.khr. 1500, I, 1–6.

Document of Central State Archive of Uzbekistan TsGA (f. 2456) op. #1, ed.khr. 94, I 1.

Fathi, H. (2006) Gender, Islam, and social change in Uzbekistan. *Central Asian Survey* 25, ((3), 303–317Vol.. DOI: 10.1080/02634930601022575.

Gordlevsky, V.A. (1960) *Dervishi Akhi Evrana i tsekhi v Turtsii*. Izbrannie sochineniya, Tom 1. Moscow.

Hughes, A.W. (2013) *Muslim Identities: An Introduction to Islam*. Columbia University Press, New York.

Ibbotson, S. and Lovell-Hoare, M. (2016) *Uzbekistan*. Bradt Travel Guides.

Karimov, E. and Abramson, D. (2009) *Religion Made Official: A Comprehensive Collection of Documents from the State Archives of Soviet Uzbekistan, 1920s – 1960s*. Institut Francais d'Etude sur l'Asie Centrale, Almaty, Kazakhstan.

Khalid, A. (2014) *Islam after Communism. Religion and Politics in Central Asia*. University of California Press, Berkeley, California.

Liutikas, D. (2012) Experiences of valuistic journeys: motivation and behaviour. In: Sharpley, R. and Stone, P. (eds) *Contemporary Tourist Experience: Concepts and Consequences*. Routledge, New York, pp. 38–56.

Liutikas, D. (2015) In search of miracles: pilgrimage to the miraculous places. *Tourism Review* 70(3), 197–213. DOI: 10.1108/TR-08-2013-0046.

Lonely Planet (2014) *Central Asia*. Lonely Planet Publications.

Louw, E.M. (2007) *Everyday Islam in Post-Soviet Central Asia*. Routledge.

Mendikulova, G. and Nadezhuk, E. (2018) Religious migrations in contemporary central Asia. In: Yasuda, S., Raj, R. and Griffin, K. (eds) *Religious Tourism in Asia. Tradition and Change through Case Studies and Narratives*. CAB International, Wallingford, UK, pp. 110–117.

Navruz-zoda, B. and Navruz-zoda, Z. (2018) The destination marketing development of religious tourism in Uzbekistan: a case study. In: Yasuda, S., Raj, R. and Griffin, K. (eds) *Religious Tourism in Asia. Tradition and Change through Case Studies and Narratives*. CAB International, Wallingford, UK, pp. 149–157.

Ro'i, Y. and Wainer, A. (2018) Uzbekness and Islam: a survey-based analysis of identity in Uzbekistan. In: Laruelle, M. (ed.) *Being Muslim in Central Asia. Practices, Politics, and Identities*. Brill, Leiden, The Netherlands, pp. 62–84.

Shterenshis, M. (2002) *Tamerlane and the Jews*. Routledge.

Sultanova, R. (2011) *From Shamanism to Sufism. Women, Islam and Culture in Central Asia*. I. B. Tauris.

Tomb of Imam al-Bukhari (2020). Available at: https://madainproject.com/tomb_of_imam_al_bukhari (accessed 20 February 2020).

Uzbekistan: Land of a thousand shrines (2018) BBC News. Available at: https://www.bbc.com/news/world-asia-44685414 (accessed 12 February 2020).

Wheeler, B. (2006) *Mecca and Eden. Ritual, Relics, and Territory in Islam*. The University of Chicago Press.

Yemelianova, G. (2018) How 'Muslim' are Central Asian Muslims? A historical and comparative enquiry. In: Laruelle, M. (ed.) *Being Muslim in Central Asia Practices, Politics, and Identities*. Brill, Leiden, The Netherlands, pp. 10–37.

16 Pilgrimage Tourism to Palestine: The 'Come and See' Initiative in Palestine

Rami K. Isaac*

Centre for Sustainability, Tourism and Transport, Breda University of Applied Sciences, Breda, The Netherlands

Introduction

The origins of various forms of religious tourism and pilgrimage have been discussed at length in the literature, with many scholars arguing that perhaps the practice of pilgrimage is the earliest type of tourism (Griffin and Raj, 2015). In recent years, pilgrimage and religious tourism has aroused the interest of many scholars and academics (Drule *et al.*, 2012; Raj and Griffin, 2015). Currently, many academics from different disciplines are studying the topic of religious tourism and pilgrimage – sociologists, historians, theologians, psychologists, geographers, anthropologists and economists (Vukonić, 1996) – which, at the same time, are becoming a big business. One of the areas for investigation that is absent in the tourism literature is pilgrimage tourism in the state of Palestine, the newest country, but probably with the oldest nation brand in the world (Cakmak and Isaac, 2016). This chapter aims to explore the new phenomenon of 'Come and See' that is gaining ground in Palestine targeting the pilgrimage market. Under the slogan of 'Comes and See', many initiatives – Kairos Palestine, The Alternative Tourism Group (ATG), the Holy Land Trust in Palestine – urged Christians in the whole world to continue their pilgrimage to the land of the Bible, despite the situation of continued political tensions, in order to receive 'a message of faith, hope and love' and in order to 'know the facts and the people of this land', Palestinians and Israelis alike. In addition, this study aims at exploring the motives and different experiences of these pilgrimages that are visiting Palestine and Israel. Analysing participants who went on a tour through ATG, data was collected from the online reviews of these organizations.

Pilgrimage tourism is a noteworthy and rapidly growing market segment within the tourism industry. It is also sometimes referred to it as faith-based travel, Christian travel, Muslim travel or any other religious denomination connected with the words travel or tourism (Isaac, 2016). The exact numbers of people travelling for religious purposes in this niche segment is unclear, for example the World Tourism Organization (UNWTO, 2011) estimates that there are over 600 million national and international religious trips every year, with between 300 million and 330 million people visiting a religious site. In spite of the lack of evidence and reliable statistics, religious tourism and pilgrimage have experienced considerable growth in the last 30 years, although it appears that the greatest increase has taken place over the last decade (Isaac, 2018, p. 124). According to Mintel

*Isaac.R@buas.nl

Group (2012) the number of foreign pilgrims to perform the Hajj in Saudi Arabia has increased from 1.4 million in 2010 to 1.8 million in 2011, whereas the number of pilgrims visiting the Vatican has more than doubled in the last decade, recording over 5 million for the first time.

According to Olsen (2019, p. 110), 'Pilgrimage is one of the oldest forms of travel'. Barber (1993, p. 1) states that pilgrimage is defined as 'a journey resulting from religious causes, externally to a Holy place, and internally for spiritual purposes and internal understanding'. Pilgrimage, either traditional and religious or modern and secular, is experiencing renewal around the world and ancient shrines still act as magnets for those in search of spiritual goals (Digance, 2003). Examination of the pilgrimage metaphor has resulted in many intriguing discussions in tourism, although too little discussion on pilgrimage itself (Cohen, 1979; Graburn, 1989; Vukonić, 1996; Attix, 2002) and its values and motivations.

Pilgrimage and Religious Tourism

Several scholars (Raj and Morpeth, 2007) have identified that within the eclectic range of themes that comprise religious tourism and pilgrimage studies there is a common consensus on some aspects such as the term 'religious space', referring both to the limited space within a shrine or cathedral and to the geographical space that a pilgrim traverses on his/her pilgrimage. Segmenting this market was done early from a general perspective, frequently revolving around attempts to distinguish and separate between pilgrims and tourists (see Cohen, 1992; Pfaffenberger, 1983; Graburn, 1989). For example, Smith's typology is the best-known one, where she places pilgrims and tourists at polar opposites of a continuum with a seemingly endless combination of sacred-secular motivations a person could imaginably have when deciding to travel to a sacred place (Smith, 1992). Following Smith, Santos (2003) developed a similar typology, placing tourism and religion at opposite ends of a spectrum with the grades between the two based on the religious background of travellers, the values they place on various religious spaces and the experiences they expect to have in those spaces. Other approaches include Cohen (1992) typology between concentric-formal and peripheral-popular sacred sites; Rinschede's (1992) extricating of short-term religious travel (religious tourism) and long-term religious travel (pilgrimage); and Singh (2013) contrast of pilgrimage and tourism attributes. Defining pilgrimage and religious tourism has led to several studies related to the division of religious motivations, specifically religious and non-religious motivations for travel within contemporary Christian travel (see Nolan, 1987; Nolan and Nolan, 1989, Nolan and Nolan, 1992; Fleischer and Nitzav, 1995; Fleischer, 2000; Collins-Kreiner and Kliot, 2000; Weidenfeld, 2006; Olsen, 2006; Ron, 2009). Other studies and discussions also emerged concerning the various tourist motivations relative to two dimensions, along the two axes of pious pilgrim-secular tourist and sacred pilgrimage-secular (Nolan and Nolan, 1989; Murray and Graham, 1997), while Kurlantzick (2007) claims that the religious market is homogenous and developed into much larger and more segmented markets, with niches ranging from high-end religious travel, to volunteer-oriented religious travel to modern-day pilgrimages.

Christian Pilgrimage in Palestine

Palestine plays a vital and important role in attracting and motivating tourists who find interest in religion, history, culture, tradition and environment. Palestine is sacred for the three monotheistic world religions: Islam, Christianity and Judaism. Regarding the Christian religious sites of Jerusalem, these places are regarded as the most important in the world and contain various churches and monasteries of the Christian religion such as the Church of the Holy Sepulchre (The Palestinian Association for cultural exchange PACE, 1999; Shahin, 2007). As Jerusalem is a unique place and a major consideration in religious tourism, there are a lot of villages, districts and towns considered sacred which attract tourists and pilgrims to visit and stop over in them such as Hebron, Bethlehem, Jericho, Nazareth and Nablus (Fig. 16.1).

Revenues generated by tourism in Palestine are made up of of incoming international

Fig. 16.1. Basilica of the Nativity in Bethlehem. (Photo: D. Liutikas)

tourism (50%) and domestic tourism (including Palestinians travelling from Israel) (50%). Over 85% of tourism revenues generated by international visitors come from pilgrimage. As a tourist offering, pilgrimages have special characteristics (Centre, 2014):

- In contrast with other types of tourism, the decision to undertake a pilgrimage is much less susceptible to influence by means of marketing because the main motivation is spiritual and hence deeply personal. In consequence, it appears difficult to stimulate international demand in this area.
- The pilgrim market has a fluctuating high season between March and April (due to Easter increasing the number of pilgrims) and between September and November (as a by-product of Jewish holidays which lead to high occupation rates in Israel, pushing pilgrims into Palestinian accommodation).
- Pilgrims in most cases follow a fixed programme which generally includes meals, accommodation in and tours of almost exclusively sacred places such as the Holy Sepulchre in Jerusalem or the Basilica of the Nativity in Bethlehem, together with visits to churches, monasteries and convents.
- Pilgrims usually have a relatively low budget as often the one purpose of their trip is to rediscover simplicity. Many are thus looking for simple accommodation and meals and do not intend to spend a lot of money on luxuries. This means that margins are rarely significant (except for devotional artefacts).
- Pilgrimage cannot be easily bundled with other types of tourism. Additional services that can be bundled with the traditional ones need to be 'religion-friendly'.

The tourism sector's direct contribution to GDP is approximately 4%, which is less than the immediate regions' economies (6% in Israel, 20% in Jordan, 13% in Egypt and 37% in Lebanon). This weakness also applies to the contribution of the tourism sector to total employment (World Travel and Tourism Council, 2013). The tourism sector contributes approximately 2% to total Palestinian employment. This is the lowest rate compared to the immediate regions' economies (8% in Israel, 19% in Jordan, 11% in Egypt, and about 4% in Lebanon) (Al-Falah, 2012). However, the limited availability of reliable statistics makes it difficult to describe with accuracy the size and dynamics of the state of Palestine's tourism sector both in terms of volumes and value.

The lack of control over borders (air, land and sea); the vulnerability to regular invasion and consequent physical damage to tourism infrastructure; the lack of freedom of movement for Palestinians and tourists; the regular closures of Palestinian areas; and the Separation Wall, which cuts deep into Palestinian areas, are only some of the problems resulting directly from the Israeli occupation. Furthermore, as a consequence of this occupation, Palestinians have been unable to fully develop their tourism potential and, more importantly, have been unable to plan for future development without a clear indication of when the conflicts will end and how Palestine will look once a final status agreement is reached (Isaac, 2014). Israeli measures in holy places and throughout occupied Palestine explicitly target religious and cultural heritage, including damaging and demolition of mosques and ancient sites. Constant military incursions and unilateral measures taken in occupied Palestine severely damage tourism through creating threats to any developments or investments in the sector (see Isaac, 2010).

'Come and See': The Palestinian Christian Initiative

In December 2009, Palestinian Christian leaders launched the so-called Kairos document, a statement that shares their daily realities of life under military occupation and calls on Christian sisters and brothers and churches worldwide to be witnesses to these realities, to be in solidarity and to take action. A guideline was developed in 2010 at a consultative meeting in Geneva to promote justice tourism for pilgrimage to Palestine-Israel. The meeting was represented by 14 countries. A group of 27 theologians, Palestinian Christian activists and professionals in the tourism industry called on Christian pilgrims to live their faith as they visit the Holy Land, going beyond homage to ancient sites and to follow the steps of Jesus in showing concern for the Palestinian people living there whose lives are severely constricted by the Israeli occupation of their lands. The meeting was organized by the Alternative Tourism Group (ATG) in collaboration with the Ecumenical Coalition on Tourism (ECOT), Kairos Palestine, the World Council of Churches (WCC) and the Palestine-Israel Ecumenical Forum (PIEF). The ATG is a Palestinian NGO specializing in tours and pilgrimage that combines critical examinations of the Holy Land's history, culture and politics (Kairos Palestine, 2014).

These guidelines have been written for visionary pilgrims and visitors seeking an authentic face-to-face human encounter in the Holy Land, who wish to connect with Palestinian Christians – the 'living Stones' – who share their faith. In order to understand the reality of Palestine, they say to the Churches: 'Come and See'. They will fulfil their role to make known to visitors the truth of their reality, receiving you as a pilgrimage coming to Palestine to pray, carrying a message of peace, love and reconciliation. Visitors and pilgrims will know the facts and the people of this land, Palestinians and Israelis alike. At the same time, they call on you as a pilgrim to say a word of truth and to take a position of truth regarding Israel's occupation of Palestinian land. The Come and See Initiative addresses churches, tourists and pilgrims of all faiths and ethnicities in the hope that they will see the injustice happening in Palestine and walk and work in solidarity with the Palestinians along the path to justice. Kairos Palestine, in addition, believes in facilitating the 'sight' of the many people who visit Palestine, whether as a tourist or a pilgrim. Joining the 'Come and See' initiative is a wave that spreads the message as well as draws in more alternative-justice tourists and pilgrims.

According to Kassis, director of the ATG (2009) transformational pilgrimage to Palestine, there are also justice tourists, seeking to understand and make a positive difference in the lives of people whose lands they visit. Meeting Palestinians who are living under occupation is an act of solidarity that brings hope to the people

Fig. 16.2. Saint George Monastery of Choziba in Palestine. (Photo: D. Liutikas)

and, in addition, contributes to their economic development. Like the rest of the Palestinian economy, tourism faces extraordinary difficulties caused by the Israeli occupation. Israel controls all entrances into Palestine and controls the flow of tourists, issuing visas, using Israeli touring buses and favouring its own tourism industry. In addition, Israel severely restricts business in Palestine. Therefore, engaging churches, social movements and faith-based organizations needs to promote pilgrimage for transformation. It is hoped that pilgrims will be inspired by and will work for justice-based peace and reconciliation for the Palestinians and Israelis. While more tourists and pilgrims visit Israel, the number of visitors to Palestine has been increasing annually (Fig. 16.2). Justice tourism to Palestine has its goal: 'promoting peace with justice for the people in the Holy Land'. According to MOTA (2019), the number of overnight stays of inbound and domestic tourists in the first lf of 2019 reached 954,217, whereas the number of inbound and domestic visitors to Palestine reached 3,174,781. Table 16.1 shows the top 10 nationalities of overnights for inbound tourism in Palestine.

The 'Come and See' initiative is also related to sharing the experiences with friends and relatives. In this context, values are fundamentally abstract beliefs that are culturally defined and assist as guidelines for worldviews, principles and behaviours, particularly in Palestine. For example, Muller (1991, p. 57) stated that values 'govern a person's lifestyle and provide a direct and useful explanation of the multitude of interests, outlooks on life (and death), consumption practices and activities that define a lifestyle'. In addition, Kahle (1983) states values could be defined as strategies to adapt to one's environment and to one's situation, which is the case in Palestine.

Liutikas (2017, p. 218) states that 'identity helps establish a person as an independent individual and is usually constructed in relation to difference'; although there are different streams that help tourists (re)construct their personal

Table 16.1. The top 10 nationalities of overnights for inbound tourism in Palestine.

Nationality	Number of overnights for inbound tourism
Poland	139,633
Romania	111,033
USA	53,154
Arab 48	49,907
India	48,194
Russia	43,494
Italy	40,421
Germany	32,976
Greece	31,506
Egypt	25,237
Indonesia	24,325
China	23,905
Korea	20,625
Others	236,377

identity. For example, Higham and Hinch (2009) argued that tourism is an instrument that manifests or constructs one's identity through contact with others. Tourism influences the values and identity of both hosts and guests and destinations serve as reference points in the activities related to the manifestation of identities and values. Through the 'Come and See' initiative, pilgrims' identity and values could be reconstructed when understanding the current political situation, the 'see it for yourself experience', which are facts on the ground regarding the segregation wall and the current Israeli occupation. In this case, motivations for pilgrimage for travelling are related to the construction or manifestation of personal and/or social identity and self-fulfilment. These provide a framework for understanding values and identity impacts on motivation and behaviour of pilgrims.

Nolan and Nolan (1989) stated that the term pilgrimage is best-known for religion-motivated travel to sacred places. However, many scholars argue new secular forms of pilgrimage that go beyond traditional religious practices: political or patriotic, musical, sport, nostalgic, and others (Hall, 2006). One of the concepts is holistic tourism. Smith (2003) defined holistic tourism as directed towards inner harmony underpinned by the pursuit of personal enhancement or enlightenment in congruence with nature. The main value of a pilgrimage is, arguably, the satisfaction of physical, mental and spiritual well-being needs (Smith and Kelly, 2006). In this vein, pilgrims heading to Palestine through 'Come and See' initiatives are partially motivated to satisfy their personal enhancement in conjunction with the political situation in Palestine.

Methodology

Netnography is a recently developed research methodology. The method was first used in 1995 by Robert Kozinets as an online marketing research technique, this in order to provide consumer insights (Kozinets, 2010). Netnography has some advantages in comparison to more traditional research methods – for instance, it is less time-consuming and less expensive than other methods such as interviews, focus groups and questionnaires (Kozinets, 2002).

Netnography mainly uses consumer-to-consumer communication on the internet. In order to understand needs, emotions, opinions and motivations, netnography uses information that is available on online forums and blogs (Kozinets, 2002). According to Kozinets, it is useful to reveal interaction styles, personal narratives, communal exchanges, online rules, practices, rituals and discursive styles across academic fields (Kozinets, 2015). The data researched will be analysed using quantitative data analysis techniques such as word frequency and semantic network analysis.

Tripadvisor is the largest travel review website in the world and is therefore the most logical tool to analyse reviews. Furthermore, the decision was made to make use of reviews from seven different organizations and tours that offer some kind of alternative tourism activities within the West Bank. The tours and trips were found via Tripadvisor by searching terms such as alternative tours, alternative trips, etc. The reviews of these tours (a total of 652), ranging from a few words to multiple paragraphs, were put together in a word document. To avoid getting lost in translation, only reviews in the English language were used. The following seven Tripadvisor pages were chosen: Alternative Tourism Group (ATG), Free Bethlehem Walking Tours, Banksurfing Alternative Tours, Beyond Borders Tours, Murad Tours, Private Tours by Michael Jackaman, and Hebron Alternative Tour.

The decision was made to analyse the data with the help of software program NVivo 12, of which a free trial version was downloaded. NVivo 12 is a software tool that helps to organize, store and retrieve data.

First, all the reviews were collected from the different Tripadvisor pages and put together in a word document; thereafter the document was added in NVivo 12. The data with the reviews that were collected in a separate Word document were put in NVivo and a word count was run that selected the 1500 most frequent words used. On NVivo, the word count includes stemmed words (e.g. talk, talking) to gain a more realistic overview of the most used words. Furthermore, certain 'stop' words such as 'the', 'I', 'is', 'he', 'a' etc. are excluded from the word count. Also, names that are not relevant are filtered out of the overview, for example the names of the guides.

Additionally, several semantic network figures were constructed in order to provide a clear overview on the impressions that the review writers had of the tours they undertook in the West Bank.

The in-depth expert interviews were semi-structured and conducted on location in Bethlehem. For this research there have been eight interviews conducted. The interviews were based on the principles of motivations, their vision and engaging with the initiative with 'Come and See'. Interviews have been conducted with representatives of the following stakeholders of alternative-justice tourism in Bethlehem:

ATG

To counter the lack of contact with the local communities and to give tourists the chance to understand more of the political conditions, alternative tourism has emerged in Palestine with the creation of the Alternative Tourism Group (ATG) (Alternative Tourism Group, 2012). Alternative tourism is an attempt to build the Palestinian economy through tourism without Israeli co-operation. Alternative tourism should be a pillar of tourism development for Palestinians, but not the only means of tourism development. Since its establishment in 1995, ATG has welcomed thousands of people from all over the world in Palestine. ATG would accompany visitors and pilgrims on a journey with a purpose and with intent to both learn and unlearn through gaining accurate insights into Palestine and the struggle for justice as viewed through the eyes of Palestinians and based on their aspirations (Kassis *et al.*, 2016). Interviewee was Mr Daffar Kassis (Advocacy officer).

Siraj Center

The Siraj Center is a non-profit organization based in Beit Sahour, Bethlehem Governorate. The goal of the organization is to make connections between Palestinian people and people from all around the world. It also organizes trips across Palestine to allow foreigners to witness first-hand the ongoing Israeli occupation by meeting with both Palestinians and Israelis. Furthermore, the Siraj Center organizes home stays with local Palestinian families as well as other forms of intercultural interaction. The Siraj Center believes that by encouraging visitors and pilgrims to stay with local families and hotels, it will help to boost the local economy. Moreover, it also organizes hiking and biking trips. Mr Michel Awad is the representative of the Siraj Center who was interviewed. Mr Awad is the Executive Director of the Center.

Holy Land Trust

The Holy Land Trust is a Palestinian NGO located in Bethlehem. For over 20 years it has crafted specialized tours in Palestine to meet the demands for both pilgrimage, individuals and organizations. Their work is based on six main points: empathy and compassion, non-violence, community and individuality, equality and justice, trust and respect, and spirituality. It distinguishes itself by not being political nor religious. The organization recognizes that the Holy Land is holy for all the people that live in it. Mr Elias D'eis is the representative of the Holy Land Trust

Table 16.2. Most frequently used words.

Word	Count	Word	Count
tour	508	Jerusalem	69
Bethlehem	269	refugee	69
guide	140	situation	68
day	129	political	61
tours	124	city	58
Palestinian	120	trip	58
West	113	much	57
Bank	110	camp	55
see	110	get	55
experience	104	like	52
Palestine	101	highly	51
Israel	98	take	51
people	95	lot	50
time	94	saw	50
recommend	91	went	49
wall	91	around	48
also	90	interesting	48
one	87	area	47
life	85	conflict	46
really	81	just	46
great	80	local	46
well	79	church	45
Hebron	77	perspective	45
visit	76	took	45
history	71	amazing	44

who was interviewed. Mr D'eis is the Executive Director of the Trust.

Results of Netnography and the Semantic Network of Reviews

The reviews on Tripadvisor were analysed by tabulating the phrases and words based on frequency on the website. Table 16.2 reveals the 50 most frequently occurring words and phrases including the frequency mentioned.

The most frequently used words were used to create a semantic network diagram to analyse the tourists' reviews and their impressions. In the three figures below, the most frequently used words are linked to a certain theme that comes up often in the frequency analysis. Frequently used words that are linked to the theme are made bold in quotes to underline how they were used by the writers of reviews.

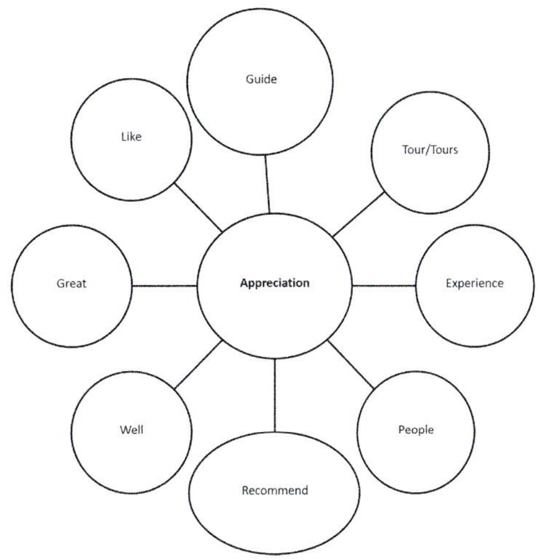

Fig. 16.3. Appreciation of the trip.

The first theme that occurs in many of the reviews was the theme of **appreciation**. Many of the reviews reveal their appreciation for the trip they undertook. The writers of the reviews do this in multiple ways; for example, they thank their guide and recommend the trip they participated in; naming their guide is something that occurs many times (Fig. 16.3).

*'I cannot **recommend** this **tour/experience** enough'* is a good example of a pilgrimage who was highly appreciative and thankful for the tour. Examples of those who were appreciative of the guide is: *'Abuud is a wonderful **guide**, friendly and professional'*, and *'Salah is a fantastic tour **guide**'*.

Other frequently used words that emphasized the positive experience that tourists had are words such 'like', 'great' and 'well'. Quotes that underline this are *'...had **great** passion to share his stories'* *'this experience is **well** worth it'*, and *'...but I **liked** the way our **tour** was'*.

The reviews also show appreciation by talking about the experience often, such as *'Thank you for the **amazing experience** Salah!'* Furthermore, the reviews often showed appreciation for the local Palestinians. For instance:

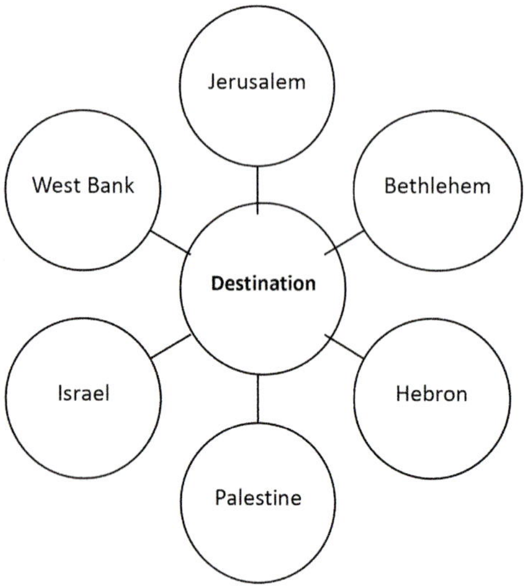

Fig. 16.4. The destination (Palestine and Israel).

Fig. 16.5. The experience of the pilgrimage.

'you will notice that ***people*** *are very friendly and helpful'* and *'The* ***people*** *we met were warm'.*

The second theme that can be hit upon often while analysing the table with frequently occurring words was the theme of **destination** (Fig. 16.4). Since one of the aims of Tripadvisor is to get reviews about locations where members of the website have been, this makes complete sense. However, the analysed trips were focusing on tours in the West Bank. '**Israel**' was also a frequently mentioned word; the word was used mainly to encourage others to not only visit Israel but also Palestine; for example: *'I highly encourage any traveler in* ***Israel*** *to visit Bethlehem'* and *'Any holiday to* ***Israel*** *would not be complete without a visit to the West Bank'*. It was also common to mention the conflict: *'we now understand a lot more of the ongoing conflict between* ***Israel*** *and* ***Palestine****'* and *'it will educate you on the entire historical, cultural and political context of the* ***Israeli Palestinian*** *conflict'.*

The other words occurring frequently are locations that the analysed tours visit regularly, such as the cities of **Bethlehem**, **Jerusalem** and **Hebron**. Also, mentioning the **West Bank** was something frequently done in the reviews. This makes complete sense since the tours usually go to this destination. These destinations were often mentioned in ways like: *'... took a tour around greater* ***Bethlehem****'*, *'we did the Ramallah/****Hebron*** *Tour with Salah'* and *'I came from* ***Jerusalem*** *by myself'.*

The third theme that has been created was the **experience**. It is noticed that the reviews frequently talked about the experience that the tourists had (Fig. 16.5). The first phrase that came up was '**see**', which was often used to talk about how they experienced the destination; for instance, *'I would recommend anyone who wants to* ***see*** *and understand Bethlehem beyond the normal tourist attractions to go on this tour!'* and *'You will get to* ***see*** *first-hand some of the issues Palestine is experiencing'*. Furthermore, '**went**' was a word that was mentioned often in constructions like *'I* ***went*** *on a full day tour'.*

Many of the writers mentioned how they experienced the current situation in the West Bank, with words such as '**situation**', '**political**' and '**conflict**'; for instance: *'...which gives really good insights into the Palestinian* ***situation****'*, *'but probably an even more important point of view in current* ***political*** *situation', and. 'I felt that we were given a good overview of the* ***conflict****'.*

Mentioning the past of the destination, '**history**' was a word that was used often to describe the experience in the destination: *'it was an excellent introduction to the* ***history*** *of such an interesting country'* and *'Yamen knows a lot about country,* ***history*** *and current situation'*. Furthermore, another word that came back in many of the reviews is '**interesting**'. Many of the reviews talk about how interesting they found the **tour/experience**. Examples of reviews that talk about this are: *'It was very **interesting** to see the terrible situation at Hebron, horrible but you should see it'*, and, *'This was one of the most **interesting** trips we had lately'*.

Results of interviews

> 'I can see that the people want to get more of people and culture and mostly they come for political reasons **they want to understand the conflict**. And usually they come up with a lot of impressions and they feel like it's too much maybe. To many impressions. So, and even if they come for religious purposes then they still want to know about the conflict and then they are ready to **hear different perspectives**'.

When asking Ms Norma of VisitPalestine.ps what the motivations are of those who contact her organization to visit Palestine, she mentioned mainly that people are interested in the conflict and want to learn about it – *'see it for yourself experience'*. She underlined that tourists who come for religious purposes want to learn more about the conflict and want to hear more perspectives.

Another interviewee from the Holy Land Trust states:

> We have these activists and that actually they come from activist **background they want to come here to do action**.
>
> The other part actually they have mainly come from churches they're actually Christians and they want to come and live the way Jesus live. Like we always say like if Jesus is here today what would it be, what would his reaction would be? So, as a Christian you need to **take action for justice and for peace and justice**. So those Christian they come **here just to learn**. They are like curious to know what's going on here. How can they influence their governments when they go back or the decision maker.

On the question of what drives people to visit Palestine, Mr D'eis of the Holy Land Trust said that there is a difference of motivations. On the one hand there are the people with an activist background who come to the West Bank to do some kind of action such as volunteering. Then there are the more religious travellers who come to Palestine to learn, they are curious about the destination and the situation (Fig. 16.6).

One of the staff of the ATG states:

> My opinion would be that there are three reasons. Some people are you know in the solidarity ideology of you know coming and **supporting and helping**. Others are **political or activists** are saying let's not say political but other activists who come here knowing the background and want to help and understand more. And others are just **eager to tease their interest**. So, you know you hear a lot about the Palestinian policy of the Israelis situation, but you don't have enough information, so you come to tease those ideas.

For Mr Kassis said there are three main reasons why people contact the ATG to travel to Palestine. First, it is out of solidary, particularly due to the 'Come and See' initiative with the Palestinians and those want to support and help out; secondly, there is a group that considers themselves activists and want to help out and understand more; thirdly, there are travellers who want to tease their interest; this is the group that hears a lot about the conflict but wants to learn more about it.

According to Mr Awad of the Siraj Center, the people that contact his organization are the travellers that want to explore the Palestinian culture and want to understand the social and political situation: *'to explore the Palestinian cultural and to **understand the situation**'*.

A touristic journey can be a turning-point (Hutchinson, 2014) event for the traveller, in the context of Palestine. The turning point could be a time when a major change occurs in the person's life. It may involve transformations in how the person views themself in relation to the world (Palestine) and/or transformation of values and identity. In general, pilgrimage and tourism are quite often considered as a ritual of transformation (Franklin, 2003). Many scholars (Turner and Turner, 1978; Graburn, 1989; Burns, 1999) empathize with the ritualized function of tourism which closely resembles rites of passage. The transformative potential of journeys was examined by Morgan (2010) who highlights the importance of the inner journey to transformative learning. This learning aspect comes through the 'Come and See' initiative that is focusing on pilgrimage in Palestine. The process of the construction of identity involves the internalization of knowledge, beliefs and skills. According to Wiederkehr (2001), a pilgrimage is a transformational journey during

Fig. 16.6. Panorama of Bethlehem. (Photo: D. Liutikas)

which significant change takes place, new insights are given (through 'Come and See') and deeper understanding is attained.

Conclusion

This chapter aims to explore the new phenomenon 'Come and See' that is gaining ground in Palestine, targeting the pilgrimage market. Under the slogan of 'Comes and See', many initiatives such as Kairos Palestine, the Alternative Tourism group (ATG), the Holy Land Trust and Siraj Center in Palestine urge Christians in the whole world to continue their pilgrimage to the land of the Bible, despite the situation of continued political tension, in order to receive 'a message of faith, hope and love' and in order to 'know the facts and the people of this land', Palestinians and Israeli alike. In addition, this chapter aims to explore the motives and different experiences of these pilgrims visiting Palestine and Israel. This study revealed that alternative tours can be seen as tours that have interaction with the local community and teach about the historical and political situation of a certain destination. In the West Bank, the demand for and supply of these themed tours is growing, due to the various initiatives that also included the 'Come and See' experience. Several organizations offer tours which show tourists and pilgrims alike the political situation and conflict on different locations such as the town of Hebron and Bethlehem and the Separation Wall in Bethlehem. From the different interviews it can be concluded that there are multiple reasons why pilgrimage and tourists travel to the West Bank. For example, one of the ATG staff stated the main reasons why these pilgrims travel to the West Bank and take part in their tours, namely:

- out of solidarity with the Palestinians' 'Come and See' initiative;
- to learn more about the conflict; and
- to tease their interests.

References

Al-Falah, B. (2012) *Tourism in the Palestinian Territory: Analysis if Significance and Impact*. Palestine Economic Policy Research Institute, Ramallah.

Alternative Tourism Group (2012) *Palestine & Palestinians*. ATG Publications, Bethlehem.

Attix, S.A. (2002) New Age-oriented special interest travel: an exploratory study. *Tourism Recreation Research* 27(2), 51–58. DOI: 10.1080/02508281.2002.11081220.

Barber, R. (1993) *Pilgrimage*. The Boydell Press, London.

Burns, M.P. (1999) *An Introduction to Tourism and Anthropology*. Routledge, London, New York.

Cakmak, E. and Isaac, R.K. (2016) The state of Palestine: the newest country probably the oldest nation brand in the world. In: Isaac, R.K., Hall, C.M. and Higgins-Desbiolles, F. (eds) *The Politics and Power of Tourism in Palestine*. Routledge, London.

Centre, P.T. (2014) *Tourism Sector Export Strategy 2014–2018*. Palestine Trade Centre, Ramallah.

Cohen, E. (1979) A phenomenology of tourist experiences. *Sociology* 13(2), 179–201. DOI: 10.1177/003803857901300203.

Cohen, E. (1992) Pilgrimage and tourism: convergence and divergence. In: Morinis, A. (ed.) *Sacred Journeys: The Anthropology of Pilgrimage*. Greenwood Press, Westport, Connecticut.

Collins-Kreiner, N. and Kliot, N. (2000) Pilgrimage tourism in te Holy land: the behavioral characteristics of Christian pilgrims. *GeoJournal* 50(1), 55–67. DOI: 10.1023/A:1007154929681.

Digance, J. (2003) Pilgrimage at contested sites. *Annals of Tourism Research* 30(1), 143–159. DOI: 10.1016/S0160-7383(02)00028-2.

Drule, A.M., Chiş, A., Băcilă, M.F. and Ciornea, R. (2012) A new perspective of non-religious motivations of visitors to sacred sites: evidence from Romania. *Procedia - Social and Behavioral Sciences* 62, 431–435. DOI: 10.1016/j.sbspro.2012.09.070.

Fleischer, A. (2000) The tourist behind the pilgrim in the Holy land. *International Journal of Hospitality Management* 19(3), 311–326. DOI: 10.1016/S0278-4319(00)00026-8.

Fleischer, A. and Nitzav, Y. (1995) *Christian Pilgrims: The Tourism Potential for Peripheral Regions in Israel*. Centre for Development Studies, Rehovot.

Franklin, A. (2003) *Tourism: An Introduction*. Sage, London.

Graburn, N. (1989) Tourism: the sacred journey. In: Smith, V.L. (ed.) *Hosts and Guests: The Anthropology of Tourism*. University of Pennsylvania Press, Philadelphia, Pennsylvania.

Griffin, K. and Raj, R. (2015) The globalization of pilgrimage tourism? some thoughts from Ireland. In: Raj, R. and Griffin, K. (eds) *Religious Tourism and Pilgrimage Management: An International Perspective*, 2nd edition. CABI International, Wallingford, UK.

Hall, C.M. (2006) Travel and journeying on the sea of faith (perspectives from religious humanism). In: Timothy, J.D. and Olsen, H.D. (eds) *Tourism, Religion and Spiritual Journeys*. Routledge, pp. 64–77.

Higham, J. and Hinch, T. (2009) *Sport and Tourism: Globaliztion, Mobility and Identity*. Elsevier.

Hutchinson, D.E. (2014) *Dimensions of Human Behavior: The Changing Life Course*. Sage, London.

Isaac, R.K. (2010) Alternative tourism: new forms of tourism in Bethlehem for the Palestinian tourism industry. *Current Issues in Tourism* 13(1), 21–36. DOI: 10.1080/13683500802495677.

Isaac, R.K. (2014) Palestine: tourism under occupation. In: Butler, R. and Suntikul, W. (eds) *Tourism and War*. Routledge, London, pp. 143–158.

Isaac, R.K. (2016) Pilgrimage tourism in Palestine. In: Isaac, R.K., Hall, C.M. and Higgins-Desbiolles, F. (eds) *The Politics and Power of Tourism in Palestine*. Routledge, London.

Isaac, R.K. (2018) Religious tourism in Palestine: challenges and opportunities. In: Butler, R. and Suntikul, W. (eds) *Tourism and Religion: Issues and Implications*. Channel View Publications, Clevedon, UK, pp. 143–160.

Kahle, R.L1983) *Social Values and Social Change*. Praeger, New York.

Kairos Palestine (2014) *Come and See: A Call from Palestinian Christians: A Journey for Peace with Justice: Guidelines for Christians Contemplating a Pilgrimage to the Holy Land*. Kairos Palestine, Bethlehem.

Kassis, R., Solomon, R. and Higgins-Desbiolles, F. (2016) Solidarity tourism in Palestine: the alternative tourism group of Palestine as a catalyzing instrument of resistance. In: Isaac, R.K., Hall, C.M. and Higgins-Desbiolles, F. (eds) *The Politics and Power of Tourism in Palestine*. Routledge, London and New York, pp. 37–52.

Kozinets, R.V. (2002) The field behind the screen: using netnography for marketing research in online communities. *Journal of Marketing Research* 39(1), 61–72. DOI: 10.1509/jmkr.39.1.61.18935.

Kozinets, R. V. (2010) *Netnography: Doing Ethnographic Research Online*. Sage Publications, Thousand Oaks, California.

Kozinets, R.V. (2015) *Netnography: Redefined*, 2nd ed. Sage Publications Ltd.

Kurlantzick, J. (2007) 21st century religious travel: leave the sackcloth at home. Available at: http://nytimes.com/2007/04/29/travel/29religion.html?_r=0 (accessed 12 March 2020).

Liutikas, D. (2017) The manifestation of values and identity in travelling: the social engagement of pilgrimage. *Tourism Management Perspectives* 24, 217–224. DOI: 10.1016/j.tmp.2017.07.014.

Mintel Group (2012) *Religious and Pilgrimage Tourism*. Mintel.

Morgan, D.A. (2010) Journeys into transformation: travel to an “other” place as a vehicle for transformative learning. *Journal of Transformative Education* 8(4), 246–268.

MOTA (2019) *Palestine 2019 Tourist Statistics*. MOTA, Bethlehem.

Muller, T.E. (1991) Using personal values to define segments in an international tourism market. *International Marketing Review* 8(1), 57–70. DOI: 10.1108/02651339110003952.

Murray, M. and Graham, B. (1997) Exploring the dialectics of route-based tourism: the Camino de Santiago. *Tourism Management* 18(8), 513–524. DOI: 10.1016/S0261-5177(97)00075-7.

Nolan, M. (1987) Roman Catholic pilgrimage in the new world. In: Eliade, M. (ed.) *Encyclopedia of Religion*. The Macmillan Publishing Company, New York, pp. 332–335.

Nolan, M. and Nolan, S. (1989) *Christian Pilgrimage in Modern Western Europe*. University of North Carolina Press, Chapel Hill, North Carolina.

Nolan, M. and Nolan, S. (1992) Religious sites as tourism attractions in Europe. *Annals of Tourism Research* 19(1), 68–78. DOI: 10.1016/0160-7383(92)90107-Z.

Olsen, D.H. (2006) Tourism and informal pilgrimage among the Latter-Day Saints. In: Timothy, D. and Olsen, D.H. (eds) *Tourism, Religion and Journeys*. Routledge, London/New York, pp. 254–270.

Olsen, D.H. (2019) Religion, pilgrimage and tourism in te middle East. In: Timothy, D. (ed.) *Routledge Handbook on Tourism in the Middle East and North Africa*. Routledge, London and New York.

Pfaffenberger, B. (1983) Serious pilgrims and frivolous tourists the chimera of tourism in the pilgrimages of Sri Lanka. *Annals of Tourism Research* 10(1), 57–74. DOI: 10.1016/0160-7383(83)90115-9.

Raj, R. and Griffin, K. (2015) *Religious Tourism and Pilgrimage Management: An International Perspective*, 2nd edition. CAB International, Wallingford, UK.

Raj, R. and Morpeth, N. (2007) *Religious Tourism and Pilgrimage Management: An International Perspective*. CAB International, Wallingford, UK.

Rinschede, G. (1992) Forms of religious tourism. *Annals of Tourism Research* 19(1), 51–67. DOI: 10.1016/0160-7383(92)90106-Y.

Ron, A.S. (2009) Towards a typological model of contemporary Christian travel. *Journal of Heritage Tourism* 4(4), 287–297. DOI: 10.1080/17438730903045548.

Santos, M. (2003) Religious tourism: contributions towards a clarification of concepts. In: Fernandes, C., McGettigan, F. and Edwards, J. (eds) *Religious Tourism and Pilgrimage: ATLAS Special Interest Group*, 1st Expert Meeting. Tourism Board of Leiria/Fátima, Fátima, pp. 27–42.

Shahin, M. (2007) *Palestine: A Guide*. Arab Scientific Publisher, Ramallah.

Singh, R.B. (2013) Pilgrimage-tourism: perspective and vision. In: Singh, R.B. (ed.) *Hindu Tradition of Pilgrimage: Sacred Space and System*. Dev Publishers, New Delhi, pp. 305–332.

Smith, V.L. (1992) Introduction: the quest in guest. *Annals of Tourism Research* 19, 1–7.

Smith, M. (2003) Holistic holidays: tourism and the reconciliation of body, mind and spirit. *Tourism Recreation Research* 28(1), 103–108. DOI: 10.1080/02508281.2003.11081392.

Smith, M.K. and Kelly, C. (2006) Journeys of the self: the rise of holistic tourism. *Journal of Tourism Recreation Research* 31(1), 15–24.

The Palestinian Association for cultural exchange PACE (1999) *PACE tour guide of the West Bank & Gaza Strip Palestine: Historical & Archaeological Guide*. PACE, Ramallah.

Turner, V. and Turner, E. (1978) *Image and Pilgrimage in Christian Culture*. Columbia University Press, New York.

UNWTO (2011) *Religious Tourism in Asia and the Pacific*. World Tourism Organization, Madrid.

Vukonić, B. (1996) *Tourism ad Religion*. Elsevier Science Ltd, London.

Weidenfeld, A. (2006) Religious needs in the hospitality industry. *Tourism and Hospitality Research* 6(2), 143–159. DOI: 10.1057/palgrave.thr.6040052.

Wiederkehr, M. (2001) *Behold Your Life: A Pilgrimage through Your Memories*. Ave Maria Press, Paris.

World Travel and Tourism Council (2013) *Travel & Tourism Economic Impact 2013: Mediterranean*.

17 Tourists Versus Pilgrims: Dichotomy of Visits to Auschwitz-Birkenau

Derek Dalton*
Flinders University, Adelaide, Australia

'We are pilgrims whatever we do, and there is little we can do about it even if we wished.'

'In the tourist's world, the strange is tame, domesticated, and no longer frightens; shocks come in a package deal with safety.'

(Zygmunt Bauman, from *Pilgrim to Tourist – or a Short History of Identity* (Bauman, 1996, pp. 20, 29–30, respectively)

Introduction

This chapter seeks to explore how a recent upsurge in tourism at Auschwitz-Birkenau has produced a problematic and unhelpful distinction between two types of visitors to the site that supposedly sit at distinct ends of the ethical spectrum of worthiness. The 'pilgrim' and the 'tourist' are the two archetypes produced in academic and media discussion about visitations to the former camp, now a memorial museum.

It is my contention that the distinction between tourists and pilgrims occasions a type of representational violence. The former – the pilgrim – is avowed as deferential, respectful and prepared (via education or through religious affiliation) to do justice to the site during their visit. The latter – the tourist – is disavowed as coming to the camp as a naïve visitor possessed of a voyeuristic motive; a mere 'consumer' of the space lured by guidebooks that frame Auschwitz-Birkenau as 'a must if you're in striking distance' when in Krakow (Cole, 2013, p. 233). Strange and Kempa (2003, p. 387) frame the better class of tourist as someone who will pay their respects, whereas the worst type of tourist will 'gawp in morbid fascination at the aftermath of tragedy'. However, Reynolds (2018, p. 33) sees distinctions between different types of visitors as distinctly unhelpful, arguing 'dichotomies between the tourist and the pilgrim, the tourist and the educator, or other modulations of the "they are tourists, I am not" formula are inadequate to capture the motives, identities and experiences of visitors to the site'.

In this chapter I seek to do four things. First, I wish to ground this chapter by outlining the complexities of Auschwitz-Birkenau as a space that must accommodate lots of different purposes and a variety of visitors. Here, the controversy about the commercialization of Auschwitz will be touched upon. Secondly, I wish to explore the concepts of the pilgrim, and provide an overview of the different types of pilgrimages that take place at Auschwitz. Thirdly, I wish to explore tourism and tourist practices at Auschwitz. Eschewing a discussion on visitor numbers, this section seeks to touch upon tourist practices and experiences. To that end, this section reflects on the controversies that attend the practice of tourist photography at Auschwitz. In the fourth section I seek to ruminate on the problems that confound *both* pilgrims and tourists when they

*derek.dalton@flinders.edu.au

encounter the contemporary camp that is, of course, temporally located out-of-wartime. The chapter concludes by challenging the dichotomy between pilgrims and tourists at Auschwitz, for it argues that both types of visitors come here to bear witness and that privileging one over the other sets up an unhelpful form of classificatory antagonism.

In this chapter I will deliberately refrain from describing all tourists who visit Auschwitz as 'dark tourists'. Whilst Lennon and Foley (2000) use the term 'dark tourism' as a general descriptor for travel to places associated with mass destruction and death that resonate with us because we are familiar with them through the media (Reynolds, 2016, p. 336), ascribing the descriptor 'dark tourist' to all tourists who visit Auschwitz is extremely problematic. This is because we are not privy to their core motivations. As Pedersen (2017, p. 132) states, 'Scholars have argued that the framework of dark tourism falls short for accounting for the nuanced motives behind travelling to sites of former suffering and the subsequent encounters that tourists have with these sites.' This chapter prefers to just use the simple descriptor of tourist to describe Holocaust/Shoah tourists who visit Auschwitz-Birkenau.

Before proceeding, it is worth noting that the vast current literature that looks at the boundaries between pilgrimage and tourism (see Chapter 14) 'reflects a somewhat uneasy relationship between [the] two scholarly camps' (Kaelber, 2006, p. 284). This chapter is mindful of these tensions, but, in the interests of brevity, must largely leave these wider issues uncanvassed.

Auschwitz-Birkenau: A Complicated Space

In trying to grapple with the complexity of Auschwitz, Pettitt makes a profound observation:

> Auschwitz is a site of mass atrocity, a museum, a cemetery, a focal point of Holocaust memory, a place of education, a town in south west Poland,[1] a tourist 'must see', and a place where complex negotiations of identity and morality take place.
>
> (2019, p. 6)

Therefore, even as we juxtapose the relationship between tourism and pilgrimage at Auschwitz, we need to do so while acknowledging that the place that is Auschwitz-Birkenau is multi-faceted, and employing tropes like pilgrim and tourist can occlude the sheer diversity of visitor types and motives to this sacred and profane place. Indeed, Eade (2015, p. 127) astutely points out that:

> Given the interweaving of (religious and secular) pilgrimages and tourism, the managers of popular sites face the challenge of catering for a complex mix of visitors, whose motives may be very different, ambivalent and variable according to the context.

Auschwitz-Birkenau faces such a challenge and manages to placate both tourists and pilgrims by managing to accommodate their distinctive wishes. However, some theorists conceive of the former camps as being overly commercialized. For example, Tim Cole (2000) claims that tourists at Auschwitz consume the site in the manner of a theme park; employing the somewhat cynical term 'Auschwitz-land' as though Auschwitz-Birkenau might be juxtaposed with Disneyland. In Coles's eyes, they are not so much looking to be informed and educated by the site but seeking a thrilling experience through representations of the past. This notion of the past is worth dwelling on. Reynolds (2018, p. 31) argues that a dualism is at play:

> [T]ourists conflate the place of Auschwitz as it currently exists with its operation as an extermination camp roughly seventy years ago; legitimate visitors, on the other hand, somehow appreciate that the current place is a representation of the past, not the past itself.

Returning to Coles's concerns that Auschwitz is part of a Holocaust industry that neatly packages the former camps – replete with restrung barbed wire and restored guard towers – for a ready-made tourist encounter, Norman Finkelstein (in Reynolds, 2016, p. 344) has argued that the collective memory of the Nazi genocide has led to the establishment of a 'Holocaust Industry' which consumes the Holocaust rather than advances knowledge about the past. Summing up his position, Reynolds notes, 'Finkelstein portrays Holocaust tourists as false pilgrims on a junket' (2016, p. 344). So it is apparent that even before we consider what tourists might do when they come to Auschwitz, their very presence is constructed as problematic.

Pilgrims at Auschwitz

The figure of the pilgrim is age-old, whereas the tourist is an invention of modernity. Writing in the context of Graeco-Roman and early Christian antiquity, Williamson (2007, p. 220) frames the figure of the pilgrim in direct opposition to the tourist:

> Whatever they [the pilgrim] might take in en route, their primary motivation is religious: to visit a site of special religious or numinous significance. By contrast, tourism is seen as either mere recreational activity, or at best as visiting beautiful buildings, landscapes, or works of art to evoke an aesthetic response to wonder.

So how are we to understand the figure of the pilgrim as it relates to Holocaust sites, and Auschwitz-Birkenau in particular? Since 1947, when the former camps were inaugurated as memorial museums, Jews from all over the world have made a solemn pilgrimage to Auschwitz to discover the place where the Shoah claimed so many lives (Pettitt, 2019). Kidron (2013)) sums up the impetus for these pilgrimages, writing:

> A growing number of survivors of mass violence are accompanying their children and/or grandchildren on voyages to sites of personal suffering and survival.

Some term these visits 'descendant pilgrimages'. Indeed, Cole (2013, p. 235) points out that the survivors who return to Auschwitz 'rarely visit alone, but often return accompanied by an audience ranging from family members through groups of Jewish teenagers to film-makers'.

The category of Jewish teenagers that Cole refers to above warrants exploration; for when we think of modern-day pilgrims, these teenagers (primarily from Israel, the USA and the UK) undertake what Kidron (2013)) terms 'a sacred commemorative quest'. Walking defiantly but solemnly through the camps – often draped in huge Israeli flags as makeshift capes (Wellner, 2008) – these young people are the archetypal Holocaust pilgrims (Kugelmass, 1996; Feldman, 2008).

All over Auschwitz-Birkenau visitors see signs that pilgrims have been here. Memorial candles are left on rail tracks (Fig. 17.1), small plastic Israeli flags are lodged into brick crevices, and blue-and-white ribbons are tied to wooden bunks and adorned to fixtures. A group on a pilgrimage from Japan left a beautiful hanging garland of origami cranes hanging from a restored bunk at Birkenau (Fig. 17.2). Some memorials are hasty and rudimentary like the giant star of David I saw during my visit that had been carved into the gravel pathway near the sauna building (Fig. 17.3). As Dalton notes (2015, p. 31), 'These remnants are reassuring reminders of the countless acts of mourning and commemoration performed to honour the dead'.

Some pilgrims come to participate in the March of the Living, a Holocaust educational tour that has a profound effect on adolescent Jewish identity (Nager *et al.*, 2013). Started in 1988, Cole (2013, p. 236) argues that the March 'brings Jewish teenagers to Auschwitz to 'enter into history' and be changed in the process. Teenagers march some two miles from Auschwitz I to Auschwitz II (Birkenau) accompanied by a parent or perhaps an uncle or aunt. They typically march to honour departed relatives who managed to survive the camps. They are, of course, also honouring other relatives murdered in the camps. The March attracts significant media attention as families from 45 different countries participate (Cole, 2013). This chapter does not argue that their pilgrimage is more important than others who come in much smaller groups on days untethered to this special event. Indeed, it is helpful to perceive these other pilgrims as ostensibly belonging to the same category of visitor.

Pettitt notes (2019, p. 5) 'Cole's distinction between what the Holocaust tourist and the Holocaust pilgrim – who, he claims visits the site with 'loftier intentions' – might also be reflected in the double-ness of the museum site: Auschwitz I caters for the tourists while Auschwitz-Birkenau provides a space of quiet reflection and mourning for the pilgrims'. She elaborates: 'discourses like these do – problematically in my view – suggest a hierarchy of visitor types'. I concur, and think it is even more complicated than her schema suggests. Few people – pilgrims or tourists – can find a private place to contemplate the horrors of the camps

Fig. 17.1. Memorial candle on track at Birkenau. (Photo: the author)

at Auschwitz I; there are just too many people passing through the site. However, to suggest that tourists just visit Auschwitz I and dispense with a visit to Birkenau is problematic. Certainly, tourists on a tight time schedule may well fail to spend time quietly reflecting at Birkenau, but many others increasingly see a visit to the larger extermination camp as an essential component of their visit. To frame every tourist as engaging in a whistle-stop itinerary where they get in and out as quick as possible and only bother seeing Auschwitz I is an assumption challenged by the enormous increase in tourists visiting Birkenau over the past ten years.

It should be noted that many Catholics in Poland come to Auschwitz primarily to pay respects to their countrymen murdered by the Nazis. Many come as an act of devotion to Saint Maximilian Kolbe, a Polish priest who famously asked that an SS officer allow him to take the place of a prisoner with a family who was about to be interred in the punishment/starvation Block 11 at Auschwitz I. The SS officer granted his request and the ten men were immediately transferred to the no-food-or-water Block 11, a place of extra terror in the Auschwitz universe. Kolbe eventually died from an injection of carbolic acid administered by the SS because they were annoyed that – after two weeks – four prisoners remained alive (Dolce, 2018). Kolbe's act of martyrdom lures many Poles to make a pilgrimage to Auschwitz (Charlesworth, 1994). His former cell is constantly adorned with floral wreaths and notes of adoration to prisoner 16,670.

Writing in *Pilgrimage to Auschwitz*, Vishvapani (2013) explores how people from very diverse backgrounds make pilgrimages to Auschwitz. As a Buddhist, his aim was to bear witness by joining others on a spiritual retreat. He spoke of another person on the retreat as a caring person who had worked with dying women for many years. For her, 'coming to Auschwitz was a further encounter with death' (Vishvapani, 2013, n.p.). But it is not merely ordinary people or descendants of survivors

Fig. 17.2. Origami cranes hanging in barracks at Birkenau. (Photo: the author)

that make pilgrimages to Auschwitz. Pope John Paul II visited Auschwitz in 1979 as part of the Vatican's efforts to reconcile with Jews. John Paul, coming from Poland, had witnessed the horrific suffering inflicted on his nation during the German occupation. Pope Benedict XVI visited in 2006, his reflections no doubt coloured by the time he spent in the Hitler Youth as a teen. Pope Francis, unlike previous pontiffs, did not live through the brutality of World War II on European soil. His visit in 2016 was one of quite prayer and meditation.

It is tempting to argue that by virtue of having had a relative who survived or perished in Auschwitz the pilgrim somehow has stronger insights into the camps. We must be cautious about stereotyping the pilgrim as possessing such clarity. While some may possess a conduit of understanding far superior to a tourist, for others the remnants of the camp resist an easy connection. Furthermore, it is dangerous to construct the figure of the pilgrim as well-educated and all-knowing. Hamrin-Dahl (2010), in reflecting on her visit to many of Poland's Shoah sites, observed: 'some of the pilgrims had deep historical knowledge, and were familiar with political and cultural circumstances; other pilgrims were in the dark about these matters'.

In summary, Greenia (2018, p. 9) best sums up our contemporary understandings of pilgrimages when he writes, 'Scholars now appreciate how pilgrimage embraces many sorts of journeys freighted with special values for one's people, nation and eternal soul'.

Tourists at Auschwitz

It bears emphasizing from the outset that many tourists to Auschwitz liken themselves to pilgrims. Keil (2005, p. 479-80) observes that many tourists characterize their experience as 'a pilgrimage, as a journey of commemoration and witness'. This is even though Hodgkinson notes that most tourists who visit Auschwitz do not have any direct connection to the atrocities that took place there (Hodgkinson, 2013, p. 23). The fact that some tourists perceive themselves as pilgrims may seem an affront to those who would like to differentiate visitors from pilgrims, but it reveals to us that tourists are not the flippant, vacuous visitors that many stereotypes suggest they are. Irrespective of how tourists classify their visit, Reynolds observes that 'one travels to Auschwitz or to Belzec to see the very ground on which the crime was committed' (2016, p. 348). The pilgrim and the tourist are therefore both custodians of the memory work that memorial sites sponsor (Reynolds, 2016).

Pedersen (2017, p. 139) argues, 'Before the tourist even arrives at Auschwitz, they have already been exposed to a myriad of visual and textual interpretations. These representations motivate tourists to visit Auschwitz and tremendously affect tourist's encounter with the site.' Dalton (2015) concurs with this view, noting that representations of Auschwitz 'lure' visitors from all over the world. There can be no doubt that the reconstructed barracks and guard houses, the restrung barbed wire and the many other restored physical attributes attract tourists who yearn to see the camps; but Reynolds points to a deeper quest for authenticity when he writes: 'Holocaust is likewise a search for authenticity beyond the pleasure-oriented package tour. The frequent invocation of the term "pilgrimage" by some visitors marks this intent, though one need

Fig. 17.3. Star of David carved into gravel path near sauna building at Birkenau. (Photo: the author)

not claim the label to engage in ethical reflection' (2016, p. 346).

Reynolds (2016, p. 337) observes:

> Tourists, especially those that travel to such grim destinations, are active participants in the market, they know how to exercise agency, and many are more skilled at discerning inauthentic from authentic representations than their critics assume.

So to employ Cole's analogy that tourists traipse around Auschwitz-land in a naïve stupor is problematic. During my visit I witnessed individuals persistently questioning tour guides, quietly discussing exhibits and engaging with the space in a manner that was far from passive. Perhaps the compulsory need to join a formal tour group at Auschwitz I, and the spectacle of tourists following their assigned guide, is a factor that allows the ready imposition of a critical frame on their mass behaviour. And while many theorists have pointed out the worst forms of tourist behaviour (e.g. talking in places that warrant silence, eating within the camp confines), these tend to overshadow the solemnity of quiet contemplation and reflection that most tourists practise. If there is one behaviour that is vexatious and attracts much disapprobation at Auschwitz it is the practice of photography. My attention in the next section will tease out the complicated relationship between the camera and the camps.

Tourists – as is their universal practice – take photographs at Auschwitz. It would be easy to deride photography as a purely voyeuristic practice. Certainly – to the dismay of tour guides and ethical tourists – many tourists take photographs of the trauma exhibits (Zyklon B canisters, piles of spectacles, luggage etc.) despite signs and verbal instructions prohibiting such photography. This is profoundly regrettable. However, most tourists photograph wooden barracks, the *judenrampe*, the archetypal gates with the slogan *Arbeit macht frei*, the red-brick entrance at Birkenau and other outside extant features of the camps. Pedersen (2017, p. 143) argues that 'the photography taken by tourists during their visit tremendously affects the tourist's encounter with the environment'. Indeed, Crawshaw and Urry (1997) argue that tourists construct their own images through photography. Photography thus achieves two purposes: first, it furnishes proof that they have visited a specific location; second – to borrow Pedersen's understanding of their argument – it 'gives shape to their encounters with their surroundings' (2017, p. 143). Pedersen was disturbed during her numerous visits to Auschwitz that tourists seemed more attuned to 'anticipating the next photo opportunity' than simply being in the moment. She elaborates on her concern:

> Many tourists seemed more concerned with capturing photos of themselves among the notorious imagery countlessly depicted in popular culture than of the prospect of being unsettled by their surroundings.
>
> (Pedersen, 2017, p. 147)

Pedersen argues that the act of photography 'reduces tourists' encounters with Auschwitz to a quest for souvenirs and evidence-furnishing proof that they had visited the infamous site' (2017, p. 147). However, I wonder if Pedersen's is too quick to dismiss photography and the labour of framing the sites in the camera's viewfinder as an act that can only lead to an impoverished experience of the camp. Perhaps many tourists return to their photographs – at a quieter and less harried time – and silently ruminate on what they have seen. They may share their photographs with others upon their return home. Like the *stories* of the pilgrim, the tourist shares his/her photographs as proof that they sought to bear witness to the Holocaust by touring its remnants. The tourist's photograph is not merely a facile souvenir but a testament to a difficult and harrowing encounter.

If there is one type of photograph that is particularly maligned at Auschwitz, it is the ubiquitous selfie. In this section I wish to explore an infamous example of a selfie-taking tourist. Pettitt argues that, 'There is little evidence that visiting Auschwitz will automatically stir empathy, create better citizens, or prove in any way transformative' (2019, p. 5). Indeed, the trope of insensitive tourist, lacking in respect and decorum, has emerged recently to cast some tourists in a very poor light. If there was a poster child for this figure of the inappropriate tourist, then we might posit that role was unwittingly performed by a tourist who used the moniker Princess Breanna (Koziol, 2014). In 2014, her smiling selfie at Auschwitz was bounced all over the world in news feeds that framed her, and those of her ilk, as vacuous young people just using the camps as backdrops for narcissistic photo opportunities. And yet the scandalous news accounts about Princess Breanna's supposedly frivolous selfie 'moment' obscured a motive that had nothing to do with narcissism. She had recently lost her father. They had planned to visit Auschwitz together, but his death thwarted that plan for father and daughter to embark on a journey to Poland together. Her smile captured by the phone camera was a personal affirmation – a 'Look, Dad, I finally made it!' testament. This poignant context got lost or was wilfully obscured by media that were poised to heap indignation on the selfie-taking youth who visited Auschwitz. And while many selfies may be intrinsically narcissistic and impelled only by a desire to populate Instagram, perhaps we are too quick to admonish the young who have been acculturated to capture every aspect of their daily lives on a mobile phone affixed with a selfie stick. Some pouting-face, sultry selfies may well be indefensible, but other selfies may mask the pride in having made a long journey of discovery to see what has only been experienced hitherto in books and on the cinema screen.

Problems that Confound Both Pilgrims and Tourists to Appreciate the Contemporary Camp Complex

Irrespective of the way one approaches a visit to Auschwitz, one must acknowledge that the camp complex and its presentation of museum exhibits presents some challenges for both pilgrims and tourists alike. For example, Griffiths (2019, p. 187) notes that the visual aesthetic is distinctly different from wartime:

> The visual aesthetic is ... troublesome: red brick buildings, light brown gravel pathways, green grass, blues skies and a golden sun. Today's Auschwitz is not the filthy, disease ridden hell of period photographs.

The mud that history tells us caused many prisoners to slip and sometimes be shot for failing to rise from in it is gone. And despite the muted colour palatte of modern cinematic representations of Auschwitz, the colourful nature of the space surprises those who envisioned the camps to mimic this cinematic convention (Griffiths, 2019).

The second problem, aside from discordant aesthetics, is that we must ponder the question of how the 'carefully tended, unlovely remains of a former concentration camp' can possibly communicate something tangible about the experience of being incarcerated there (Griffiths, 2019, p. 183). And yet tourists and pilgrims willing to employ imagination to engage with the exhibits can use what they see as a conduit to empathy. Dominick LaCapra (2001) worries about the problem of vicarious victimhood; and while he is right to decree that no pilgrim or tourist can rightfully claim victimhood merely by vicariously experiencing Auschwitz-Birkenau, surely we must applaud any attempt to imagine being imprisoned here. The very space demands that an ethical visitor imagine the plight of the prisoner. Contemplating the *judenrampe* (the rail siding where selections took place) at Birkenau, Dalton (2015, p. 27) ruminated:

> I pondered how many people had walked this very pathway towards a known or unknown fate? What of the sounds of pandemonium (wailing and weeping) as people were herded towards the gas chambers and crematoria? What of the acrid stench of burning flesh wafting over the entire camp? What of the taste of fear? To be here on a typical day or night and subject to the selection process must have been terrifying in a way that no 'non-survivor' visitor can fully fathom.

Many other tourists perform similar imaginative feats when visiting the camps, feats that are respectful of the space's legacy as a place of torture, suffering and murder. By doing so, tourists conjure 'ghosts of place' (Mayerfeld Bell, 1997). Furthermore, Griffiths (2019, p. 184) argues that, irrespective of how we might seek to classify visitors to Auschwitz by assigning a label like pilgrim, thrill-seeker or dark tourist, 'their response is not always discernible or quantifiable – or even assimilated – until long after their Auschwitz encounter is complete'.

Conclusion

In her blog entitled 'The Auschwitz Pilgrim', Wellner (2008) bravely poses the question: 'Is it a desecration of memory to visit Auschwitz as a pilgrim, and tell yourself that you have survived anything other than visiting a museum?' This is in keeping with Pettitt's (2019, p. 2) observation that 'The "real place" accessible today is, however, not the same as the "real place" of history.' A similar criticism often levelled at tourists is that they have done little more than spend a few hours traipsing around Auschwitz on a tightly controlled museum tour itinerary. Might we be less willing to judge either party – the self-identifying 'pilgrim' and the 'tourist' – if we understand that both people might leave the site moved to their very core, having shed tears and been left stupefied when confronted with the apparatus of genocide and the artefacts of murder? It matters not what label we want to assign to visitors to Auschwitz-Birkenau; rather, we should acknowledge that every respectful visitor plays a vital role in witnessing the Holocaust. Reynolds argues that Holocaust tourists 'become stewards of collective memory' (2016, p. 334). Like pilgrims they carry the burden of their visit home with them and embrace opportunities to share impressions of their experience with family and friends.

Reflecting on the atrocity exhibits that one encounters at Auschwitz-Birkenau, Hamrin-Dahl wrote:

> The pilgrims were confronted with stimuli that made thoughts take on an active, embodied quality. Ultimately, however, the subject remains the facilitator of her own experience; she must actively engage her imagination, set the 'stage' for the idea to have impact.
>
> (2010, p. 133).

The tourist – when confronted with stimuli – must do precisely the *same* thing as the pilgrim. She/he must employ their imagination to comprehend what they are seeing.

Hamrin-Dahl writes:

> A *pilgrimage* can be a collective affirmation of values and the journey is both physical and psychological. In fact, visiting sites which are connected to death, have become a significant part of *tourist* experiences in many societies. (2010, p. 144, emphasis added)

This remark is revealing because its logic connects the two hitherto disconnected figures of the pilgrim and the tourist. The physical and psychological journey that two figures embark on is not so dissimilar.

Garry Weissman (2004) employs the term 'fantasies of witnessing' to describe what visitors see when they visit Auschwitz. This term is far too cynical. Tourists aren't naïve. When they stand near the dynamited remains of the gas chamber ruins, they don't fantasize that they have seen an active gas chamber; rather, by peering at the piles of bricks and pondering the ominous steps leading downwards they are able to imagine the enormity of standing in the place where countless people were murdered. Their witnessing is a type of secondary witnessing and it is profoundly valuable. Cole (2013) recalls that the members of the *Sonderkommando* buried their notebooks and diaries under the ashes in the hope that they would one day be discovered and read long after their deaths – a wish that, thankfully, came to pass. Cole cites *Sonderkommando* member Zalma Gradowski, who entreats us about their gesture of testifying:

> We have done as much as we could. And you, searching for the truth, you who have lived to see justice and liberty, what will you do?
>
> (cited in Cole, 2013, p. 343)

It is a powerful question, one that does not privilege its recipient. One can speculate that those who died here would not really care if those who visit are pilgrims or tourists – only that they come prepared to witness what remains and leave prepared to speak of it. During his visit with students from the UK, Cole remarked that the appointed educator reminded his group, 'Gradowski is talking to us since we all came here to bear witness' (2013, p. 243).

We all came here to bear witness. These seven simple words should divest us of the desire to employ a hierarchy of worthiness to Auschwitz visitations. Such a hierarchy is not productive and sets up a tension that is not helpful. As the renowned Holocaust memorial scholar James Young reminds us in his classic text *The Texture of Memory*, 'the site retains its symbolic power over us partly because we don't live here, because we must make this pilgrimage to memory' (Young, 1993, p. 144). Auschwitz-Birkenau issues a type of siren song, a wailful, sad lament that echoes across the planet. Does it really matter who heeds its call? Writing in *Tourism as Pilgrimage*, Busatta (2018, p. 5) poses the question, 'Can cultural tourism practices legitimately be considered a form of pilgrimage?' I venture to answer that question in the affirmative. The pilgrim and tourist tread the same journey of disquieting discovery. Some forty-odd years ago – and long before Auschwitz became a tourist draw card impelled by the digital age – Victor and Edith Turner (1978, p. 20) infuriated purists by decreeing that 'a tourist is half a pilgrim, if a pilgrim is half a tourist'. Their refusal to differentiate the two is even more compelling in our contemporary world of Twitter and Facebook. A hybrid subjectivity need not be threatening. The common goal of coming to the *site* to see the *sights* for themselves reveals another dualism at play in the complex phenomenon that we understand as a visitation to Auschwitz-Birkenau. As Busatta (2018, p. 14) argues: 'social scientists have been debating the relationship between the tourist and the pilgrim for a number of decades'. Indeed, Damari and Mansfeld (2016, p. 200) have pointed out that recently a 'fuzziness of boundaries between pilgrimage and tourism, as well as between pilgrims and tourists' has emerged. Furthermore, this 'has sometimes made them indistinguishable' (Damari and Mansfeld, 2016). Rather than perceive the Auschwitz pilgrim as pious and the Auschwitz tourist as frivolous, might we – to employ Busatta's metaphor– seek to connect 'pilgrimage and tourism as two sides of the same *viaticus*[2] coin' (2018, p. 14, original emphasis).

Notes

[1] For a discussion of the fraught relationship Oświęcim has with the former concentration/extermination camp within its environs, see Russell Manning (2010).

[2] Here Busatta is employing the Latin term *viaticus* as provisions/money for a journey (from via – 'road, journey, travel'). The metaphor seems apt as most pilgrims and tourists typically travel a long way to visit Auschwitz-Birkenau.

References

Bauman, Z. (1996) From pilgrim to tourist – or a short history of identity. In: Hall, Stuart. and Gay, Pauldu. (eds) *Questions of Cultural Identity*. Sage, London.

Busatta, S. (2018) Tourism as pilgrimage? Antrocom: *Online Journal of Anthropology* 14(1). Available at: http://www.antrocom.net/upload/sub/antrocom/140118/Antrocom_14-1.pdf#page=5 (accessed 17 April 2019).

Charlesworth, A. (1994) Contesting places of memory: the case of Auschwitz. *Environment and Planning D: Society and Space* 12(5), 579–593. DOI: 10.1068/d120579.

Cole, T. (2000) *Selling the Holocaust: From Auschwitz to Schindler: How History Is Bought, Packaged and Sold*. Routledge, New York.

Cole, T. (2013) (Re)visiting Auschwitz: (Re)encountering the Holocaust in its Landscapes. *Cultural History* 2(2), 232–246. DOI: 10.3366/cult.2013.0049.

Crawshaw, C. and Urry, J. (1997) Tourism and the photographic eye. In: Rojek, C. and Urry, J. (eds) *Touring Cultures: Transformations of Travel and Theory*. Routledge, London.

Dalton, D. (2015) *Dark Tourism and Crime*. Routledge, London.

Damari, C. and Mansfeld, Y. (2016) Reflections on pilgrims' identity, role and interplay with the pilgrimage environment. *Current Issues in Tourism* 19(3), 199–222. DOI: 10.1080/13683500.2014.957660.

Dolce, J. (2018) Only love is creative: Maximilian Kolbe in Auschwitz. *Quadrant* 62(4).

Eade, J. (2015) Pilgrimage studies an expanding field. *Material Religion* 11(1), 127–129. DOI: 10.2752/174322015X14259910631833.

Feldman, J. (2008) *Above the Deathpits and Beneath the Flag: Youth Voyages to Poland and the Performance of Israeli National Identity*. Berghahn, New York.

Greenia, G.D. (2018) What is pilgrimage? *International Journal of Religious Tourism and Pilgrimage* 63

Griffiths, C. (2019) Encountering Auschwitz: touring the Auschwitz-Birkenau state museum. *Holocaust Studies* 25(1-2), 182–200. DOI: 10.1080/17504902.2018.1472881.

Hamrin-Dahl, T. (2010) This-worldly and other-worldly: a holocaust pilgrimage. *Scripta Instituti Donneriani Aboensis* 22, 122–162. DOI: 10.30674/scripta.67365.

Hodgkinson, S. (2013) The concentration camp as a site of 'dark tourism'. *Témoigner. Entre histoire et mémoire* 116, 22–32. DOI: 10.4000/temoigner.272.

Kaelber, L. (2006) Place and pilgrimage: real and imagined. In: Swatos, W.H. (ed.) *On the Road to Being There: Studies in Pilgrimage and Tourism*. Brill, Leiden, The Netherlands, pp. 277–295.

Keil, C. (2005) 'Sightseeing in the mansions of the dead. *Social & Cultural Geography* 6(4), 479–494. DOI: 10.1080/14649360500200197.

Kidron, C.A. (2013) Being there together: dark family tourism and the emotive experience of co-presence in the holocaust past. *Annals of Tourism Research* 41, 175–194. DOI: 10.1016/j.annals.2012.12.009.

Koziol, M. (2014) Smiling selfie in the Auschwitz leaves a poor taste. *The Age* online. Available at: https://www.smh.com.au/technology/smiling-selfie-at-auschwitz-leaves-a-poor-taste-20140722-zvjo1.html (accessed 22 August 2014).

Kugelmass, J. (1996) Missions to the past: Poland in contemporary Jewish thought and deed. In: Antze, P. and Lambek, M. (eds) *Tense Past: Cultural Essays in Trauma and Memory*. Routledge, New York.

LaCapra, D. (2001) *Writing History, Writing Trauma*. Johns Hopkins University Press, Baltimore, Maryland.

Lennon, J. and Foley, M. (2000) *Dark Tourism: The Attraction of Death and Disaster*. Continuum, London.

Mayerfeld Bell, M. (1997). *Theory and Society* 26(6), 813–836. DOI: 10.1023/A:1006888230610.

Nager, A.L., Pham, P. and Gold, J.I. (2013) March of the living, a holocaust educational tour: effect on adolescent Jewish identity. *Journal of Religion and Health* 52(4), 1402–1414. DOI: 10.1007/s10943-013-9749-3.

Pedersen, C. (2017) Screening tourist encounters: penal spectatorship and the visual cultures of Auschwitz. In: *The Palgrave Handbook of Prison Tourism*. Palgrave Macmillan, London, pp. 131–151.

Pettitt, J. (2019) Introduction: new perspectives on Auschwitz. *Holocaust Studies* 20(1), 1–11. DOI: 10.1080/17504902.2019.1625110.
Reynolds, D. (2016) Consumers or witnesses? Holocaust tourists and the problem of authenticity *Journal of Consumer Culture* 6(2), 334–353.
Reynolds, D. (2018) *Postcards from Auschwitz: Holocaust Tourism and the Meaning of Remembrance*. New York University Press, New York.
Russell Manning, J. (2010) The palimpsest of memory: Auschwitz and Oświęcim. *Holocaust Studies* 16(1-2), 229–256. DOI: 10.1080/17504902.2010.11087252.
Strange, C. and Kempa, M. (2003) Shades of dark tourism: Alcatraz and Robben island. *Annals of Tourism Research* 30(2), 386–405.
Turner, V. and Turner, E. (1978) *Image and Pilgrimage in Christian Culture: Anthropological Perspectives*. Columbia University Press, New York.
Vishvapani (2013) Pilgrimage to Auschwitz; Buddhism in the West. Available at: https://www.wiseattention.org/blog/2013/01/01/pilgrimage-to-auschwitz/ (accessed 4 December 2019).
Weissman, G. (2004) *Fantasies of Witnessing: Postwar Efforts to Experience the Holocaust*. Cornell University Press, Ithaca, New York.
Wellner, A. (2008) The Auschwitz pilgrim? The Blog. Available at: https://www.huffpost.com/entry/the-auschwitz-pilgrim_b_100573 (accessed 22 November 2019).
Williamson, G. (2007) Mucianus and a touch of the miraculous: pilgrimage and tourism in Roman Asia minor. In: Elsner, J. and Rutherford, I. (eds) *Pilgrimage in Graeco-Roman and Early Christian Antiquity: Seeing the Gods*, pp. 219–252.
Young, J.E. (1993) *The Texture of Memory: Holocaust Memorials and Meaning*. Yale University Press, New Haven, Connecticut.

18 Tracking to a Pilgrim's Beat: Insights from a Sabbatical Journey

Josephine Pryce*
James Cook University, Cairns, Queensland, Australia

Introduction

Pilgrim journeys are associated with spiritual enlightenment, strong emotional embodiment and transformative experiences. Contemporary pilgrim journeys encompass a range of values-rich journeys that impact on an individual's personal and social identity. One such journey that is often overlooked is the sabbatical journey. Equally, little attention is given to the mode of transport and how this dimension impacts on the journey. Increasingly, pilgrims engage with ways beyond walking or combinations of travel modes, including boat (Braun, 2017), camel (Dauphin *et al.*, 2015) and train (Blake and King, 1972). This chapter brings together the sabbatical journey and train travel to present my lived experiences as a scholar on sabbatical who chose to circumnavigate the USA via train and, in the process, was catapulted into a journey of self-discovery and healing.

On Thursday 25 July 2019, I began my two-month sabbatical journey to the USA. I was accompanied by my husband on what we anticipated to be a journey of a lifetime. After some rigorous research on my part and armed with a multitude of diverse reviews, we opted to travel throughout the USA via the long-distance national railway network company known as Amtrak. We commenced the round trip in San Francisco, travelling on the *California Zephyr* to Chicago, from where we travelled on the local train to Woodstock to visit the Illinois Railway Museum. After five days in Woodstock we returned to Chicago and boarded an Amtrak train heading for Boston, from where we backtracked to Syracuse, and then on to New York, Orlando, Atlanta, New Orleans, Los Angeles and back to San Francisco. We generally spent a minimum of five days in each city. During our time on the train we experienced different services provided by Amtrak, such as coach, sleeper, dining car, café car, observation car and station lounges, each adding to the range of spaces on the train that afforded opportunities to roam, socialize, converse with each other and fellow travellers, or simply contemplate another world passing by through the window.

Initially, the train travel was an alternative and immersive approach to expand on a current research project. As I undertook the journey, I found that it became a means to gain time to 'slow down', and to find space to muse and re-energize. The external passing landscapes promoted internal meanderings of the mind and cues to marvel and reflect on a range of occurrences and phenomena. Long hours of rhythmic movement on the train encouraged time and space for deep reflection, regrounding, renewal and personal growth. I felt like a pilgrim, with a quintessence that has endured

*josephine.pryce@jcu.edu.au

and continues to metamorphose, even upon my return home.

Costa (2018, p. 66) writes of 'sabbatical as a pilgrimage to meaning and wholeness'. From the outset I knew that my journey would impact on me both personally and professionally. The USA was my father's homeland and the place where I had spent a year of my teenage life. I knew this homecoming would stir memories and emotions. Pearce (2012) writes of 'returning home' and notes that there are 'a suite of interrelated concepts in the areas of emotion, human memory, cognition, identity and the perception of time' (p. 1024). He contends that experiential consumption is triggered by sensory and environmental cues that activate cognitive activity, generating higher levels of consciousness that can challenge one's identity and, in turn, activate modifications to our being. Pearce (2012) made specific mention of the resultant responses of melancholy, nostalgia and gratitude and suggested that travellers' positive outlooks could 'assist in affirming identity and promote well-being', leading to further exploration of associated meanings relating to time and place (p. 1042). Schmitt (2003), similarly, emphasized the sensory, affective, cognitive, behavioural and relationship components of experiential consumption. Taken together with Pearce's (2012) consideration of the impact of travel experiences on identity, well-being and learning, it was inevitable that as my journey unfolded I found myself being confronted and changed in ways that I had not anticipated. The exposure to 'other ways of knowing and being' challenged my own discourses, values and identity. The intensity of the experiences made it impossible to escape self scrutiny and the construction of 'new ways of knowing and being'.

This chapter adopts a heuristic phenomenology and utilizes a qualitative narrative approach, combined with critical autoethnography. The iterative nature of the heuristic process allows for the unfolding, explication and openness that is required for capturing transformation and embodied experiences. Narrative inquiry and reflective analysis were applied to observations and my field notes, allowing for themes to be extracted. The themes highlighted professional and personal experiences that led to the realization that the physical sabbatical journey projected travel experiences into liberated time and liminal spaces that were powerful, transformative and persistent.

Sabbatical as a form of Pilgrimage

Pilgrimage has been defined as 'a journey undertaken in quest of spiritual enlightenment . . . [that] is considered an extraordinary consumption experience . . . [and] is often personal, sacred, and transformational in nature' (Van Laer and Izberk-Bilgin, 2019, p. 387). Researchers agree that as individuals on a pilgrimage slow down and separate themselves from mundane aspects of everyday life, the pilgrim's experiences are highly emotive, enriching and transformative (Higgins and Hamilton, 2019; Husemann and Eckhardt, 2019). Pilgrims traditionally travelled on foot as it was thought that this activity served to slow down the pace of engagement and immersion in the environment, thus extending the physical challenge. There is increasing recognition that such existential parallels can be similarly obtained through other means of travel. It has been thought that the definition of pilgrimage may be better defined by the place or activities in which 'mundane' existential frames of reference or 'habits of mind' are disrupted by some form of challenge or 'disorienting dilemma' (Morgan, 2010). In a similar fashion, Smelser (2009) posits the 'odyssey experience' as being a conceptual framework that captures the essence of meaningful journeys that are transformative. He defines the 'odyssey experience' as 'a finite period of disengagement from the routines of life and immersion in a simpler, transitory, often collective and often intense period of involvement that often culminates in some kind of regeneration' (p. xi). He identifies the elements of the 'odyssey experience' as follows: (i) it is finite with a beginning and end; (ii) it involves a disengagement from normal, everyday roles and obligations (social destructuring); (iii) it involves an intermediate or liminal phase (the point of withdrawal into an unstructured phase); (iv) it is likely to involve travail (physical and/or psychological uncertainty and challenge); and (v) it has a *specialness* (whereby the 'traveller' has a sense of having *left his/her world*), and a *restructuring* as the individual subsequently returns.

Smelser (2009) acknowledges that the odyssey experience can be aligned to the traditional practice of 'sabbatical' where the academic is granted 'a period of rest or respite from routine responsibilities' (p. 125). He extrapolates the defining features of the odyssey experience against the sabbatical period as follows: (i) it is of a fixed term, e.g. a semester or year; (ii) it is an example of destructuring where the academic is released from their normal university-related roles, such as teaching, administration and service; (iii) there is a liminal period during which the academic has flexibility to engage with creative activities; (iv) it includes a period of travail whereby the academic experiences self-improvement, fulfilment and regeneration through activities such as focused research and writing 'a book or a series of articles' (p. 126); (v) the academic returns from the 'period of freedom' with achievements that showcase performance during the sabbatical and enable restructuring with a newfound experience of 'one's scholarly worth' (p. 127). Using his own sabbatical experiences, Smelser illustrates the parallels between the odyssey experience and sabbatical, highlighting the importance of destructuring and how this can best be achieved by taking the sabbatical away from one's place of residence. It is only when one is away that it becomes easier to avoid being called upon or pressured by staff at the home institution. In discussions of pilgrimage, Van Gennep (1960) introduced the notion of 'liminality', a term that was later refined by Turner. Van Gennep (1960) spoke of 'limen' as an integral component of pilgrimage, as a spatial metaphor acknowledging that 'in between' space that is on a 'threshold' and is 'neither here nor there'. Turner and Turner (1978) extended this understanding to capture the importance of liminality in accessing the transformative power of the pilgrimage. It is a place where 'two worlds meet . . . [a place] of learning . . . the in-between state . . . a place where learning liminal wisdom is communicated' (p. 44). The liminal place is where the pilgrim has, both literally and symbolically, 'left home' and severed the ties with 'general norms' but has not yet assimilated or embodied the 'new state of being' (Ridsdale, 2014). The liminal space exposes the pilgrim to intense and ambiguous experiences (Turner and Turner, 1978) and 'disorienting dilemmas' (Mezirow, 1991) that are liberating and empowering. The isolation during the liminal period is a key contributing factor for making the sabbatical an efficacious odyssey (Smelser, 2009).

While I became familiar with Smelser's writing upon my return from sabbatical, I reflected on my own notes long before that time, the time of my departure from Australia. The sabbatical was a pleasing welcome, almost a reward, after close to three years of heavy administrative duties. For the last year of that time, I had felt my physical and mental fatigue increasing dramatically; so I was excited to have this opportunity to be free from the quotidian demands of university operational work and to be able to focus on my own research. I knew that it was going to be a precious time to be able to work solely on building my scholarly profile. I was excited about going out into the field and to have some space and time to reconnect with myself, to reflect and to rebalance, especially as it included a period of time overseas. I anticipated that this venture abroad would bring novel and enlivening experiences that would generate a refreshed outlook to my work. I also thought that it would be more difficult for colleagues to call upon me to engage with impending issues and that I could focus on my research. I had envisaged the sabbatical as a treasured time for personal and professional development, but had underestimated the space, time and wealth of experiences that would lead to those transformative learnings.

One of the earliest revelations came when, one week into the trip, we arrived in Woodstock (Illinois) close to ten at night – tired and hungry – only to find that the station was closed and the taxi numbers found in a Google search were not answering. We could hear laughter and merriment in the distance. For a moment I thought of walking to the place from which the light and noise was emanating, but not knowing the area, the fear of being mugged stopped me in my tracks. I rang the hotel to seek assistance with taxi numbers but the young lady just repeated numbers I already had. I was at my wits' end and my husband was growing more impatient by the minute, wondering all the while what this place was that I had insisted we visit. 'It may have the largest railway museum in the USA, but it definitely is not visitor-friendly,' he said. Then again, I thought of some of Australia's railway museums that are in isolated locations. The NSW Rail Museum in Thirlmere was one that came to mind; a good one and half-hour drive south-west of the Sydney Harbour Bridge. My mind was wandering homeward when a voice on the other end of my husband's phone replied. We were in luck. Half an hour later we were comfortably settled in our hotel room and I reflected on how ambitious we had been to venture from Chicago in the evening into unfamiliar territory. I realized that, despite us being on the older side of 50, we were fortunate to be able to have found the time to share such an adventure together. Thus, absorbed in my contemplation, I slipped into a respectful appreciation of the opportunity to integrate and balance professional and personal life in meaningful and holistic ways.

Costa (2018) writes of 'sabbatical as a sacred time for renewal and wholeness' and as a time for 'contemplative practice' (p. 65). She talks enthusiastically about life as an academic in 'the neoliberal academy' and how she transcended her negative experiences in such a regime during a transformative sabbatical in which she walked the Camino de Santiago, allowing herself to 'slow down' and engage in 'self-care' and 'deep reflection'. She says:

> . . . in academe . . .[there] is little time or patience for sitting in solitude, for assessing where we have been and what we have contributed, for thinking deeply and carefully about what to do next or even the *meaning* of the work we do. Yet I yearned for such moments with my whole being (p. 66) . . . I yearned to slow down and to reclaim a sense of purpose (p. 72) . . . For me, slowing down was the necessary foundation for a creative exploration which led to the development of new projects and course offerings and the cultivation of new relationships with other scholars and activists – precisely the intellectual work I was meant to be doing during my sabbatical. Through this redefinition of time and space I discovered that 'the field' to which I traveled was no longer a distant land or a community, but rather my own inner life. And this inner life was one in which I was compelled to grapple with additional ways of knowing that are not typically valued or even recognized in the academy, including the somatic, the intuitive, and the spiritual.
>
> (Costa, 2018, p. 77)

In her essay, Costa (2018) challenges the extant neoliberal capitalist agenda permeating academe and sets out to find the balance between work and life where there is time for

Fig. 18.1. Word-frequency cloud for articles relating to pilgrimage, sabbatical, and transformative travel.

self-care and contemplative practice and in which the whole person is respected. She emphasizes the need for it to be understood that people come to work as a 'whole person' with physical, emotional and spiritual elements to their being as well as the cognitive. In Fig. 18.1, the key words that Costa (2018) used in her narrative stand out: time, contemplative, work, sabbatical, sacred, and faculty. She concludes by noting that, 'Imagining and enacting sabbatical as sacred time . . . has been profoundly healing in the way it has reenergized my scholarship and teaching' (Costa, 2018, p. 80).

There is no denying that prior to the sabbatical I felt fragmented and weary. Like Costa (2018), I sought time to slow down and the space to engage with deep reflection to allow me to reconnect with all that makes me a 'whole person' and a better academic. Costa (2018) sets out with the view that sabbatical is sacred time and grants herself the 'gift' to 'go slow'. I, on the other hand, was wanting to pack as much as I could into my sabbatical. I was relieved to be away from the operational, administrative and service work of university life and looked forward with eager anticipation to uninterrupted time for researching, writing and reflecting. I wanted to be as productive as possible. These thoughts accompanied me as we set out for the USA. It was only while we were on the train that I realized it was madness to be in a constant state of frenzied activity. It was during those long stretches that I gave in to opportunities to gaze out the window and allow my thoughts to wonder that the 'inner voice' surfaced. Costa (2018) also talks of the 'inner voice' and how the silence and movement during her walking days created 'space for the inner voice to emerge' (p. 66). In my case, the stark reality was that things wouldn't always work out in the way I had planned but I had the luxury of the opportunity of finding a relevance in the present that would advance my research and writing. Ridsdale (2014) describes the 'ritual of pilgrimage' as being defined by the 'rhythm of walking day after day . . . [linking] space and time in non-habitual ways for the person' (p. 23). She elaborates, highlighting the pilgrim's arrival in new places as enacting 'an unfolding embodied encounter' with the 'here and now' (p. 23). At these places, the pilgrim engages in actions that constitute 'travail' (e.g. attending mass, staying in hostels, enduring physical hardships) while purposefully practising attributes such as mindfulness, sharing, courtesy, respect, generosity, patience and gratitude. Ridsdale (2014, p. 29) sums up the experience as follows:

> To walk the Camino is to experience contradictions and extremes much more powerfully and vividly than in everyday life. One lives the whole range of experience between: communitas and contestation; elation and despair; the ephemeral and the durable; physical empowerment and physical fragility; loneliness and communion; deep silence of nature and noisy, chaotic environments; and one's identity being simultaneously disrupted and deeply affirmed . . . on the Camino, everyone is simply a pilgrim.

Ridsdale adopted a 'ritual autoethnography' that enabled her to walk the Camino as a 'ritual practice' and determine 'the spiritual resonance between pilgrimage and singing' (p. 16). She, like I, felt that the autoethnographic methodology was well suited to the self-reflective nature of the fieldwork of our respective projects. Ellis (1999) comments that, 'Autoethnography is an autobiographical genre of writing and research that displays multiple layers of consciousness' (p. 673). This statement is relevant to both Ridsdale and myself. This approach allows for the collection, analysis and representation of data that align with qualitative methods and afford interpretive orientations. As is the practice with autoethnography, reflective journaling, photography, and audio and video recordings are commonly utilized. Coupled with heuristic phenomenology, the methodology allows for framing of the project as an exploratory one. Moustakas (1990) explains heuristic research as a journey where, 'whatever presents itself in the consciousness of the investigator... represents an invitation for further elucidation . . . [involving] inner awareness' (pp. 10–11). It is an unfolding process wherein the researcher begins, not knowing where or how the journey will transpire. This approach is conducive to research relating to transformation as it provides scope for openness and discovery.

For Ridsdale (2014), this research protocol accommodated the trials and tribulations throughout her four-week journey. In the end, it granted her the courage to accept that her 'plan for how the singing should unfold was completely undone' and that instead she could 'surrender to a flow' (p. 33). Ridsdale's experience as a researcher brought to mind my own internal struggles. From the time we visited the Illinios Railway Museum in Woodstock, I knew that I, too, would meet with obstacles and would have to realize and accept that my plans for understanding the role of volunteers in heritage museums may be compromised. In Woodstock, I realized that gaining physical access to the museums was going to be more difficult and costly than I had anticipated. People drove everywhere. Public transport was limited in places like Woodstock and so we had to rely on taxis to get to the museum, which was about 20 minutes out of town. So fare and tips had to be accommodated. Such obstacles were compounded by the experience of arriving in Penn Station, New York, and being overwhelmed by the struggle to get out of the station. We were caught in the mass of commuters that carried us with them in the direction opposite to the one we needed to go. Eventually surfacing up on West 31st Street, we were able to quickly locate a taxi that took us to our accommodation across the road from the American Museum of Natural History. The intention was to spend the days visiting places of research interest near Midtown Manhattan (such as the United Nations Headquarters). I also wanted to spend some time at the New York Historical Society, which was close to where we were staying. However, upon visiting the society on our first full day in New York, I was faced with an unwelcoming attitude that made me rethink the value I would draw from that institution. And again, unexpected costs were involved. As a way to buy some time, I marched straight out and headed across the road and into Central Park. At this point I was feeling a little deflated. Traumatized by the experience at Penn Station, I was not ready to tackle the crowds again; and now I was on the verge of dismissing the research with the Historical Society. As I walked through Central Park and immersed myself in the landscape and observed the people coming and going, I became aware of an alternative plan that was forming in my head. I began to think that I could shift the focus of my research to an examination of the utility of green spaces in cities. I was already involved in similar work back home and this park would present a valuable contrast. So, with some doubts and more than a little trepidation, I opted to change my plans with the result that the week in New York was spent with daily visits to Central Park. This allowed me to indulge my interests in nature and to pursue a rich alternative vein of research. I reflected on how inspiring, educational, enriching and therapeutic the time in Central Park was. As a bonus, I was able to share more time with my husband as we explored the natural landscape of the park, observing its visitors, workers and inhabitants. We even spent time investigating the geological history of the area. In the end, I had a more relaxed approach to our stay in New York and a realistic view of what could be achieved. I am reminded of Friedman's reflections on his sabbatical year.

While Friedman (2018) did not specifically refer to the sabbatical as a pilgrimage, he commented on 'the meaningful and transformative' nature of the sabbatical year, both personally and professionally. Interestingly, he referred to it as 'a priceless gift' that allows one to 'spend quality time in sharing memorable experiences with one's family (p. 656). He supports this comment with reference to a study of 258 faculty members who had reported an increase in their 'sense of wellbeing' as a result of having taken a sabbatical. Friedman (2018) stresses that while it is important to plan appropriately and to have a sound idea of the goals and timeline, sabbaticants should 'remain flexible and open to new ideas' (p. 657). He elaborated on the importance of taking the sabbatical away from the institution (preferably 'offshore') as an essential ingredient for a transformative and productive sabbatical. Such an exposure to new environments presents opportunities for meaningful experiences that 're-define one's professional identity and interests' (p. 657). Friedman also highlighted the continued enrichment from sabbatical experiences upon return, and the value in sharing knowledge and insight gained with colleagues, friends and family.

Sabbatical, Pilgrimage and Transformative Travel

In 2017, Professor Ann Gallagher (2017) of Surrey University wrote in her blog that 'sabbatical provides a time and space to read and think and listen and learn from others' (n.p.). Charney (2019) concurs and adds that sabbatical allowed her to, 'Step back. Breathe . . . Travel in liminal mind spaces . . . identifying possibilities. Focus . . . Interact with public library colleagues' (p. 2). For me, the sabbatical did provide the space and time to engage with meaningful and enriching activities, and the travel added a dimension that made me think of the sabbatical as a pilgrimage. It is widely understood that travel presents opportunities to 'broaden the mind' (e.g. Morgan, 2010; Lean, 2012). This notion is evident in the intentions of travel as early as the mid-17th century when the Grand Tour was traditionally taken by privileged young men to travel around Europe as a way to enhance their education. Today, the same value is realized in study-abroad programmes, study tours and the 'gap year', serving to build academic, social and personal benefits. Traditionally, much of this developmental travel is focused on engendering encounters with other cultures in order to expand intercultural awareness. Increasingly, there has been promotion of 'an intimate engagement with . . . the "more-than-human" world of nature and landscape' (e.g. Morgan, 2010) and a realization that transformative travel experiences can include both encounters with other cultures and engagement with environments in a holistic and connected way. Morgan (2010) writes at length about how 'travel represents a potentially fruitful vehicle for transformative education and learning' (p. 247). There is an emerging school of thought that centres on 'transformative travel'. Morgan (2010) elucidates by defining travel as 'journeying' and 'actual physical movement through space over time', and transformation as 'psychological development or growth' of the 'inner' self. My time in Boston was exemplary of the transformation that is possible through travel. As I looked outward into the novel and different environments, I simultaneously found myself gazing inwards and reflecting on myself. For example, I considered the prominence in my life of my work as an academic countered against my family, and the lack of balance there.

I had not set out with the intention or any anticipation of being transformed. Rather, I thought that my sabbatical journey would be one of networking with other academics and learning about places and people from another country. I didn't expect that the trip would present anything that would be overly unfamiliar or foreign. After all, I had already met and come to know the 'other' during my visit to the USA in my youth. Patterson (2015, p. 347) talks of 'habits of mind' that condition travellers' interpretations of their experiences. He presents the example of a participant in his study who visited China, and, in her mind, she expected to see 'a China of the past, untouched by globalization and modernity' (p. 363). Instead she was met by a China that was 'modern and in many ways similar' to the environment back in the USA. In my case, my 'habits of mind' were expecting a culture very similar to the one in Australia – there were similarities that made the travel easier, e.g. the language; but there were often subtleties that presented unease or a sense of discomfort. At other times, the difference(s) was palpable. It was these moments in my travels that infused and shaped my experiences and influenced and reshaped my own 'habits of mind'; and so it was that as we progressed through the USA I began to realize that this journey was deeply personal, enlightening, enriching and emotive.

It was a journey of dynamic transformation. For example, I repeatedly found myself thinking about 'excessive consumption' compared with minimalism. Wingfield (2017) wrote that sabbaticals were a time to lose some of one's 'creature comforts' and that being 'minimalist' is part of the challenge. Adopting such an outlook emancipates sabbaticants to then 'focus on research and writing' (p. 1). It also often means having to 'live frugally'. This may sound harsh, but I, like Wingfield (2017), learned to appreciate, 'to make do with minimal home facilities'. I began to realize that even the clothing I needed for travelling could be severely curtailed. I deliberated upon my level of consumerism and resolved to further liberate myself from over-consumption. Not that I am overly materialistic, but during my journey I found myself thinking at length of the imbalance in consumerism across society, the excesses of our consumption and the need for a more ethical outlook. These are issues that do concern me every day, but I found that the journey shifted my thinking and practices to being even more mindful of unnecessary consumption. So it was that on many levels I found myself beginning to think differently about various aspects of my life. Most noticeable was the shift in my perspective on the importance of each aspect. For example, I realized that I needed to spend more time on myself and my fitness and not feel guilty about it. Walking is important to me, and I began to think that I should make time to achieve my set goal of daily steps and not compromise this goal for what often seem urgent work matters. Being healthy and fit were key to achieving the work expectations. The enlightenment was gradual and subtle, but it was transformative and enduring. Even now I make time every day for 'my steps'. As I became conscious of this extraordinarily enriching change, I began to think of my sabbatical as 'My USA Pilgrimage'.

The kaleidoscope of experiences created a melting-pot that led to self-discovery and healing. These outcomes are best captured by our stay in Boston. By the time we reached Boston, I was ready to immerse myself in some indulgences. Attending the Academy of Management (AOM) conference was the anticipated highlight of the sabbatical for me. I had waited a long time, over a decade, to have the opportunity to attend an AOM conference and at last the time had arrived. The conference exceeded my expectations. It was a watershed of engaging debates, spirited conversations, judicious networking and meaningful learning. Beyond the conference, Boston itself grounded me. While attending the conference, I loved my daily walk through the Boston Common and the Boston Public Garden as I made my way from the hotel to the Hynes Convention Center. In the afternoons, my husband would meet me on the Boston Common and we would explore the area, walking off in different directions each day. We also had several days before and after the conference to explore Boston. Our wanderings led us to discover some amazing parts of Boston. The highlight for me was a fortuitous finding of the Italian District in the North End. My taste buds were titillated by the myriad restaurants, cafés, bakeries and delicatessens. Conversations caught in passing brought me back to my childhood spent in an Italian enclave in Melbourne. The narrow streets, laid with cobblestones, added to the nostalgic ambience. As we strolled through the neighbourhood, I was enthralled by the vibrancy, authenticity, and history. As a finale to our stay in Boston, we met a First Nation American gentleman on our last day. He was part of the concierge team at the hotel we stayed in and we were fortunate to have a lengthy discussion with him about his memories of Boston, how it had transformed, and proposed future changes. This conversation segued into our departure from Boston and back on the train as we journeyed to Syracuse.

Space and Time: Travelling Along on the Train

As mentioned, travelling on the train provided time and space to 'slow down', ponder and reenergize. The limited internet access for long stretches of track meant that connectivity to the outer world and my dependence on it was often lost; and so the confined space of the carriage encouraged contemplation. Both Morgan (2010) and Lean (2012) reflect on what it is about places that triggers transformation. Lean (2012) says: 'While a traveller may physically remove their body from a specific geographic location, contemporary and historic flows of people, ideas, information, objects, memories and symbols create mobile spaces, places, landscapes and identities, where both familiarity and difference abound' (p. 151). He adds:

> All elements within place (both mobile and static), create symbolic representations, any of which could trigger memories, reflection, imagination, conversations and performances that may maintain and/or transform a particular reality during physical travel . . . Even when visiting destinations perceived as vastly different from their own, travellers often frequent the same locations (especially through tourism), creating 'enclaves' around particular attractions that host the bulk of accommodation and traveller services, and in which travellers may find themselves in 'familiar' surrounds . . . These zones are often separate from the reality of local inhabitants outside of them, or at least offer complex hybridisations. Some participants believed that their transformation was only possible through making a conscious effort to move outside of these 'bubbles' and what they believed to be a familiar performance of 'the tourist'.
>
> (Lean, 2012, p. 165)

When travelling throughout the USA, there were times when the familiarity of certain things (e.g. Starbucks, foods, the English language) made me feel like I was back home. At other times I felt that I was a tourist among a mass of visitors that were truly removed from 'the authentic'. For example, the French Quarter in New Orleans was so packed with tourists and offerings for tourists that it felt like there were exhibitions and performances on show that were not truly genuine. It was only when we went off exploring on our own to some of the areas less frequented by tourists that I felt like I was getting glimpses of the real New Orleans. For example, we visited the quaint Beckham's Bookshop, which has been operating since 1967 and is one of the oldest bookshops in New Orleans. It sold antiquarian and rare books as well as books by local authors. The bookshop is locally owned and operated and has a resident cat. Bookshops were interesting because they often exposed us to what lay beneath the city's tourist veneer.

Likewise, travelling on the train provided greater opportunity to meet people other than tourists. We were all travellers – some international travellers and many domestic travellers, the latter mostly commuting from one area of the USA to another and not necessarily on holiday. On one occasion we met two teenage sisters who were travelling from their grandparents' town back home. They shared with us many insights into their everyday lives, including that their father had recently passed away. Thus, the bounded space of the train opened up opportunities for interesting conversations that enriched our experiences and deepened our understanding of American culture.

Cavanaugh (2008), p. 346) spoke of the quest for transcendence and 'the spiritual quest that lies behind attempts to transcend everyday life' as being as fundamental to the modern authentic traveller as it was to earlier pilgrims. He notes that medieval pilgrimage was the forerunner of modern travel and that both traditional pilgrims and modern travellers seek 'transformation of the self' (p. 349). Cazaux (2011) also explores the history of pilgrimage, but with his focus on the Camino de Santiago he points out the journey is one of self-discovery, of learning and of gaining freedom. In this sense, pilgrims undertake a physical journey that simultaneously parallels their inner spiritual journey.

As we physically circumnavigated the USA via train, an inner parallel journey was underway. For example, travelling across the Rockies was an extraordinary experience. The spectacular panoramic views stirred admiration and respect for the environment and heralded thoughts and emotions of our obligations as global citizens of this planet. I thought of my own endeavours to minimize my own footprint and what other practical steps I could take, especially when I returned home.

Perhaps one of the starkest and most revealing moments was on our last night on Amtrak's Sunset Limited as we travelled from New Orleans to Los Angeles. During the night, as we were crossing the south-western deserts in Arizona, I woke up and realized that the train was stationary. I had a look out the window into the partial darkness. Some illumination was coming from what appeared to be a farmhouse in the distance. I checked the time and went back to sleep. Sometime later, I woke again and noticed that the train was again not moving. I looked out the window and realized that we were stationary in the same spot as previously. I checked the time and noted that almost two hours had passed. I lay back down and pondered on my satisfaction with our epic journey. Two words came to mind: self-efficacy and action. I knew that I was coming away from this trip with a greater belief in my own ability to take action for the things I care about (e.g. minimizing waste, work-life balance, and equity in education). As my mind wandered, I realized that I had reconnected with my previous practices of mindfulness. It was delightful to be able to be present in the moment, without any pressures of deadlines or urgency. I reminded myself that I should engage daily with such practices of mindfulness. A good start would be returning to my daily yoga. Being in the train carriages had been physically restricting but, even so, intensely liberating and empowering (Fig. 18.2).

Fig. 18.2. Inward, outward, tracking through the landscape. (Photo: the author)

Sabbatical, Pilgrimage and Transformative Travel: Analysis of the Literature

In order to better understand the themes extant in the literature, a 'word-frequency cloud' was generated for literature pertaining to each of the respective topics: pilgrimage, sabbatical and transformative travel. Fig. 18.1 shows the resultant word-frequency cloud.

The most frequent words emerging from the literature included, in order of descending occurrence: pilgrimage, experience, pilgrim, transforms, research, travel, place, sacred, time, life, cultures, learning, traditions, journey, meanings, spirituality and changing. Many of these themes have appeared in the narrative of this chapter as it sought to understand how 'the sabbatical' is a pilgrimage, especially in relation to transformation of the traveller.

The literature showed that travel, including that for pilgrimage and/or sabbatical, has educational benefits. It is said that, 'Traversing a foreign space, and transforming one's understanding of self and world as a result' has immense transformative potential (Patterson, 2015, p. 502). It can result in individuals shifting their worldviews, expanding their personal horizons, and increasing their awareness of the world in which we live. Pratt (2008) contended that learning through travel is a 'dynamic process of potentialities produced at the nexus of multiple places and spaces, presents and pasts, and individual and collective histories' (p. 109). Patterson (2015) suggested that travel resulted in a complex array of outcomes. Travel has as the potential to precipitate transformation and international travel pushes individuals outside of their comfort zone. Wingfield (2017) referred to the motivational quote: 'Outside of the comfort zone is where the magic happens' (p. 2). My sabbatical journey took me to places that were outside my comfort zone. In doing so, I was rewarded with enriching and enduring experiences that made me rethink so many aspects of myself. Each city we visited, each place we occupied, each person we met – all were a rich source of inspiration and learning that formed threads in the web of transformation that was created as we journeyed through the USA.

Conclusion

This chapter has shown that the sabbatical provides a lens for understanding the pilgrimage experience. It has clarified that, in tune with the nature of pilgrimage, the sabbatical allows for transformative potential in travellers as they cross borders and come to know other people, environments and societies. It has demonstrated that it can lead to transformative learnings that build aspects of our being such as self-awareness, self-efficacy, open-mindedness and mindful consumerism. Such changes are powerful and enduring as authentic travellers on sabbatical learn about other worlds and peoples, themselves and life, in ways that precipitate transformations in personal values and identity. The experiences '[seep] into the marrow of one's bones' as Ridsdale (2014, p. 68).

My sabbatical sojourn parallelled a pilgrimage because I felt intensely, markedly changed by my travels during this time. I noticed differences in my thinking, my perspectives and my being. It was subtle and protracted but noticeable by the time we were in New York. It is difficult to pinpoint when the transformation began. If I were to reflect concertedly, I would say that it happened the minute we landed in Honolulu and were immersed into a welcoming hospitable culture with a different and endearing way of doing things. The stopover in Honolulu was unplanned and the result of our flight from Sydney to San Francisco being cancelled after check-in. So the detour had not held any lengthy preconceived expectations and we actually welcomed the opportunity to break up the long flight from Australia to San Francisco. We stayed within the airport but already I began to experience an internal stirring to learn about peoples and environments that were unfamiliar.

This heightened state of awareness and receptivity accompanied me throughout the trip, but it remains a challenge to pinpoint what it is that has changed within me. I know that it was more palpable as the journey unfolded. Each city brought new experiences and immersed us in 'new ways of knowing and being'. Most notable times of deep reflection were when I was on the train. In my mind, the slow and rhythmic pace of the train travel created space and time to think, to reflect, to dream and to be me.

As we travelled from west to east, a diversity of landscapes unfolded, exposing us to contrasting environments that titillated and tested our emotions, captured minds and wrestled with our thoughts. The natural landscapes presented rich and contradictory images that engraved striking memories of the USA. The built environments added dimensions that constructed cameos of a complex world composed of opportunity and wealth on the one hand, and poverty, disrepair and neglect on the other. I have seen such contrast in Australia but the stark reality of its presence in the USA shook my being. My worldview was under siege.

In the train, we gained access to the backyards of homes and cities. We travelled along landscapes that are not generally marketed in brochures and websites. We were exposed to the heartbeat of the land and its people – 'the good, the bad, and the ugly'. It was confronting. It was humbling.

The passing scenery enthralled me. I didn't want to take my eyes away for a second in case I missed something. At the same time, I was consumed by trying to capture every aspect of it on my camera or in my notes. But at times the scenery was monotonous. An endless wall of tree growth obscured the world beyond the tracks, directing my attention inward. Within the confines of the train, another world pulsated, a world which we explored and engaged with intermittently as we ventured to the observation or dining cars or wandered in the passageways to stretch our legs. Mostly we retreated to our own room. At times, conversation enveloped us; at other times we revelled in our own thoughts.

It was in these moments of granted silence that the solitude of my thoughts entwined with the passing landscape and I revelled in contrasting what I saw out the window with what I knew of 'Life, the Universe, and Everything', as Douglas Adams (1979) sought in *The Hitchhiker's Guide to the Galaxy*. The '42' I knew was becoming obscured and elusive as I sought a new world order internally.

References

Adams, D. (1979) *The Hitchhiker's Guide to the Galaxy*. Pan Books, London, UK.

Blake, G. and King, R. (1972) The Hijaz Railway and the pilgrimage to Mecca. *Asian Affairs* 3(3), 317–325. DOI: 10.1080/03068377208729639.

Braun, I. (2017) Barristers on the move: slow boat sabbatical. *Victorian Bar News* 162, 50–51.

Cavanaugh, W.T. (2008) Migrant, tourist, pilgrim, monk: mobility and identity in a global age. *Theological Studies* 69(2), 340–356. DOI: 10.1177/004056390806900206.

Cazaux, F. (2011) To be a pilgrim: A contested identity on Saint James' Way. *Tourism* 59(3), 353–367.

Charney, M. (2019) Sabbatical report 2018 – climate change conversations in libraries. Available at: https://works.bepress.com/charney_madeleine/100/(accessed 5 August 2020).

Costa, L.M. (2018) Sabbatical as sacred time: Contemplative practice and meaning in the neoliberal Academy. *Gender & Women's Studies Faculty Scholarship* 5.

Dauphin, C., Ben Jeddou, M. and Castex, J.-M. (2015) To Mecca on Pilgrimage on Foot and Camel-back: The Jordanian *Darb al-Hajj*. *Bulletin for the Council for British Research in the Levant* 10(1), 23–36. DOI: 10.1179/1752726015Z.00000000029.

Ellis, C. (1999) Heartful Autoethnography. *Qualitative Health Research* 9(5), 669–683. DOI: 10.1177/104973299129122153.

Friedman, S.L. (2018) A sabbatical: the gift that keeps on giving. *Cellular and Molecular Gastroenterology and Hepatology* 5(4), 656–658. DOI: 10.1016/j.jcmgh.2018.01.010.

Gallagher, A. (2017) Arrival in Alabama: sabbatical week 1. Available at: https://blogs.surrey.ac.uk/ethics-in-care/2017/09/30/arrival-in-alabama-sabbatical-week-1/ (accessed 15 November 2019).

Higgins, L. and Hamilton, K. (2019) Therapeutic servicescapes and market-mediated performances of emotional suffering. *Journal of Consumer Research* 45(6), 1230–1253. DOI: 10.1093/jcr/ucy046.

Husemann, K.C. and Eckhardt, G.M. (2019) Consumer deceleration. *Journal of Consumer Research* 45(6), 1142–1163. DOI: 10.1093/jcr/ucy047.

Lean, G.L. (2012) Transformative travel: a mobilities perspective. *Tourist Studies* 12(2), 151–172. DOI: 10.1177/1468797612454624.

Mezirow, J. (1991) *Transformative dimensions of adult learning*. Jossey-Bass, San Francisco, USA.
Morgan, A. (2010) Journeys into transformation: Travel to an 'other' place as a vehicle for transformative learning. *Journal of Transformative Education* 8(4), 246–268.
Moustakas, C. (1990) *Heuristic Research: Design, Methodology and Applications*. Sage, Newbury Park, California.
Patterson, T. (2015) The transformative power of travel? Four social studies teachers reflect on their international professional development. *Theory & Research in Social Education* 43(3), 345–371. DOI: 10.1080/00933104.2015.1064504.
Pearce, P.L. (2012) The experience of visiting home and familiar places. *Annals of Tourism Research* 39(2), 1024–1047. DOI: 10.1016/j.annals.2011.11.018.
Pratt, M.L. (2008) *Imperial Eyes: Travel Writing and Transculturation*, 2nd edn. Routledge Press, London, UK.
Ridsdale, L. (2014) *Finding My Feet, Finding My Voice: An Autoethnography of Pilgrimage and Song on the Camino de Santiago. MA thesis*. University of Limerick, Limerick, Ireland.
Schmitt, X. (2003) *Customer Experience Management*. John Wiley & Sons, Hoboken, New Jersey.
Smelser, N.J. (2009) *The Odyssey Experience*. University of California Press, London, UK.
Turner, V. and Turner, E. (1978) *Image and Pilgrimage in Christian Culture*. Columbia University Press, New York.
Van Gennep, A. (1960) *The Rights of Passage*. University of Chicago Press, Chicago, Illinois.
Van Laer, T. and Izberk-Bilgin, E. (2019) A discourse analysis of pilgrimage reviews. *Journal of Marketing Management* 35(5-6), 586–604. DOI: 10.1080/0267257X.2018.1550434.
Wingfield, B. (2017) The privileges and opportunities of a research sabbatical. *South African Journal of Science* 113(9/10), 1–2. DOI: 10.17159/sajs.2017/a0234.

19 Contemporary Perspectives of Pilgrimage

Daniel H. Olsen[1]* and Dallen J. Timothy[2]
[1]*Brigham Young University;* [2]*Arizona State University*

Introduction

Spirituality and religion have been key indicators of cultural identity in every society since the beginning of the Anthropocene. As a result, sacred sites and rituals are among the most ubiquitous cultural expressions throughout the world. Pilgrimage and religious tourism to these sites today is a clear manifestation of the broader phenomenon of heritage tourism, which, in part, comprises the tangible and intangible faithscapes and practices of pilgrimage that people inherit from the past and utilize and value in the present (Graham *et al.*, 2000; Timothy, 2021). Belief systems, religious doctrines, music traditions, rituals and celebrations, holy scriptures, tangible artifacts, and sacred buildings are therefore important elements of faith heritage at different scales, ranging from the global to the personal.

Although pilgrimage to sacred sites has a long and glorified history, in recent years, post-modern views of pilgrimage have radically changed the types of travel that are considered pilgrimage, as well as what constitutes a sacred site. Rather than disappearing in a post-modern, secularized world, pilgrimages have seen a resurgence in terms of the sheer number of people who engage in religiously motivated travel (Feldman, 2017). Indeed, pilgrimage is viewed as a subset of religious tourism (Timothy and Iverson, 2006), which niche market has greatly expanded over the past three decades, leading to a splintering of this niche market as tour agencies begin to specialize in catering to the specific needs of pilgrims and religious tourists from a variety of faith traditions (Olsen and Ron, 2013). However, as found in other tourism niche markets, there are several negative impacts that can occur, ranging from commodification to overtourism to conflicts over ownership, interpretation and management of sacred sites.

The purpose of this chapter is to highlight contemporary perspectives of pilgrimage in the (post-)modern world, particularly focusing on changes that have taken place in pilgrimage travel due to the 'democratisation of travel' (Richter, 2003, p. 340) and the use of *pilgrimage* and *pilgrim* as a metaphor to describe other types of mobilities and experiences. After discussing the changing definitions and meanings of pilgrimage in a post-modern and post-secular era, different types of pilgrimage destinations and motivations for and uses of pilgrimage travel, attention turns to several issues that affect pilgrimage today, including the politics of sacred-site management, over-/undertourism, and heritage designation.

*Corresponding author: dholsen@byu.edu

Defining Pilgrimage in a Globalizing and Post-secular World

Stemming from the Latin word *peregrinatio*, pilgrimage entails, at a basic level, a 'journey to a distant sacred goal' (Barber, 1991, p. 1). Traditionally, pilgrimage has been performed by religiously motivated persons who, as part of a religious tradition, travelled to places deemed sacred by their faith to participate in worship rituals either for initiatory purposes, to ensure a happier afterlife, for educational purposes, or to petition for divine intervention in their daily lives (Morinis, 1992). These sacred spaces ranged from single nodal sites such as cathedrals and temples to entire cities and archaeological sites. Other sacred locales revolve around features of the natural landscape, such as islands, mountains, rivers, lakes and groves of trees (Timothy, 2012; Olsen, Forthcoming). Sacred sites are also linked together through organic and purposive pilgrimage routes (Timothy and Boyd, 2015), such as the Camino de Santiago de Compostela in western Europe and the Kumano Kodo in Japan, which link several religious sites together while, at times, being anchored by a final spiritual destination (Jimura, 2016; Olsen and Trono, 2018; Timothy and Olsen, 2018). These religious nodes and lines have, historically, led to entire economic circuits and cities being built and organized to cater to the wants and needs of religious pilgrims. Many of these cities and routes continue to serve as pilgrim paths and pilgrimage centres today (Greif, 2006; Leppäkari and Griffin, 2017; Timothy and Olsen, 2018).

Pilgrimage, like tourism, is an identity-building exercise. In many ways, pilgrimage is a very personal activity that helps congeal one's individual identity and solidify one's place in a faith-based society. It also encourages participants to develop their spiritual selves in partnership with deity, even when undertaken in the company of others. The enlightenment received, the forgiveness sought, and the intimacy with deity that people seek are extremely private, and through the act of pilgrimage become part of the personal spiritual heritage and identity of the pilgrim (Timothy, 1997; Belhassen *et al.*, 2008; Oviedo *et al.*, 2014; Collins-Kreiner and Wall, 2015; Amaro *et al.*, 2018; Rahimi and Eshaghi, 2019). Pilgrimage, however, is also a part of community building, particularly the faith community, where leaders of religious organizations encourage believers to visit sacred sites, using pilgrimage as a tool to revitalize and maintain religious and spiritual group values and identities (Olsen, 2019a). Indeed, these pilgrimage-cum-revitalization efforts are structure-affirming (Olsen, 2012a, b; Albera and Eade, 2015) and are designed to solidify both individual and collective religious identities (Olsen, 2019a).

While exact numbers of people who engage in pilgrimage are unknown, some scholars have suggested that between 200 and 600 million people visit a religious site each year (UNWTO, 2011; Timothy, 2021). Some pilgrimages, such as the Kumbh Mela in India and the Hajj in Saudi Arabia, are some of the largest religious gatherings in the world, with millions of people participating in these events. The number of Kumbh Mela participants, for example, continues to grow. In 1954, approximately 5 million pilgrims attended. In 1989, some 15 million people participated; some 70 million attended in 2001, and in 2013, a record-breaking 120 million pilgrims were estimated to have participated in the Allahabad gathering (Economic Times, 2018). Between 15 January and 4 March 2019, Mela celebrations in Allahabad were estimated to have attracted approximately 150 million Hindu worshippers and other visitors from around the world (Hacohen, 2019).

Changes in transportation, access and human mobility, as well as the development of more tourism-oriented services for pilgrims (e.g. accommodations in luxury tents, portable toilets, and mobile phone apps to help guide the pilgrims in the destination and through the rituals), have driven an expansive growth in Kumbh Mela attendance. Also, owing to its role as the keeper of the holy places in Islam, Saudi Arabia draws approximately 18–20 million religious travellers each year, 74% of which are *umrah* pilgrims and 16% Hajj pilgrims. Approximately 2–3 million Hajj pilgrims descend on Mecca each year during the holy month *zul-Hijjah* to participate in sacred rituals, which each Muslim is required to undertake at least once in his/her lifetime (Timothy and Iverson, 2006; Zamani-Farahani *et al.*, 2019).

The increase in the practice of pilgrimage might be surprising to some, considering that for decades scholars have argued that globalization and industrial capitalism would serve as the death knell of religion and, by extension, religious pilgrimage (Luckmann, 1963; Berger, 1967). This 'secularization thesis' held that as globalization unfolded it would be accompanied by the irrelevance of religion in public and political life. Bauman, 1996 also argues that the 'chronological succession' or genealogical shift from pilgrimage (as emblematic of a previous epoch) to tourism would lead to the disappearance of pilgrimage as a type of mobility.

However, this 'thesis', which came to prominence in the 1960s, has not come to full fruition, as the decline of the importance of religion in the public sphere and in public life seems to have been geographically limited to western Europe (Turner, 2005). As well, there has been a 'post-secular' turn in many fields of scholarship, including international relations, where scholars are noting the resurgence of religion in public and political life (e.g. May *et al.*, 2014). Indeed, globalization has in many ways transformed and intensified the importance of religion around the world (Gauthier *et al.*, 2013), as manifested in religiously motivated terrorism (Stump, 2000), the rise of spiritual movements (Olsen, 2019a), diasporic transplanting of ethnic religious traditions (Prorok, 2003; Olsen, 2016, 2019a), and increasing commodification of religion (Kitiarsa, 2010; Ornella, 2013).

In addition, in the same way as the relationship between politics and religion has in many cases become blurred (May *et al.*, 2014), the relationship between pilgrimage and tourism has also become blurred (Kaelber, 2006). This is because of the democratization of travel and advances in transportation and telecommunication technologies, leading to increased opportunities for people of many socio-economic levels to travel for religious purposes. As such, pilgrimage has seen a resurgence of popularity, with increasing numbers of people travelling to religious sites (Bremer, 2005; Feldman, 2017; Olsen, 2019a; Timothy, 2021). Because of its use of tourism infrastructure and because of its economic potential, pilgrimage travel has become strongly intertwined with the tourism industry, in part because of the economic potential of pilgrimage, which, like the overall number of pilgrims, is not known (Olsen, 2019b; Ron and Timothy, 2019). This has led some tourism scholars to note the structural similarities between pilgrimage and tourism, wherein pilgrims and tourists leave home, use similar travel infrastructure, experience a sacred 'high' or state of hyper-reality, and then return home (MacCannell, 1976; Eco, 1983; Jafari, 1987; Graburn, 1989; Holmberg, 1993; Gupta, 1999; Fleischer, 2000). This has led, in part, to tourism industry leaders viewing pilgrims as a subniche of the tourism market (Timothy and Iverson, 2006). As such, religion has become a stronger, not a weaker, motivation for travel in the 21st century.

The Effects of Globalization on Pilgrimage

The contemporary world is largely a product of globalization, where people, goods, services, money and technologies flow across an increasingly borderless world. As a part of these increasing flows, pilgrimage, as an ancient form of mobility, has been modified in part because of new conceptualizations of religion's place in the new global economy. Olsen (2019a, pp. 275-279), for example, highlights three of these changes and challenges. First, globalization and the rise of increasing mobilities has led to the diasporic transplanting of religious and spiritual traditions. As members of cultural groups migrate to different regions of the world, many of them have a need to adapt and recover their religious traditions. This is done in several ways, including recreating sacred spaces, establishing new sacred spaces and pilgrimage traditions, and engaging in pilgrimage or 'sacred remittances' back to their religious homelands (Garbin, 2019; Prorok, 2003; Olsen, 2016).

Second, increasing numbers of religious site visitors have led to several management issues, particularly where religious and tourism spaces overlap. These issues include managing visitor flows, interpreting sacred sites in relation to the wide range of visitors' expectations and activities, and maintaining the physical infrastructure of sacred places.

Third, globalization has led to increased commodification and hybridization of religion and spirituality. While religious groups have long commodified religious rituals, sites and ceremonies to maintain a strong economic-material base from which to survive (Olsen, 2003), globalization has resulted in 'cultural deregulation' (Gauthier *et al.*, 2013), where high and low culture have merged, and where everything can be viewed as a commodity. This is evidenced by the rise of 'spiritualized progressivism' (McLennan, 2010), as seen in various New Age movements, new hybridized forms of religion and spirituality, and the spiritual tourism niche market, all of which 'allow consumers greater freedom and choice in how they form their religious and spiritual selves' (Olsen, 2019a, p. 278). This has led to pilgrimage being redefined as any journey 'undertaken by a person in quest of a place or a state that he or she believes to embody a *valued ideal*' (Morinis, 1992, p. 4; italics added). As such, travel to a baseball game, to Elvis Presley's Graceland mansion, or to a patriotic monument, might be considered a pilgrimage if it is deemed as such by an individual or group (Campo, 1998; Alderman, 2002; Gammon, 2004). This hybridization of religion and spirituality has also arisen due to increased 'techno-spiritual practices' (Bell, 2006), where social media, cell phones, the internet and other technological advances heighten the pilgrim experience and how these experiences are shared.

In addition to these three changes and challenges to pilgrimage, here the authors highlight other issues and challenges that arise from the processes of globalization, including pilgrimage as heritage and World Heritage status, overtourism and forced undertourism.

Pilgrimage sites as world heritage

From a broad perspective, religion and spirituality are the foundation of much global heritage. Some religious structures and sacred locales have become popular tourist attractions beyond the spiritual needs of pilgrims, such as cathedrals in Europe, Islamic mosques in north Africa, sacred mountains and shrines in Japan, sacrosanct rivers in south Asia, and historic temples in Thailand. Religious archaeology and other faith-based built environments provide some of the most valued and visited heritage resources in the world (Koren-Lawrence and Collins-Kreiner, 2019; Farra-Haddad, 2020). In fact, many of the cultural properties on UNESCO's World Heritage List are either religious sites or are somehow associated with a faith, a religion, or some form of spirituality (see Table 19.1). Similarly, many intangible elements of religious heritage are important cultural resources that are commemorated by communities throughout the world. UNESCO's Representative List of the Intangible Cultural Heritage of Humanity is replete with examples of living culture that faith groups, communities, states and global society at large value as important and deem worthy of protecting and promoting (see Table 19.2).

Although it has received substantial criticism for procedural political biases and nepotism (Meskell, 2015; Meskell *et al.*, 2015), the UNESCO 'brand' is now widely recognized as a trademark that ensures a certain level of heritage authenticity, sustainable management practices and global exceptionality (Poria *et al.*, 2011; Adie, 2017; Timothy, 2021). Heritage branding is likewise widespread within the realm of 'religiouscapes' (McAlister, 2005, p. 251) and faith practices. Faith-based heritage is so important in some countries that they utilize it as a tourism brand (Fig. 19.1). The Taj Mahal in India, for instance, which has strong religious connections to Islamic history and Muslim architecture, has become an iconic 'brand' for that country's tourism industry. The same is true with the rock-hewn churches of Lalibela, Ethiopia, the Angkor temples of Cambodia, and Machu Picchu in Peru.

Many developing countries seek to inscribe as many cultural properties and elements of living heritage as possible on UNESCO's heritage lists in the ill-conceived belief that such a brand will increase tourist arrivals. While some studies have suggested that the UNESCO brand can be successful in increasing visitation (Geary, 2008; Yan and Morrison, 2008), other studies find that while the brand may increase the visibility of a certain cultural location on the global tourism map, it does not necessarily increase visitation rates (Hardiman and Burgin, 2013; Adie *et al.*, 2018; Timothy, 2021). In the religious context, Geary (2008) and Fyall *et al.* (2020) present evidence that religious tourism and other pilgrimage forms at Bodh Gaya, India, have grown since the city's holy sites were placed on the World Heritage List. As such, there are conflicting findings or mixed results with regard to the UNESCO brand increasing visitor numbers at WHSs. Overall, however, heritage is

Table 19.1. Examples of religious sites and structures on the World Heritage List. (From UNESCO, 2020b)

Country	Religious World Heritage Site	Year inscribed	Religious importance
Argentina	Jesuit Missions of the Guaranis	1983	Remains of Jesuit missions among the Guarani Indigenous people
Bangladesh	Historic Mosque City of Bagerhat	1985	Early example of Muslim architecture in Bangladesh
Bolivia	Tiwanaku Spiritual Centre of the Tiwanaku Culture	2000	Spiritual and political centre of pre-Hispanic culture
Cambodia	Temple of Preah Vihear	2008	Ancient Hindu temple and example of Khmer architecture
China	Temple and Cemetery of Confucius	1994	The cemetery and family home of Confucius, the great spiritual thinker
Ethiopia	Rock-hewn Churches of Lalibela	1978	Thirteen-century monolithic cave churches in the 'New Jerusalem'
India	Buddhist Monuments at Sanchi	1989	Oldest Buddhist sanctuary on earth dating from the 2nd century BC
Israel	Bahá'i Holy Places	2008	Holy shrines and most sacred sites in the Bahá'i faith
Japan	Itsukushima Shinto Shrine	1996	Sacred Shinto locale since at least the sixth century AD
Lebanon	Baalbek	1984	Ancient Phoenician city and centre of worship for Phoenicians, Greeks and Romans
Nepal	Lumbini	1997	Birthplace of Buddha, commemorated with many temples and shrines
Palestine	Birthplace of Jesus	2012	Church of Nativity built over the traditional birthplace of Jesus
Spain	Santiago de Compostela	1985	One of Europe's most important Catholic shrines and pilgrimage centres
Turkey	Ephesus	2015	Greek and Roman settlement that played an important role in the early spread of Christianity
USA	San Antonio Missions	2015	Spanish missions built in the 1700s to help colonize the area for Spain and convert the natives to Christianity

becoming an increasingly popular asset for global tourism, and countries such as China, Italy and Mexico are keen to utilize their UNESCO-branded World Heritage Sites (WHSs) as key tourist attractions. Whether or not the WHS brand automatically increases visits, there are increasing numbers of tourists at UNESCO-branded attractions. With the inscription of additional religious properties and spiritual practices on UNESCO's heritage lists, what was once in many cases local faith heritage has now become global religious heritage, which is now consumed by the masses on an international scale.

Overtourism

Whether on the UNESCO World Heritage List or not, evidence suggests that visitor numbers to religious sites are growing. This is part-and-parcel of continuing growth rates of both

Table 19.2. Examples of religious practices on the Representative List of the Intangible Cultural Heritage of Humanity. (From UNESCO, 2020a)

Country	Intangible religious heritage	Year inscribed	Type of religious heritage
Algeria	Annual Pilgrimage to the Mausoleum of Sidi 'Abd el-Qader Ben Mohammed	2013	Pilgrimage
Algeria	Sbuâ, Annual Pilgrimage to the Zawiya of Sidi El Hadj Belkacem in Gourara	2015	Pilgrimage
Belarus	Celebration in honour of the Budslaŭ icon of Our Lady	2018	Pilgrimage
Bolivia	Festival of the Santisima Trinidad del Señor Jesús del Gran Poder in La Paz	2019	Religious festival
Brazil	Círio de Nazaré in the City of Belém, Pará	2013	Pilgrimage & procession
China	Mazu belief and customs	2009	Religious ceremonies & folk practices
Colombia	Holy Week processions in Popayán	2009	Pilgrimage & procession
India	Kumbh Mela	2017	Pilgrimage
Iran	Qālišuyān Rituals of Mašhad-e Ardehāl in Kāšān	2012	Pilgrimage & celebratory ritual
Iraq	Provision of services and hospitality during the Arba'in Visitation	2019	Pilgrimage & pilgrim hospitality
Japan	Nachi no Dengaku	2012	Religious performing art
Kenya	Practices of the Kayas in the sacred forests of the Mijikenda	2009	Performing arts in sacred forests
Luxembourg	Hopping procession of Echternach	2010	Procession & pilgrimage
Mexico	La Romería: Ritual Cycle of the Carrying of the Virgin of Zapopan	2018	Pilgrimage
Mongolia	Mongolian traditional practices of worshipping sacred sites	2017	Nomadic people worshipping sacred sites
North Macedonia	Feast of the Holy Forty Martyrs in Štip	2013	Pilgrimage & feast
Peru	Pilgrimage to the Sanctuary of the Lord of Qoyllurit'i	2011	Pilgrimage
Philippines	Buklog, thanksgiving ritual system of the Subanen	2019	Rituals of thanksgiving to the spirits
Vietnam	Worship of Hùng Kings in Phú Thọ	2012	Pilgrimage & celebration

domestic and international tourism in general. Increasing numbers of tourists have led to what many scholars label 'overtourism', which is one of the most vexing tourism issues of our day. Since the 1970s, scholars have been concerned about the negative impacts of mass tourism (Mathieson and Wall, 1982), suggesting that overcrowded destinations and the unmitigated growth of tourism will have negative ecological and social outcomes. These impacts are manifold and range from environmental damage to community dissatisfaction with tourism and outrage toward tourists' misbehaviour. The term 'overtourism' was recently coined to bring currency to the debate and highlight these negative impacts (e.g. Dodds and Butler, 2019; Milano *et al.*, 2019; Pechlaner *et al.*, 2020).

Religious tourism and pilgrimage, like other forms of tourism, have become a mass human experience in many contexts and locations.

Fig. 19.1. Varanasi (India) – sacred town. (Photo: D. Liutikas)

Many religious localities are now global-scale mass destinations for both pilgrims and experience-seeking secular tourists. Prominent examples include St Peter's Basilica in the Vatican City, the sacred Catholic pilgrimage cities of Lourdes and Santiago de Compostela, and the Old City of Jerusalem. Even sanctified places that are consumed primarily by pilgrims are now becoming increasingly popular among non-pilgrim tourists. Research shows that some popular pilgrimage routes are now so crowded that they are becoming de-sanctified and their religious role is eroding (Timothy and Boyd, 2015; Ron and Timothy, 2019; Clancy, 2020). In many major cities, religious edifices are some of the most popular tourist attractions. However, with the experience of overtourism, these sacred localities are now over-run by mass tourists. The churches and shrines of Prague, Venice and Barcelona are particularly well-documented examples of collateral damage from the over-tourism currently being experienced in those European cities (Navarro-Ruiz *et al.*, 2019).

At the time of writing, there were 53 WHSs on UNESCO's List of World Heritage in Danger. Two of these were in danger as a result of overtourism. The first is the historic centre of Shakhrisyabz, Uzbekistan, which is located along the Silk Road. This site is endangered because of the excessive construction of tourism services and infrastructure (e.g. hotels) within the World Heritage area. The second is the historic city of Jerusalem, which is sacred to all three Abrahamic faiths: Judaism, Islam and Christianity (Fig. 19.2). The city is so over-run by religious tourists and other visitors that it has been on UNESCO's List of World Heritage in Danger since 1982. Overtourism in Jerusalem has physically altered the structural and spatial integrity of that holy city, and the pressures of religious tourism have contributed to increased inter-sectoral conflict at some of the holiest sites there (Olsen and Ron, 2013; Timothy and Emmett, 2014).

With the resurgence of the Way of St James (Camino de Santiago) in the late 1980s

Fig. 19.2. Old City of Jerusalem. Western Wall. (Photo: D. Liutikas)

as a pilgrimage and cultural route, tourism in Santiago de Compostela, Spain, grew tremendously, and visitor numbers swelled during holy years and at other celebratory times (Timothy and Olsen, 2018, Ron and Timothy, 2019). Since the popularity of 'the Camino', the city and cathedral have become far too crowded with both pilgrims and non-pilgrim tourists, creating significant congestion and safety hazards, which also reduce the quality of the spiritual experience for pilgrims (Fig. 19.3). City planners, tourism representatives and Church officials have recently undertaken major visitor flow management steps and planning efforts to dissipate congestion and help protect the physical urban and intangible spiritual environment of the city (Santos Solla and Pena Cabrera, 2014; Pérez Guilarte and Lois González, 2018). The over-commercialization of religious tourism in Lourdes, France, has also seen the sacred space in that city quickly becoming profane as pilgrimage has become nearly indistinguishable from leisure tourism (Thomas *et al.*, 2018; Ron and Timothy, 2019).

Even in sacred localities that remain more geared towards a pilgrimage market than a general tourism market, over-visitation has become a salient problem and has begun to reach a critical breaking point. The world's largest tourism gatherings are of a religious nature, the most famous being, as noted above, the Kumbh Mela and the Hajj. Although it is unknown whether the Kumbh Mela's inscription on UNESCO's list of intangible heritage in 2017 has had any impact on increasing pilgrims' travels or non-pilgrim travellers' propensity to visit the sacred sites and celebratory times of the pilgrimage, the fact remains that it is one of the largest spiritual gatherings on the planet. In the case of the Hajj, Mecca would likely see increasing numbers of pilgrimage if Saudi Arabia were to eliminate its quota system, which limits the number of pilgrims each year. The quotas set by the government act as a means of managing

Fig. 19.3. Tourists and pilgrims waiting in a queue to enter the Cathedral of Santiago de Compostela through the Holy Door, 2010. (Photo: D. Liutikas)

the pilgrim crowds at Mecca's holy sites, yet the popularity of this pilgrimage has led to a severe over-commodification of the place and its rituals. As Qurashi's (2017) research illustrates, Mecca's traditional orientation towards humble pilgrims has now been replaced by a hyper-commercialized, leisured Hajj consumerscape where five-star hotels, resorts, recreation facilities, elegant restaurants, surrogate ritual guides and up-market transportation have replaced the tents, camping areas, pedestrian paths and food vendors of yesteryear, so much so that most Muslims can no longer afford to undertake the requisite Hajj experience because they are essentially required to purchase expensive package tours.

The extravagant and over-commercialized nature of many contemporary pilgrimages is only one manifestation of the 'over-touristification' of the sacred. The extreme crowded conditions in Lourdes, Santiago de Compostela and Jerusalem have begun to raise significant security concerns for the destinations involved. In the largest and most crowded pilgrimage events – the Kumbh Mela and the Hajj –millions of pilgrims in a relatively small space contribute to urban degradation and over-development (Shinde, 2012; Maroufi and Rosina, 2017; Pérez Guilarte and Lois González, 2018), as well as the spread of infectious diseases, trampling deaths and threats of terrorism that frequently accompany large gatherings (Sridhar *et al.*, 2015; Gautret and Steffen, 2016; Qurashi, 2019).

Current Undertourism

While overtourism has received much attention in the last few years, undertourism can also be a problem for religious destinations but is often circumstantial or situational. This is evidenced with the present COVID-19 (coronavirus) pandemic which, at the time of writing, had devastated the tourism industry worldwide. Efforts to increase distances between people during this pandemic have led to restrictions on religious ceremonies, gatherings, rituals and events, as well as the closure of places of worship, including major religious sites around the world, for both pilgrims and leisure tourists. For example, the government of Saudi Arabia has suspended the 2020 *umrah* pilgrimage and Hajj visas over coronavirus concerns (Al-Arshani, 2020; Chopra, 2020; Sabah and Nereim, 2020). The Ministry of Hajj and Umrah also closed the Al-Haram Mosque in Mecca and the Al-Masjid al Nabawy Mosque in Medina for sterilization. Likewise, four key Shiite religious sites in Iran have been closed (AFP, 2020). The Church of the Nativity in Bethlehem was closed after employees at a nearby hotel were diagnosed with the coronavirus (Qawasma, 2020; Sudilovsky, 2020b), as have the pilgrimage centres of Lourdes, Medjugorje and Santiago de Compostela. Most pilgrimage destinations throughout the world have been affected by the 2019–2020 pandemic because of their propensity to attract large crowds.

Nonetheless, some religious sites remain open and continue to function as much as possible given the dire situation and in light of government restrictions on assemblies. For example, even though pilgrims from South Korea and Greece contracted the coronavirus while visiting Israel (Sudilovsky, 2020a), religious sites in Jerusalem were running 'business as usual' at the end of March 2020. However, Jerusalem's religious leaders have grave concerns regarding the impact of the coronavirus on the number of people who will be attending Lent and Easter pilgrimages in the city. Likewise, while some religious sites in India remain open, Muslims in Agra have postponed their pilgrimage to Ajmer Sharif, and many Hindu worshippers are avoiding visiting large, crowded temples (Qureshi, 2020).

To accommodate pilgrimage needs during this volatile time, some religious site managers are turning to techno-spiritual practices to engage homebound pilgrims, such as the Tower of David Initiative, which uses virtual and augmented reality to guide pilgrims and curiosity-seekers around Jerusalem's holy sites. While not a true substitute for being there, the project allows pilgrims who cannot physically attend special religious rituals to feel a part of 'being there' (Teracin, 2020). It is impossible to quantify the economic impacts of COVID-19 on religious attractions, which, in many cases, are entirely dependent on donations or entrance fees for maintenance and upkeep. While governments around the world are setting up bailouts for major industries, so far, nothing has been considered regarding assisting religious groups that have had to shutter their sacred sites, most likely because religious sites and spiritual practices are not considered a public good or public service to be funded by the state.

Conclusion

Rather than declining in importance as a form of mobility, pilgrimage has seen a resurgence in the increasingly globalized world. Every year, millions of people travel to religious sites to perform sacred rituals and engage in individual and group identity-making. While pilgrimage is still an act of individual and group identity formation, what constitutes pilgrimage has changed in recent decades. According to some observers, any form of travel that embodies a valued idea constitutes post-modern pilgrimage. This expansion of the definition of pilgrimage, as well as cultural deregulation brought about by globalization, has enabled the growth of religious mobilities to include diasporic religious travel, New Age and spiritual tourism niche markets, and new ways of commodifying religion.

Religious rituals, celebrations and sacred sites are an omnipresent part of the heritage landscapes of places. Intangible and tangible manifestations of faith and spirituality are increasingly acknowledged as the universal heritage of all humankind, not just the patrimony of believers or faith organizations. More sacred sites and practices are becoming mainstream heritage tourist attractions, and some now brand the tourism industries of many countries. This worldwide enlargement of hallowed spaces and the heritagization of sacrosanct practices for broader tourism consumption have significant implications for site managers, religious leaders, tourism intermediaries, destination management organizations and pilgrims. Of particular concern are the ways in which mainstream tourism's commoditization of the sacred may depreciate the other-worldly encounters that pilgrims seek to experience, as well as the pressures of overtourism on the physical and social environments of consecrated places as pilgrimage and mainstream tourism continue to grow in these localities.

Although in normal conditions overtourism has led to deep concerns over the long-term sustainability and viability of many religious sites, the 2019–2020 COVID-19 pandemic has led to situational undertourism in these same places. This is a double-edged sword; on the one hand, the pandemic provides relief from the pressures of over-visitation in pilgrimage destinations; on the other hand, it prevents people from carrying out the corporeal manifestations of their faith, decreases employment and tax revenues in pilgrimage destinations, and empties the coffers of sites that desperately need maintenance and care. In a period of uncertainty, new interpretations and manifestations of pilgrimage, and globalized and commoditized religious heritage, it is likely that technology (e.g. augmented and virtual reality) will play an increasingly important part in pilgrimage, religious tourism and other travel modalities.

References

Adie, B.A. (2017) Franchising our heritage: the UNESCO world heritage brand. *Tourism Management Perspectives* 24, 48–53. DOI: 10.1016/j.tmp.2017.07.002.

Adie, B.A., Hall, C.M. and Prayag, G. (2018) World heritage as a placebo brand: a comparative analysis of three sites and marketing implications. *Journal of Sustainable Tourism* 26(3), 399–415. DOI: 10.1080/09669582.2017.1359277.

AFP (2020) Virus-hit Iran closes four key religious sites. Yahoo! News. Available at: https://news.yahoo.com/virus-hit-iran-closes-four-key-religious-sites-172419064.html (accessed 24 March 2020).

Al-Arshani, S. (2020) Before and after photos show how coronavirus fears have emptied out some the busiest holy sites. *Business Insider*. Available at: https://www.businessinsider.com/these-photos-show-how-coronavirus-fears-left-religious-sites-empty-2020-3 (accessed 13 March 2020).

Albera, D. and Eade, J. (2015) International perspectives on pilgrimage studies: putting the anglophone contribution in its place. In: Eade, J. and Albera, D. (eds) *International Perspectives on Pilgrimage Studies: Itineraries, Gaps and Obstacles*. Routledge, London, pp. 1–22.

Alderman, D.H. (2002) Writing on the Graceland wall: on the importance of authorship in pilgrimage landscapes. *Tourism Recreation Research* 27(2), 27–33. DOI: 10.1080/02508281.2002.11081217.

Amaro, S., Antunes, A. and Henriques, C. (2018) A closer look at Santiago de Compostela's pilgrims through the lens of motivations. *Tourism Management* 64, 271–280. DOI: 10.1016/j.tourman.2017.09.007.

Barber, R. (1991) *Pilgrimages*. Boydell Press, Woodbridge, UK.

Bauman, Z. (1996) From Pilgrim to Tourist - Or a Short History of Identity. In: Hall, S. and du Gay, P. (eds) *Questions of Cultural Identity*. Sage, London, pp. 18–36.

Belhassen, Y., Caton, K. and Stewart, W.P. (2008) The search for authenticity in the pilgrim experience. *Annals of Tourism Research* 35(3), 668–689. DOI: 10.1016/j.annals.2008.03.007.

Bell, G. (2006) No more SMS from Jesus: Ubicomp, religion and techno-spiritual practices. In: Dourish, P. and Friday, A. (eds) *International Conference on Ubiquitous Computing*. Springer Verlag, Berlin, pp. 141–158.

Berger, P.L. (1967) *The Sacred Canopy: Elements of a Sociological Theory of Religion*. Anchor Books, New York.

Bremer, T.S. (2005) Tourism and religion. In: Jones, L. (ed.) *Encyclopedia of Religion*. Thomas Gale, Detroit, Michigan, pp. 9260–9264.

Campo, J.E. (1998) American pilgrimage landscapes. *The ANNALS of the American Academy of Political and Social Science* 558(1), 40–56. DOI: 10.1177/0002716298558001005.

Chopra, A. (2020) Saudi suspends 'umrah' pilgrimage over coronavirus fears. *The Jakarta Post*. Available at: https://www.thejakartapost.com/news/2020/03/05/saudi-suspends-umrah-pilgrimage-over-coronavirus-fears.html (accessed 24 March 2020).

Clancy, M. (2020) Overtourism and resistance: today's anti-tourist movement in context. In: Pechlaner, H., Innerhofer, E. and Erschbamer, G. (eds) *Overtourism: Tourism Management and Solutions*. Routledge, London, pp. 14–24.

Collins-Kreiner, N. and Wall, G. (2015) Tourism and religion: spiritual journeys and their consequences. In: Brunn, S.D. (ed.) *The Changing World Religion Map*. Springer, Dordrecht, the Netherlands, pp. 689–707.

Dodds, R. and Butler, R. (eds) (2019) *Overtourism: Issues, Realities and Solutions*. Walter de Gruyter, Berlin and Boston.

Eco, U. (1983) *Travels in Hyperreality*. Harcourt Brace Jovanovich Publishers, San Diego, California.

Economic Times (2018) Kumbh Mela: how up will manage one of the world's biggest religious festivals. Available at: https://economictimes.indiatimes.com/news/politics-and-nation/kumbh-mela-how-up-will-manage-one-of-the-worlds-biggest-religious-festival/articleshow/67177421.cms (accessed 11 March 2020).

Farra-Haddad, N. (2020) Archaeology and religious tourism: sacred sites, rituals, sharing the Baraka and tourism development. In: Timothy, D.J. and Tahan, L.G. (eds) *Archaeology and Tourism: Touring the Past*. Channel View Publications, Bristol, UK, pp. 106–120.

Feldman, J. (2017) Key figure of mobility: the pilgrim. *Social Anthropology* 25(1), 69–82. DOI: 10.1111/1469-8676.12378.

Fleischer, A. (2000) The tourist behind the pilgrim in the Holy land. *International Journal of Hospitality Management* 19(3), 311–326.

Fyall, A., Leask, A. and Barber, S. (2020) Marketing archaeological heritage for tourism. In: Timothy, D.J. and Tahan, L.G. (eds) *Archaeology and Tourism: Touring the Past*. Channel View Publications, Bristol, UK, pp. 69–86.

Gammon, S. (2004) Secular pilgrimage and sport tourism. In: Ritchie, B.W. and Adair, D. (eds) *Sport Tourism: Interrelationships, Impacts and Issues*. Channel View Publications, Bristol, UK and New York, pp. 30–45.

Garbin, D. (2019) Sacred remittances: money, migration and the moral economy of development in a transnational African church. *Journal of Ethnic and Migration Studies* 45(11), 2045–2061. DOI: 10.1080/1369183X.2018.1433528.

Gauthier, F., Martikainen, T. and Woodhead, L. (2013) Acknowledging a global shift: a primer for thinking about religion in consumer societies. *Implicit Religion* 16(3), 261–276. DOI: 10.1558/imre.v16i3.261.

Gautret, P. and Steffen, R. (2016) Communicable diseases as health risks at mass gatherings other than hajj: what is the evidence? *International Journal of Infectious Diseases* 47, 46–52. DOI: 10.1016/j.ijid.2016.03.007.

Geary, D. (2008) Destination enlightenment: branding Buddhism and spiritual tourism in Bodhgaya, Bihar. *Anthropology Today* 24(3), 11–14. DOI: 10.1111/j.1467-8322.2008.00584.x.

Graburn, N.H.H. (1989) Tourism: the sacred journey. In: Smith, V.L. (ed.) *Hosts and Guests: The Anthropology of Tourism*. University of Pennsylvania Press, Philadelphia, pp. 21–36.

Graham, B., Ashworth, G.J. and Tunbridge, J.E. (2000) *A Geography of Heritage: Power, Culture and Economy*. Arnold, London.

Greif, A. (2006) *Institutions and the Path to the Modern Economy: Lessons from Medieval Trade*. Cambridge University Press, Cambridge, UK.

Gupta, V. (1999) Sustainable tourism: learning from Indian religious traditions. *International Journal of Contemporary Hospitality Management* 11(2/3), 91–95.

Hacohen, H. (2019) Hindu holy festival Kumbh Mela attracts 150 million visitors. *Jerusalem Post*. Available at: https://www.jpost.com/International/Hindu-holy-festival-Kumbh-Mela-attracts-150-million-visitors-582197 (accessed 3 March 2020).

Hardiman, N. and Burgin, S. (2013) World heritage area listing of the greater blue mountains — did it make a difference to visitation? *Tourism Management Perspectives* 6, 63–64. DOI: 10.1016/j.tmp.2012.12.002.

Holmberg, C.B. (1993) Spiritual pilgrimages: traditional and hyperreal motivations for travel and tourism. *Visions of Leisure and Business* 12(2), 18–27.

Jafari, J. (1987) Tourism models: the sociocultural aspects. *Tourism Management* 8(2), 151–159.

Jimura, T. (2016) World heritage site management: a case study of sacred sites and pilgrimage routes in the Kii mountain range, Japan. *Journal of Heritage Tourism* 11(4), 382–394. DOI: 10.1080/1743873X.2016.1146287.

Kaelber, L. (2006) Paradigms of travel: from medieval pilgrimage to postmodern tour. In: Timothy, D.J. and Olsen, D.H. (eds) *Tourism, Religion and Spiritual Journeys*. Routledge, London, pp. 49–63.

Kitiarsa, P. (2010) Towards a sociology of religious commodification. In: Turner, B.S. (ed.) *The New Blackwell Companion to the Sociology of Religion*. Blackwell, Chichester, UK, pp. 563–583.

Koren-Lawrence, N. and Collins-Kreiner, N. (2019) Visitors with their 'backs to the archaeology': religious tourism and archaeology. *Journal of Heritage Tourism* 14(2), 138–149. DOI: 10.1080/1743873X.2018.1503279.

Leppäkari, M. and Griffin, K. (eds) (2017) *Pilgrimage and Tourism to Holy Cities*. CAB International, Wallingford, UK.

Luckmann, T. (1963) *The Invisible Religion: The Problems of Religion in Modern Society*. Macmillan, New York.

MacCannell, D. (1976) *The Tourist: A New Theory of the Leisure Class*. Schocken, New York.

Maroufi, H. and Rosina, E. (2017) Cities hosting holy shrines: the impact of pilgrimage on urban form. *Iconarp International Journal of Architecture and Planning* 5(Special Issue), 30–44. DOI: 10.15320/ICONARP.2017.24.

Mathieson, A. and Wall, G. (1982) *Tourism: Economic, Physical and Social Impacts*. Longman, London.

May, S., Wilson, E.K., Baumgart-Ochse, C. and Sheikh, F. (2014) The religious as political and the political as religious: globalisation, post-secularism and the shifting boundaries of the sacred. *Politics, Religion & Ideology* 15(3), 331–346. DOI: 10.1080/21567689.2014.948526.

McAlister, E. (2005) Globalization and the religious production of space. *Journal for the Scientific Study of Religion* 44(3), 249–255.

McLennan, G. (2010) Space of postsecularism. In: Molendijk, A., Beaumont, J. and Jedan, C. (eds) *Exploring the Postsecular: The Religious, the Political and the Urban*. Brill, Leiden, The Netherlands, pp. 39–62.

Meskell, L. (2015) Transacting UNESCO world heritage: gifts and exchanges on a global stage. *Social Anthropology* 23(1), 3–21. DOI: 10.1111/1469-8676.12100.

Meskell, L., Liuzza, C., Bertacchini, E. and Saccone, D. (2015) Multilateralism and UNESCO world heritage: decision-making, states parties and political processes. *International Journal of Heritage Studies* 21(5), 423–440. DOI: 10.1080/13527258.2014.945614.

Milano, C., Cheer, J.M. and Novelli, M. (2019) *Overtourism: Excesses, Discontents and Measures in Travel and Tourism*. CAB International, Wallingford, UK.

Morinis, A. (1992) Introduction: the territory of the anthropology of pilgrimage. In: Morinis, A. (ed.) *Sacred Journeys: The Anthropology of Pilgrimage*. Greenwood Press, Westport, Connecticut, pp. 1–28.

Navarro-Ruiz, S., Casado-Díaz, A.B. and Ivars-Baidal, J. (2019) Cruise tourism: the role of shore excursions in the overcrowding of cities. *International Journal of Tourism Cities* 6(1), 197–214. DOI: 10.1108/IJTC-04-2018-0029.

Olsen, D.H. (2003) Heritage, tourism, and the Commodification of religion. *Tourism Recreation Research* 28(3), 99–104. DOI: 10.1080/02508281.2003.11081422.

Olsen, D.H. (2012a) Teaching truth in third space': the use of history as a Pedagogical instrument at temple square in salt lake City, Utah. *Tourism Recreation Research* 37(3), 227–237. DOI: 10.1080/02508281.2012.11081711.

Olsen, D.H. (2012b) Negotiating religious identity at sacred sites: a management perspective. *Journal of Heritage Tourism* 7(4), 359–366.

Olsen, D.H. (2016) Ritual journeys in North America: opening religious and ritual landscapes and spaces. *International Journal of Religious Tourism and Pilgrimage* 4(1), 34–48.

Olsen, D.H. (2019a) Religion, spirituality, and pilgrimage in a globalizing world. In: Timothy, D.J. (ed.) *Handbook of Globalisation and Tourism*. Edward Elgar, London, pp. 270–283.

Olsen, D.H. (2019b) Religion, pilgrimage, and tourism in the Mena region. In: Timothy, D.J. (ed.) *Routledge Handbook on Tourism in the Middle East and North Africa*. Routledge, London, pp. 109–124.

Olsen, D.H. (Forthcoming) Pilgrimage, religious tourism, biodiversity, and natural sacred sites. In: Shinde, K.A. and Olsen, D.H. (eds) *Religious Tourism and the Environment*. CABI, Oxfordshire, UK.

Olsen, D.H. and Ron, A.S. (2013) Managing religious heritage attractions: the case of Jerusalem. In: Garrod, B. and Fyall, A. (eds) *Contemporary Cases in Heritage*. 1. Goodfellow Publishers, Oxford, pp. 51–78.

Olsen, D.H. and Trono, A. (eds) (2018) *Religious Pilgrimage Routes and Trails: Sustainable Development and Management*. CAB International, Wallingford, UK.

Ornella, A.D. (2013) Commodification of religion. In: Runehov, A.L.C. and Oviedo, L. (eds) *Encyclopedia of Sciences and Religions*. Springer, Dordrecht, The Netherlands, pp. 430–431.

Oviedo, L., de Courcier, S. and Farias, M. (2014) Rise of pilgrims on the Camino to Santiago: sign of change or religious revival? *Review of Religious Research* 56(3), 433–442. DOI: 10.1007/s13644-013-0131-4.

Pechlaner, H., Innerhofer, E. and Erschbamer, G. (eds) (2020) *Overtourism: Tourism Management and Solutions*. Routledge, London.

Pérez Guilarte, Y. and Lois González, R.C. (2018) Sustainability and visitor management in tourist historic cities: the case of Santiago de Compostela, Spain. *Journal of Heritage Tourism* 13(6), 489–505. DOI: 10.1080/1743873X.2018.1435665.

Poria, Y., Reichel, A. and Cohen, R. (2011) World heritage site: an effective brand for an archeological site? *Journal of Heritage Tourism* 6(3), 197–208. DOI: 10.1080/1743873X.2011.591932.

Prorok, C.V. (2003) Transplanting pilgrimage traditions in the Americas. *Geographical Review* 93(3), 283–307. DOI: 10.1111/j.1931-0846.2003.tb00034.x.

Qawasma, M. (2020) Palestinians declare coronavirus emergency as church of the Nativity closed. Reuters. Available at: https://www.reuters.com/article/us-health-coronavirus-palestinians-idUSKBN20S17R (accessed 13 March 2020).

Qurashi, J. (2017) Commodification of Islamic religious tourism: from spiritual to touristic experience. *International Journal of Religious Tourism and Pilgrimage* 5(1), 89–104.

Qurashi, J. (2019) The hajj: crowding and congestion problems for pilgrims and hosts. In: Dodds, R. and Butler, R.W. (eds) *Overtourism: Issues, Realities and Solutions*. de Gruyter, Berlin, pp. 185–198.

Qureshi, S. (2020) Coronavirus in India: people in Agra avoid visiting religious sites. *Microsoft News*. Available at: https://www.msn.com/en-in/news/other/coronavirus-in-india-people-in-agra-avoid-visiting-religious-sites/ar-BB11vHYh (accessed 23 March 2020).

Rahimi, B. and Eshaghi, P. (eds) (2019) *Muslim Pilgrimage in the Modern World*. University of North Carolina Press, Chapel Hill, North Carolina.

Richter, L.K. (2003) International tourism and its global public health consequences. *Journal of Travel Research* 41(4), 340–347. DOI: 10.1177/0047287503041004002.

Ron, A.S. and Timothy, D.J. (2019) *Contemporary Christian Travel: Pilgrimage, Practice, and Place*. Channel View Publications, Bristol, UK.

Sabah, Z. and Nereim, V. (2020) Saudi Arabia suspends Mecca pilgrimage visas on coronavirus. *Microsoft News*. Available at: https://www.msn.com/en-us/news/world/saudi-arabia-temporarily-suspends-all-religious-pilgrimages-over-coronavirus-fears/ar-BB10rzV3 (accessed 24 March 2020).

Santos Solla, X.M. and Pena Cabrera, L. (2014) Management of tourist flows. The Cathedral of Santiago de Compostela. *PASOS Revista de Turismo y Patrimonio Cultural* 12(4), 719–735. DOI: 10.25145/j.pasos.2014.12.053.

Shinde, K. (2012) Policy, planning, and management for religious tourism in Indian pilgrimage sites. *Journal of Policy Research in Tourism, Leisure and Events* 4(3), 277–301. DOI: 10.1080/19407963.2012.726107.

Sridhar, S., Gautret, P. and Brouqui, P. (2015) A comprehensive review of the Kumbh Mela: identifying risks for spread of infectious diseases. *Clinical Microbiology and Infection* 21(2), 128–133. DOI: 10.1016/j.cmi.2014.11.021.

Stump, R.W. (2000) *Boundaries of Faith: Geographical Perspectives on Religious Fundamentalism*. Rowman & Littlefield, Lanham, Maryland.

Sudilovsky, J. (2020a) As coronavirus spreads, Holy Land Catholics concerned, but sites still open. *Catholic Philly*. Available at: https://catholicphilly.com/2020/02/news/world-news/as-coronavirus-spreads-holy-land-catholics-concerned-but-sites-still-open/ (accessed 13 March 2020).

Sudilovsky, J. (2020b) Bethlehem holy sites to be closed after coronavirus reported at hotel. *America Magazine*. Available at: https://www.americamagazine.org/politics-society/2020/03/05/bethlehem-holy-sites-be-closed-after-coronavirus-reported-hotel (accessed 13 March 2020).

Teracin, R. (2020) At the time of coronavirus, Jerusalem comes to pilgrims in VR. Available at: https://www.jpost.com/Israel-News/At-the-time-of-coronavirus-Jerusalem-comes-to-pilgrims-in-VR-621919 (accessed 24 March 2020).

Thomas, S., White, G.R.T. and Samuel, A. (2018) To pray and to play: post-postmodern pilgrimage at Lourdes. *Tourism Management* 68, 412–422. DOI: 10.1016/j.tourman.2018.03.021.

Timothy, D.J. (1997) Tourism and the personal heritage experience. *Annals of Tourism Research* 24(3), 751–754. DOI: 10.1016/S0160-7383(97)00006-6.

Timothy, D.J. (2012) Religious views of the environment: sanctification of nature and implications for tourism. In: Holden, A. and Fennell, D.A. (eds) *The Routledge Handbook of Tourism and the Environment*. Routledge, London and New York, pp. 53–64.

Timothy, D.J. (2021) *Cultural Heritage and Tourism: An Introduction*, 2nd edn. Channel View Publications, Bristol, UK.

Timothy, D.J. and Boyd, S.W. (2015) *Tourism and Trails: Cultural, Ecological and Management Issues*. Channel View Publications, Bristol, UK.

Timothy, D.J. and Emmett, C.F. (2014) Jerusalem, tourism, and the politics of heritage. In: Adelman, M. and Elman, M.F. (eds) *Jerusalem: Conflict & Cooperation in a Contested City*. Syracuse University Press, Syracuse, New York, pp. 276–290.

Timothy, D.J. and Iverson, T. (2006) Tourism and Islam: considerations of culture and duty. In: Timothy, D.J. and Olsen, D.H. (eds) *Tourism, Religion and Spiritual Journeys*. Routledge, London, pp. 186–205.

Timothy, D.J. and Olsen, D.H. (2018) Religious routes, pilgrim trails: spiritual pathways as tourism resources. In: Butler, R. and Suntikul, W. (eds) *Tourism and Religion: Issues and Implications*. Channel View Publications, Bristol, UK, pp. 220–235.

Turner, B. (2005) The sociology of religion. In: Calhoun, C., Rojek, C. and Turner, B. (eds) *The SAGE Handbook of Sociology*. Sage, London, pp. 284–301.

UNESCO (2020a) Representative list of the intangible cultural heritage of humanity. Available at: https://ich.unesco.org/en/lists (accessed 10 March 2020).

UNESCO (2020b) World heritage list. Available at: https://whc.unesco.org/en/list/ (accessed 10 March 2020).

UNWTO (2011) *Religious Tourism in Asia and the Pacific*. World Tourism Organization, Madrid.

Yan, Cheng (Grace). and Morrison, A.M. (2008) The influence of visitors' awareness of World Heritage Listings: a case study of Huangshan, Xidi and Hongcun in southern Anhui, China. *Journal of Heritage Tourism* 2(3), 184–195. DOI: 10.2167/jht059.0.

Zamani-Farahani, H., Carboni, M., Perelli, C. and Torabi Farsani, N. (2019) Islamic tourism in the middle East. In: Timothy, D.J. (ed.) *Routledge Handbook on Tourism in the Middle East and North Africa*. Routledge, London, pp. 125–136.

20 Conclusions: Pilgrimage During and After the Pandemic Crisis

Darius Liutikas*
Lithuanian Social Research Centre

Tourism and pilgrimage flourished at the end of the 20th and the beginning of the 21st centuries. It was the time of the rise of mass transportation (e.g. cheap flights), the development of tourism services (accommodation places, individualized one-day trips to various destinations inside and outside the countries) and the creation of new tourism products (cultural and religious tourism routes, the emergence of new tourism destinations). The phenomenon of pilgrimage absorbed different trips related to nationality, culture, heritage, sport, dark tourism, roots tourism and personal identity. Pilgrims of the 21st century, having access to information, could easily organize their own trips – to choose a valuistic destination and travel route, and to find the best alternatives for transport, accommodation and travel speed.

The year 2020 was marked by the global outbreak of the COVID-19 pandemic. The coronavirus affected the majority of the countries and territories in the world. People who contacted this highly infectious disease showed symptoms like high fever, coughing and shortness of breath, which could lead to pneumonia, multi-organ failure and death. The number of confirmed infections worldwide increased rapidly. On 1 April 2020 around 1,000,000 cases and 50,000 deaths had occurred. On 2 May, according to official statistics, more than 3,500,000 coronavirus cases had been detected, and more than 250,000 deaths had been registered. On 1 June more than 6,000,000 coronavirus cases had been detected and more than 370,000 deaths had been registered (World Health Organization, 2020). Many countries implemented quarantine measures for individuals and introduced other measures of infection prevention and control. These measures included isolation (individuals with the disease were separated from others for as long as they were infectious), social distancing (large groups of people were restricted from gathering) and movement restrictions (travel prevention, closing of countries' borders, quarantine and health observation of travellers).

UNWTO (2020) reported that 100% of global destinations imposed travel restrictions, including complete bans on all travel. Tourism was among the most affected of all economic sectors. The huge decline of international tourists and the loss of thousands of jobs within the sector affected the lives of ordinary people. We must examine the adaptation and the response of the pilgrimage phenomenon to this global crisis in response to post-pandemic travel.

*darius.liutikas@gmail.com

Religious pilgrimage has more than 1000 years of history. We have to remember this fact when we discuss the adaptability or the ability to adapt to this crisis, including the challenges and changes of pilgrimage. Resilience includes such elements as buffering capacity (the size or kinds of disruptions the system can absorb or adapt to without a breakdown), flexibility (the system's ability to restructure itself in response to external pressures), margin boundaries (how precarious the system is currently, relative to one or another kind of performance boundary) and tolerance (how a system behaves near a boundary) (Woods, 2006).

Religion could serve as a healing and emotional relief in time of crisis. Pilgrimage related to personal and social identity and encompassing different values and multi-sided motivational sets could be a very flexible and adaptive system. Even one of the main conditions of pilgrimage – physical movement in the geographical space – can be changed to adapt to the situation.

Alternative Forms of Pilgrimage

Pilgrimage is not just about geographical space. Morinis (1992, p. 2) claims that the 'ideals that are the pilgrim's goal are usually – but not always – fixed in space within the perceptible world.' Coleman and Eade (2004) distinguish embodied, imagined and metaphorical forms of pilgrimage motions. Similarly, Kaelber (2006) understood pilgrimage as physical, imagined or virtual movement. Non-geographical goals of pilgrimage are related to finding ways to express identity and values and seek valuistic ideals. Valuistic functions of pilgrimage could be satisfied or achieved in the home environment. Spiritual pilgrimage and virtual pilgrimage could be practised in a very small geographical area.

Spiritual pilgrimage

Digance (2006) notices that the inner journey is the essential element of any pilgrimage, 'namely the quest and search for meaning'. In religious pilgrimage, prayers and meditations require a high concentration of the mind, so 'travel to oneself', to self-knowledge, is common among adherents of different religions. When such spiritual journeys are considered a major part of daily religious activity they become the main goal of believers. The term pilgrimage is also used as a metaphor for life's journey toward death and eternity. However, this inner notion of pilgrimage has some elements similar to the physical journey: direction towards a destination or a goal and the satisfaction of the achievement of that goal.

The absence of a geographical destination is the most important feature of allegorical pilgrimage. Allegorical pilgrimage has been known since the Middle Ages. Olin (1979) notices that life can be viewed as pilgrimage, and this notion has old and established traditions in Christian thought. This idea of spiritual pilgrimage is based on the desire to deepen and place more emphasis on the importance of personal piety. Spiritual pilgrimage often leads to fantastic and allegorical descriptions of travel to a holy place, the content of which consists more of moral and religious reflections than of a retelling of the journey geographically. The most famous example of such descriptions is John Bunyan's *Pilgrim's Progress*, first published in 1678 in England. Another example of such narrative is *The Pilgrim's Story* of Ignatius Loyola, written in the 16th century. Ignatius Loyola references himself as the pilgrim throughout his life story. In the spiritual handbook *Spiritual Exercises*, Ignatius Loyola invites readers to imagine 'synagogues, villages, and towns where Christ our Lord preached' and later to ask for grace (*The Spiritual Exercises of St Ignatius of Loyola*, 2015).

Olin (1979) extends the spiritual meaning of pilgrimage. In this case, the pilgrimage is defined as an 'essentially interior journey toward some goal or ideal. It can involve the pursuit of meaning or a mission for one's life, and it may effect inward growth or transformation' (Olin, 1979, p. 388).

Virtual pilgrimage and hyper-real reality

The post-modern concept of travel is characterized by simulation and the use of hyper-real experience. This notion also influences the perception of pilgrimage. Post-modern travellers feel

Fig. 20.1. Holy Mass on a computer screen, Easter 2020, Vilnius. (Photo: the author)

that there is no authentic experience. For them, the difference between reality and fantasy is very little. Augmented reality (computer-mediated reality) could include perceptual enriched experiences for pilgrims. A traveller could get interactive and digitally controlled information about the surroundings.

New pilgrimage trends are related to the new technologies, especially the development of the internet and hyper-real reality (Fig. 20.1). It is already possible to travel to a holy and valuistic place in a virtual way. Today it is possible to form a group of virtual pilgrims who, for different purposes – spiritual, cognitive or others – visit the virtual sacred space and the most famous pilgrimage places and events. There are also virtual route trips from one site to another. Virtual travellers can safely follow the path of medieval pilgrims and can even perform various rituals, e.g. virtual repentance. Pilgrim centres invite internet users to visit their websites and virtually visit the sanctuary itself. Pilgrims on virtual chat sites can communicate with each other and share impressions. Digance (2006) states that cyber-pilgrims, too, may routinely undertake certain rituals (e.g. making a cup of coffee, lighting a candle or incense, logging on only at a certain time of day, etc.) each time they use the internet to 'do religion' online. Cyber-piligrims can recreate the sacred site or valuistic event, while the virtual space simulates the representation of place to the point where the genuine ritual experience can occur (Helland, 2013).

In Japan, travellers can visit the graves of the dead virtually. The graves of the ancestors of family members can be seen on the websites of the agencies providing these services. When you click on the picture, the tomb is automatically decorated with flowers and the readable text of the sacred sutras is broadcast. To prove their visit, the visitors can register and leave a message in the guestbook page (Inoue, 2000).

Contemporary modern technology makes it possible to combine physical activity with virtual travel. As an example, indoor cycling systems, with the combination of relevant

Fig. 20.2. Indoor cycling to a pilgrimage place. (Photo: the author)

software, allow pilgrims to cycle to Santiago de Compostela or Fatima without leaving their living-rooms (Fig. 20.2). Such systems include Google Maps, satellite images and videos of the road together (Gholipour, 2014). The screen in front of your eyes can also show your course, information about roadside objects, your cycling speed, distance taken, ride time, the slope of the route and your heart rate.

The Roman Catholic Church, in 1998, discussed virtual pilgrimage: 'Characteristic of our days, is a particular form of pilgrimage of the human mind, the informatic or virtual pilgrimage which is diffused along the ways of telecommunications. These routes, with all the risks, deformations and deviations that they involve, can be a means of proclaiming faith and love, positive messages, fruitful and efficacious contacts. It is therefore important to set out along these roads avoiding the dispersion and destruction of true communication against the "background noise" of a Babel-like myriad of information' (Pontifical Council for the Pastoral Care of Migrants and Itinerant People, 1998).

Some authors (MacWilliams, 2004; Helland, 2013) conclude that a virtual connection with the physical place can be a way for people to navigate into a liminal space where they feel an authentic relation to the sacred site. Virtual reality could serve as a symbolic substitution, a mechanism facilitating the experience of a pilgrimage journey. Moreover, social networks and social software create new opportunities to express values and identities.

Digital media and television helped people to support various aspects of their identities during the quarantine and self-isolation of the COVID-19 pandemic in 2020. Instead of a physical journey to Rome or Jerusalem to celebrate the Easter feast, Christian pilgrims had to express their religious identity through the help of television and digital media. They celebrated the feast using digital interactions within the social network of family and friends.

New virtual products for pilgrims were also launched during the quarantine in 2020. A Jerusalem museum released a virtual reality tour of Jerusalem holy sites. The experience requires a team of two – one person wears the virtual reality glasses and the other keeps a tablet (Kauffman, 2020). Using the software, the pilgrim can visit the Church of the Holy Sepulchre, al-Aqsa Mosque, the Western Wall and other holy sites. The online version supports the 360 degree immersive experience. So, the internet and hyper-real reality enhance the possibility for pilgrims to practise religion or to express their secular values and to find meaning in everyday life.

Replaced Travellers and Actions for Pilgrims

Pilgrimage is also characterized by the possibility to ask another person to go to a holy or valuistic place instead of oneself or to send something to it. In medieval times, it was quite popular to hire a professional pilgrim. Testators indicated specific qualifications for the pilgrims who had to act on their behalf (Webb, 2000). Sometimes, such persons were neighbours or friends of the testator, but it was also possible to arrange for a vicarious pilgrimage in one's own lifetime (Webb, 2000, p. 195). At the beginning of such practice, the vicarious pilgrim had to be of equal social rank to the original, but later such a provision was abandoned.

The coronavirus crisis let people create alternatives to participate in religious life. Many initiatives included sending photographs to their own parish churches or pilgrimage centres. Such photographs, posted on the church benches, somewhat compensated for the loss pilgrims felt when they could not physically go to the holy place and participate at the religious services. Symbolically, this action marked the presence of the believers who actually stayed home. Similar initiatives have been implemented by sports fans. Images of fans made from plastic or cardboard were placed on the empty seats at the stadiums.

Pilgrim centres provide other services for pilgrims who cannot come physically. For instance, religious pilgrims can ask priests to light candles for them that will be placed in the Lourdes Basilica of Our Lady of the Rosary. The recommended offering for this service is 1 euro. 'Each flame is the sign of a presence, your presence. Each flame radiates Hope!' (Lourdes Sanctuaire, 2020).

Another practice and characteristic feature of pilgrimage is to leave something in a holy or valuable place that would give meaning and mark the journey and its motives. Ex-votos are material expressions of pilgrims' thanks or a given vow. Such practice of offering ex-votos is common to different cultures and religions. To leave ex-votos or signs of miraculous healings and experienced graces is a very popular form of thanksgiving in Catholic countries (Fig. 20.3). In the face of a crisis, pilgrims can send ex-votos via courier or by mail. Pilgrims also can send prayers to the shrines using e-mails or place them at the special internet platforms online. Other offerings, such as money donations, can be sent using online transactions.

Creation of Sacred Places in the Home Environment

A sacred place is always socially constructed (Badone, 2014). Various relics, images, other objects of the cult, elements of the natural world, geographical places and famous people become sacred through the meanings projected onto them. The sanctity of the places is reinforced through the rituals, prayers and created images of the believers. Pilgrims can create sacred spaces in their local environments or even homes. Every believer has different holy artefacts in their home environment. These could be small statues, icons or images, but also sacred spaces such as prayer rooms, small chapels or altars could be created. Believers could dedicate one or another part of the house or room to the sacred space. In this sacralized environment, the individual alone, or with the family, could express religious or other values and engage in acts of religious or valuistic devotion. Sometimes people are very involved in the creation and consecration of such places. In such places where pilgrims could keep devotionalias, sacred texts or other sacred things. The pilgrims could separate the meaning of different spatial environments and

Fig. 20.3. Votive offerings (tablets) to Saint Mary at the Lourdes Grotto in Oostakker, Flanders. (Photo: the author)

create special rules and rituals for entry into the sacred domain (special prayer, wearing of clean clothes, purification bath, etc.). If it is a separate room or a separate part of the house, it is possible to decorate or specially mark the entrance to that area. Hindu families create a family shrine in the *pooja* area, a sacral niche and praying place (Mazumdar and Mazumdar, 1993).

During quarantine and self-isolation, the idea was to move away from the secular (profanum) world to the sacrum (purified) environment. Pilgrims' bodies move only several metres, but they move towards a spiritual atmosphere and something thought of as holy and divine (Fig. 20.4). Mazumdar and Mazumdar (1993) notice that the process of the creation of such places is related to our identity and personality. Sacred acts involving sacred symbols also create place attachment and people bond with place. It is important to understand that people haven't developed sacred places, rather they have *discovered* them (Chapter 12).

Suspended international travel and the increased caution of people to travel long distances open up new opportunities for domestic tourism. One of the possible consequences of the COVID-19 outbreak is increased travel to national shrines. Some sacred places are replicas of 'true' sacred places, bringing their holiness closer to the home environment. This phenomenon is known in Japan (see Chapter 10) and other religions such as Christianity and Hinduism. These replicas of the original place are very convenient to visit for people who cannot afford to make the pilgrimage to the authentic site.

Reader (2015) noticed that various forms of labyrinth have been found across cultures as a substitute for the pilgrimage to the Holy Land. Replicas of the Grotto of Lourdes and replicas of Calvary are the main objects created as an attempt to bring sacred spaces closer to believers in Christianity. The creation of such places takes into account the natural elements of the real places and reproduces their landscape. Calvaries seeks to replicate the main elements of Christ's Way of the Cross in Jerusalem. Located in open areas, they have such features as a hilly wooded landscape, the Mount of Olives, Golgotha and the Cedron Stream. Located on the Way of the Cross, the chapels, gates and churches symbolize the suffering of Christ (Fig. 20.5). The replicas of the grotto of Lourdes (France) spread in the world after the apparition of Virgin Mary to St Bernadette Soubirous in 1858. Replicas of the Lourdes grottos are usually built on the slopes of the hills or made from the stones in flat areas. There is often a spring or a stream next to the Grottos of Lourdes.

As an example, the Catholic country Lithuania has four Calvaries (Liutikas, 2015a) located in open, natural landscape. Several more

Fig. 20.4. Holy space in the house. (Photo: the author)

chapels of the Way of the Cross are installed in churchyards. In 2013, in Lithuania, there were 33 Grottos of Lourdes, most of them in the western part of the country (27 Lourdes) (Liutikas, 2014). Lithuania also has 11 places of the apparitions of the Virgin Mary, which have traditions of visitations (Liutikas, 2015b). Ten out of 11 of these apparition places are not officially recognized by Church authorities and are related to private religiosity. It is important to mention that the apparitions of Virgin Mary usually occur during times of social and political tensions (Liutikas, 2015b). New pilgrimage sites related to apparitions of the Virgin Mary could emerge during the COVID-19 outbreak. Some believers in different places have already

Fig. 20.5. The Way of Cross in Vilnius Calvary, Lithuania. (Photo: the author)

stated that they have recently seen apparitions of the Virgin Mary, e.g. San Carlos (Argentina) (Catholic Online, 2020) or the Maracha District (Democratic Republic of Congo) (Buruga, 2020).

Sometimes a sacred object can make a 'pilgrimage' to your home. Such journeys are made by sacred items passed from hand to hand such as statues, paintings, relics or religious symbols. One example is the copy of the statue of Fatima Virgin Mary travelling around the world to the homes of believers. A similar example is the pilgrimage of the Rosa Mystica statues. These statues found their way to private homes all over the world. The home of the family that received the statue became a holy place for a week or several, attracting relatives, friends and acquaintances.

Since 1984, the World Youth Day cross and icon of the Blessed Virgin Mary have been carried around the world. It is believed that the cross and icon could help young people better prepare for the World Youth Day event (See Chapter 14). Buddhist relics also travelled around the world, e.g. relics of Buddha assembled from three countries – Thailand, Myanmar and Sri Lanka, in a five-year journey from 2000 to 2005, visiting 30 countries. Another example is the Maitreya Project Heart Shrine Relic Tour (later named the Maitreya Loving Kindness Tour). During this tour, between 2001 and 2015, the unique collection of Buddhist relics, including the remains of Buddha himself, visited six continents and 68 countries.

After making these observations we can assume that the phenomenon of pilgrimage is extremely flexible and resistant to external factors. If we recall the definition of pilgrimage given in the introduction – pilgrimage is a journey to the particular places or events of searching and expressing a traveller's values and identities – we will see that each element of this definition is adaptive and resilient to challenges. A physical journey can turn into a spiritual or a virtual one, and valuistic places and events could be created in the home environment. Even the traveller could be replaced by another person, and the intentions of the journey itself are fulfilled by using objects and giving them special meanings. Such flexibility and tolerance for travel forms and places is possible because the main element of the trip – the internal motives related to the expression of personal or social identity and values – does not change.

Main Findings and Discoveries

The authors in this book have focused on pilgrims' motivations and their values and identities. Several other themes can be highlighted from the chapters. The traditional theme of pilgrimage research is its relationship to tourism and the differences between pilgrims and tourists (see Chapters 3, 14 and 17). The difference of the pilgrim and tourist approach becomes apparent in their behaviour and engagement with rituals. However, such trends as commercialization and consumerism include both pilgrims (see Chapter 11) and tourists (see Chapter 3). New Age spirituality includes ancient temples in the process of consumption (see Chapter 13).

Some authors (Russell, Chapter 5; Thompson, Chapter 8; Pryce, Chapter 18) make arguments based on their personal travel experience. Phenomenology is an important approach in travel studies. A pilgrim journey can be experienced, and we can identify the ways we experience it and the meanings it has in our experience. By examining and reflecting on the travel experience, new meanings and ideas can be developed. The authors share their opinions and orient us on how we understand a valuistic journey.

Many chapters highlighted the fact that pilgrims' motivations are very different (see Chapters 2, 4, 6, 9 and 10). Pilgrims are very heterogeneous, and their experience is no longer bound to the religious arena (see Chapters 2, 4 and 17). Empowerment agents to manifest values and identities in travelling could be multidimensional, linking personal attitudes and cultural norms.

Another key theme that emerges is the impact of pilgrimage journey. This could be on personal lives (Chapter 6) or in a change of understanding of the political situation (Chapter 16). Spiritual fulfilment (Chapters 10 and 13) is related to the experience of identity manifestation. In the process of adapting the trip to affirm one's own or social identity, traditions also have an important role (see Chapters 12 and 15). The trend of post-secularism (Chapters 7 and 19) impacts the popularity of religious pilgrimage. Pilgrims always want to experience the *communitas* (Chapter 8) or achieve personal transformations (Chapters 12 and 16). Although pilgrim places are not fixed in space any more (Chapter 14), these sites may face such challenges as overtourism or undertourism (Chapter 19).

Post-COVID-19 Pilgrimage

The lockdown in response to the COVID-19 pandemic has generated a lot of initiatives. During the pandemic, everyday religious services took on new forms, such as drive-in Holy Mass, or fly-in Holy Mass. Some priests during the coronavirus lockdown in Italy offered Holy Mass on the church roof (Brockhaus, 2020). In this case, nearby parishioners could participate from their balconies and windows. The doors of the shrines reopened as the quarantine conditions were lifted. However, in accordance with the requirement of social distance, only a small number of people could initially present religious rites. As long as there were restrictions on international travel and border entry, pilgrims have had the opportunity to focus more on national and regional pilgrimage sites.

Pilgrimages at a local level are grounded in their local context, which provides opportunities. Local pilgrimages do not cause even a temporary break of an individual's everyday connections and social networks (Dubisch, 1995, p. 95). It is one of the forms of travelling which is often implemented in the low season, when there is less crowding. Using a car, pilgrims can travel more than a few hundred kilometres on a day trip and return on the same day. Another opportunity for pilgrims who want longer trips is to use pilgrim routes. Although such routes link religious buildings, pilgrimage sites and events, they can also include natural, cultural, historical or gastronomic elements (Liutikas, 2020). For example, the network of St James' Way is settled throughout all Europe. Such pilgrim routes link to slow tourism or slowness, which encourages travellers to spend more time in a particular place and to experience the spirit of a place.

Individual trips can become more popular compared to organized bus trips. The requirement of social distance and the wish to protect at-risk individuals may limit the organization of such religious events as processions, parades and bigger gatherings. Secular, and especially cultural pilgrimage, is associated with events. However, the COVID-19 outbreak limited opportunities for events. A lot of events such as music festivals, sport competitions, cultural exhibitions and celebrations have been cancelled or postponed.

Pilgrimages will remain an important type of travel in the changed world after the COVID-19 pandemic. The tourism and hospitality industry will recover, although it is difficult to say when. People may be more inclined to keep social distance and more cautious choosing long-distance travel. The question of what impact this pandemic will have on the behaviour of pilgrims and tourists remains open and unanswered. Their behaviour could be influenced by existing restrictions, perceived health risks or changed capacities of consumption. The first initiatives aimed at revitalizing travel have been taken at the macro-regional level. Estonia, Latvia and Lithuania launched the first travel bubble in Europe on 15 May 2020. These countries opened their borders and allowed citizens from each country to travel freely in and out. Other European countries such as Greece and Italy opened their borders in the summer; however, various limitations and restrictions were in place.

The 2020 lockdown was a good time for self-reflection related to our life values and identities. It was an opportunity for potential pilgrims to inquire about possible travel destinations and to understand in what ways it is most acceptable for them to express their values. Thus it can be stated that in contemporary society, in addition to the usual and traditional forms, pilgrimage acquired new forms.

Pilgrimages in space is a part of the tourism industry. Like other types of tourism, the pilgrimage will undergo constant changes. In the near future, these journeys, like the entire tourism sector, will be affected not only by the global health situation and management of COVID-19 consequences, but also by various technological, social, environmental and legal factors. The main technological factors that will influence the future of pilgrimage are travel digitalization and development of tourism infrastructure, including development of transport and changes in urbanization.

Travel digitalization includes virtual and augmented reality, artificial intelligence and robotics, and digital communication. Touchless travel, from going to an airport to entering the hotel room, will be the most noticeable change in the near future. Virtual communities could communicate and make journeys online. Virtual reality, combined with the already existing equipment such as indoor sport walking tracks or indoor cycling systems, could create such services as virtual trips to Santiago de Compostela or a cycling trip to the Hill of Crosses in Lithuania. Pilgrimage and tourism could get benefits from such research areas as neurolinguistics and neuromarketing, related to the research areas of the travel simulation stimulus.

The main social factors that will influence pilgrimage could be changes in lifestyle, changes of mobility, demographic shifts and ageing, migration and global health. Social and cultural attitudes, personal and community values and their change will also be important for valuistic journeys. Pilgrimages will also be influenced by such factors as the changing economic situation in the world, a region or in a particular country, and the likely increase in the share of leisure time in the structure of personal time use. Environmental factors are linked to climate change, air pollution and travel sustainability. Legal factors are related to security and risk resilience including international security and cybersecurity, and travel barriers including geopolitics and the future of global health. The main determinant of pilgrimage discourse will become safety, but also already very well-known elements such as knowledge development and the preservation and valorization of authenticity. More accessible knowledge creates new areas of activity, new interests and new values. Goeldner and Ritchie (2012) claim that travellers will be seeking more individualized experiences in a knowledge-based society. They want to enrich their lives through authentic and quality experiences.

However, physical visits to the holy places will always remain the most important element of the concept of pilgrimage as an external expression of human values and identity. The travel experience could be measured by such indicators as (i) the number of times a person has travelled (abroad and domestically); (ii) the number of different destinations a person has travelled to (abroad and domestically); and (iii) the amount of time a person has spent travelling (abroad and domestically) (Pearce, 2005; Filep and Greenacre, 2007). In this case, it is also important whether the traveller travelled in a group or individually. In the case of pilgrimage, these formal indicators must be supplemented by the elements of performed rituals or achieved personal transformations.

The time spent on the trip is sacred (Graburn, 1989), extraordinary and exceptional. Pilgrimages show not only to others but also to us who we really are and what values we cherish. Valuistic travel will remain one of the most flexible and dynamic in the travel industry. Despite various challenges, pilgrimages still offer opportunities to create or demonstrate values and identities. The search for values and identity establishes new travel destinations that pilgrims will inevitably strive to visit in order to remain themselves or to be part of their community.

References

Badone, E. (2014) Conventional and unconventional pilgrimages: conceptualizing sacred travel in the twenty-first century. In: Pazos, M.A. (ed.) *Redefining Pilgrimage. New Perspectives on Historical and Contemporary Pilgrimages*. Ashgate, Farnham, UK, pp. 7–31.

Brockhaus, H. (2020) Roman priest offers Easter mass on church roof amid coronavirus. Available at: https://www.catholicnewsagency.com/news/roman-priest-offers-easter-mass-on-church-roof-amid-coronavirus-57082 (accessed 15 May 2020).

Buruga, V. (2020) Thousands throng Maracha for suspected Virgin Mary apparition. Available at: https://www.redpepper.co.ug/2020/05/thousands-throng-maracha-for-suspected-virgin-mary-apparition/ (accessed 15 May 2020).

Catholic Online (2020) Virgin Mary appears over Argentina – is this a message to us? Available at: https://www.catholic.org/news/hf/faith/story.php?id=84645 (accessed 20 May 2020).

Coleman, S. and Eade, J. (2004) Introduction: reframing pilgrimage. In: Coleman, S. and Eade, J. (eds) *Reframing Pilgrimage: Cultures in Motion*. Routledge, London and New York, pp. 1–25.

Digance, J. (2006) Religious and secular pilgrimage: journeys redolent with meaning. In: Timothy, J.D. and Olsen, H.D. (eds) *Tourism, Religion and Spiritual Journeys*. Routledge, pp. 36–48.

Dubisch, J. (1995) *In a Different Place: Pilgrimage, Gender, and Politics at a Greek Island Shrine*. Princeton University Press, Princeton, New Jersey.

Filep, S. and Greenacre, L. (2007) Evaluating and extending the travel career patterns model. *Tourism* 55(1), 23–38.

Gholipour, B. (2014) Bike around the world, from home: CycleOps virtual training review. Available at: https://www.livescience.com/46714-virtual-training-indoor-cycling.html (accessed 20 May 2020).

Goeldner, R.C. and Ritchie, B.R.J. (2012) *Tourism. Principles. Practices. Philosophies.* Jonn Wiley and Sons, Inc.

Graburn, H.H.N. (1989) Tourism: the sacred journey. In: Smith, L.V. (ed.) *Hosts and Guests: The Anthropology of Tourism*. University of Pennsylvania Press, pp. 21–36.

Helland, C. (2013) Ritual. In: Cambell, A.H. (ed.) *Digital Religion. Understanding Religious Practice in New Media Worlds*. Routledge, pp. 25–40.

Inoue, N. (2000) From religious conformity to innovation: new ideas of religious journey and holy places. *Social Compass* 47(1), 21–32. DOI: 10.1177/003776800047001004.

Kaelber, L. (2006) Place and pilgrimage, real and imagined. In: Swatos, H.W. (ed.) *On the Road to Being There. Studies in Pilgrimage and Tourism in Late Modernity*. Brill, Leiden, The Netherlands, pp. 277–296.

Kauffman, J. (2020) Jerusalem's Old City, virtual reality edition: a game brings quarantined people around the world into the holy neighborhood. Available at: https://www.jta.org/2020/04/17/lifestyle/jerusalems-old-city-virtual-reality-edition-a-game-brings-quarantined-people-around-the-world-into-the-holy-neighborhood (accessed 15 May 2020).

Liutikas, D. (2014) Religinių objektų Lietuvoje geografinių vietų analizė. *Geografijos metraštis* 47, 103–117.

Liutikas, D. (2015a) Religious landscape and ecological ethics: pilgrimage to the Lithuanian Calvaries. *International Journal of Religious Tourism and Pilgrimage* 3(1), 12–24.

Liutikas, D. (2015b) In search of miracles: pilgrimage to the miraculous places. *Tourism Review* 70(3), 197–213. DOI: 10.1108/TR-08-2013-0046.

Liutikas, D. (2020) Management of pilgrimages in Lithuania: resources, stakeholders and projects. *International Journal of Religious Tourism and Pilgrimage* 8(2), 45–61.

Lourdes Sanctuaire (2020) A flame, a presence. Available at: https://www.lourdes-france.org/en/a-flame-a-presence/ (accessed 15 May 2020).

MacWilliams, M. (2004) Virtual pilgrimage to Ireland's Croagh Patrick. In: Dawson, L. and Cowan, D. (eds) *Religion Online: Finding Faith on the Internet*. Routledge, New York, pp. 223–238.

Mazumdar, S. and Mazumdar, S. (1993) Sacred space and place attachment. *Journal of Environmental Psychology* 13(3), 231–242. DOI: 10.1016/S0272-4944(05)80175-6.

Morinis, A. (1992) Introduction: the territory of the anthropology of pilgrimage. In: Morinis, A. (ed.) *Sacred Journeys. The Anthropology of Pilgrimage*. Greenwood Press, Westport, Connecticut and London, pp. 1–28.

Olin, J.C. (1979) The idea of pilgrimage in the experience of Ignatius Loyola. *Church History* 48(4), 387–397. DOI: 10.2307/3164532.

Pearce, P.L. (2005) *Tourist Behaviour: Themes and Conceptual Schemes*. Channel View Publications, Clevedon, UK.

Pontifical Council for the Pastoral Care of Migrants and Itinerant People (1998) The pilgrimage in the great jubilee. Available at: http://www.vatican.va/roman_curia/pontifical_councils/migrants/documents/rc_pc_migrants_doc_19980425_pilgrimage_en.htm (accessed 1 May 2020).

Reader, I. (2015) *Pilgrimage. A Very Short Introduction*. Oxford University Press.

The Spiritual Exercises of St Ignatius of Loyola (2015) Digireads.com Publishing.

UNWTO (2020) 100% of global destinations now have COVID-19 travel restrictions, UNWTO reports. Available at: https://www.unwto.org/news/covid-19-travel-restrictions (accessed 20 May 2020).

Webb, D. (2000) *Pilgrimage in Medieval England*. Hambledon and London.

Woods, D.D. (2006) Essential characteristics of resilience. In: Hollnagel, E., Woods, D. and Leveson, N. (eds) *Resilience Engineering Concepts and Precepts*. Ashgate Publishing Limited, Aldershot, UK, pp. 21–34.

World Health Organization (2020) Coronavirus disease (COVID-19) pandemics. Available at: https://www.who.int/emergencies/diseases/novel-coronavirus-2019 (accessed 1 June 2020).

Discussion Questions

Chapter 2

1. What are the important values and identities for pilgrimage motivation?
2. Do you notice people travelling in a group of pilgrims who travel in this group not for reasons related to the expression of identity?
3. What are the most common rituals of pilgrims? What do they symbolize?
4. On what basis do we create segmentation models for valuistic travellers? What are the main groups of religious and cultural travellers?

Chapter 3

1. Is it possible to draw distinctions between pilgrim and tourist when a particular traveller might engage in both practices while on the same journey, even on the same day? How can we understand the distinction between pilgrims and tourists?
2. Can sacred places continue to attract pilgrims when they seem to be profaned by tourism? If so, how?
3. Given the mythic interpretation versus a conceptual understanding of wilderness, how do contemporary hikers experience liminality on a wilderness trail?
4. How do pilgrims encounter the face of alterity?

Chapter 4

1. What is the meaning of death in the process of shaping tourist space?
2. Discuss the way death spaces are organized and the motivations to travel to places of death. Are they the result of adopted cultural and religious norms?
3. The development of pop culture led to the growth of post-modernist interest in death space. Which tourist products developed, making use of the elements of death space?
4. Geographical diversity and the historical evolution of cultures are reasons why a contemporary traveller takes interest in the heritage of death. Do you agree with this statement?

Chapter 5

1. How does a pilgrim speak of her/his source of strength if it isn't measurable?
2. What does gait and rhythm tell us about a pilgrim experience?

3. If experience isn't determined by language, how can we speak of it?
4. What is the value of a physical experience to a spiritual practice?

Chapter 6

1. Why is The Way of St James so attractive?
2. Why is The Way a post-secular pilgrimage?
3. Does the pilgrimage along The Way produce any change in pilgrims' personal lives?
4. How do pilgrims' lives change?

Chapter 7

1. What explains the success of the Way of Saint James? What are the keys to this success?
2. Reflections on the Way of Saint James often focus on the debate between religion and tourism. Is it possible to contribute other approaches that help to understand the nature of this route?
3. What is the role of the governance of the Way in relation to the values and ideals that make up the pilgrimage?
4. Is it possible to conceive of the existence and success of the Way without considering the role played by the Associations of Friends of the Way?

Chapter 8

1. How does Turner's concept of 'communitas' differ from that of Carreño?'
2. According to which characteristics do pilgrims walking the Camino de Santiago define themselves? Why might these qualities or characteristics become integral to pilgrim motivations and identities?
3. How does photography contribute, if indeed it does, to the emergence of a universal pilgrim identity along the Camino de Santiago?
4. According to this chapter, how does commensality contribute to and inform pilgrim values, motivations and identities?

Chapter 9

1. What are the main motivations that lead tourist-pilgrims to go the PIWSC?
2. Characterize the identity of Portuguese and foreign tourist-pilgrims.
3. What is the contribution and importance of groups of hikers and pilgrim associations for the valorization and maintenance of PIWSC?
4. What are the main identity aspects of PIWSC and what differentiates it from other pilgrimage paths (Fatima; other Ways of Saint James and regional pilgrimage paths)?

Chapter 10

1. Discuss the influence of contemporary spirituality and heritage tourism in other pilgrimage sites. Has it influenced visitor motivations and values? If so, how?
2. Discuss contemporary spirituality in Japan and how its content is similar or different to other countries.
3. Give examples of other pilgrimage sites that are visited by inbound visitors and discuss their influence in the site.
4. Discuss the growing presence of secular pilgrimage and how it influences the traditional concept of pilgrimage related to religion.

Chapter 11

1. Discuss the characteristics of the *goshuin* movement in Japan from the perspective of religious commodification.
2. Categorize new narratives related to the *goshuin* movement.

3. Discuss the social impact of the *goshuin* movement in Japan from the perspective of cost-effectiveness.
4. Discuss the development of cost-effectiveness in religious experiences in comparison with other religions or field areas.

Chapter 12

Discuss these issues:

1. Hindu pilgrimage faces threats from the secular political ideologies as they are least in favour of religious tourism or pilgrimage-tourism. In another way, a sense of fear is propagated that religious tourism promotes Hindu fundamentalism and societal desperateness (Singh, 2011a).
2. Since the ancient past, spirituality has been at the core of all sorts of Hindu travels and visits to sacred places. Religious tourism should be conceived and promoted in the frame of 'spiritual tourism', as illustrated in the case of Yoga-tourism and tourism based on nature-healing (McCartney, 2020).
3. Religious tourism in India is kept at the margin in development strategy as structured by the ruling political party. Their strategy is using religious tourism mostly for getting power and support from the devout masses, who are mostly illiterate and god-fearing. This is illustrated with the present government at central and state (Uttar Pradesh) levels (ruling since 2014), who are more inclined to promote the Hindu religion and its ideologies for their political strength (Singh, 2011b).
4. In Indian culture, the most common and important attribute is the Hindu belief system, which is deeply rooted in cultural traditions and the Hindu psyche; and this will continue in spite of any sort of superimposition and transformations (Srivastava *et al.*, 2013). Therefore, pilgrimage-tourism will be the better choice. However, this should integrally be linked as part of inclusive development, making pilgrimage-tourism a role model for 'spiritual tourism' and understanding global harmony.

Chapter 13

1. Critique the position that a successful pilgrimage site needs to have a certain potency of mystery which is empowered by the lingering vestige of the divine.
2. Within the scope of pilgrimage or faith-based travel, what values would you attach to originality, character, spirit of place and authenticity when looking for a relevant place to visit?
3. What would you understand spiritual fulfilment means when visiting places of relevance, such as sites of old or modern worship, events, landscapes or ancient places?
4. What kind of values would you look for, expect or demand in valuistic journeys or in experiential travel?

Chapter 14

1. Could a place not be fixed in time or space?
2. Could a pilgrimage shrine be moveable?
3. How can youth be attracted to a traditional practice as pilgrimage?
4. How can religious institutions adapt themselves to a post-modern scenario?

Chapter 15

1. What are the main pilgrimage sites in Central Asia and what are they related to?
2. What are the reasons for visiting sacred places?
3. Discuss the situation of sacred places in the Soviet era. What influence did this period have on the religious identity of the people?
4. Discuss women's pilgrimage opportunities in Central Asia. How important are religious rituals performed only by women for their personal and social identity?

Chapter 16

1. Do you think that business is able to, or should, play a role in forcing a breakthrough in the Palestinian-Israeli conflict?

2. Make an ethical argument for and against each of the following two statements: (a) The cancelling of all pilgrimage holiday excursions to Palestine and Israel will contribute to respecting human rights; (b) Palestinian business such as ATG and other alternative tourism groups should not cease co-operation with Israeli businesses.
3. Is it morally and ethically right to make profits from people living under occupation like the Palestinians?
4. Reflect on the following statements:

a. Many pilgrimages head to Palestine, supporting Palestine and advocating for peace but they do not really know that the way their tour is structured is creating injustice and is widening the gap between Israel and Palestine and between rich and poor.
b. In 2019, over 3 million Christian visitors came to the Holy Land and most of them, the overwhelming majority, came to run where Jesus walked. They are in a hurry and they run from one archaeological site or church to another, touching all these ancient stones, but actually missing the most important thing, which is the people, the culture, the stories and the daily life of the Palestinians.

Chapter 17

1. Is it helpful in the post-modern, digital age to maintain a strict classification between tourist and pilgrims in relation to Holocaust visitations?
2. Should photography at Holocaust museums preclude 'selfies', and would such a ban be enforceable?
3. Irrespective of whether one thinks of oneself as a tourist or a pilgrim, what role do you think imagination and empathy play when visiting Auschwitz in the contemporary era?
4. Victor and Edith Turner (Turner and Turner, 1978, p. 20) once decreed that 'a tourist is half a pilgrim, if a pilgrim is half a tourist'. Is this summation accurate in the context of tourism at *other* places where the Holocaust is memorialized?

Chapter 18

1. Identify the defining aspects of sabbatical that make it a pilgrimage.
2. The notion of 'transformative travel' is an emerging area of scholarship, with various models/frameworks advanced. Choose one of these models/frameworks and discuss its relevance to transformative travel.
3. The concepts of 'space' and 'time' are often associated with the literature on pilgrimage and sabbatical. Choose one of these two concepts and discuss its relevance to the sabbatical.
4. It has been said that Douglas Adams's book *A Hitchhiker's Guide to the Galaxy* is a liberating read that provides escape from our existential lives. Can we say the same for the sabbatical?

Chapter 19

1. How is pilgrimage an identity-building exercise?
2. Why has pilgrimage flourished rather than disappeared in the modern world?
3. How has pilgrimage changed due to globalization?
4. What effects do you think the COVID-19 pandemic will have on pilgrimage in theimmediate future?

References

McCartney, P. (2020) Yoga-scapes, Embodiment and imagined spiritual tourism. In: Palmer, C. and Andrews, H. (eds) *Tourism and Embodiment*. Routledge, London & New York, pp. 86–106.

Singh, R.P.B. (2011a) Pilgrimage and religious tourism in India: countering contestation and seduction. In: Singh, R.P.B. (ed.) *Holy Places and Pilgrimages: Essays on India*. Shubhi Publications, New Delhi, pp. 307–334.

Singh, R.P.B. (2011b) Politics and pilgrimage in North India: Varanasi between communitas and contestation. *Tourism* 59(3).

Srivastava, C., Dhingra, V., Srivastava, A. and Bhardwaj, A. (2013) Morality and moral development: traditional Hindu concepts. *Indian Journal of Psychiatry* 55(6), 283–287.

Turner, V. and Turner, E. (1978) *Image and Pilgrimage in Christian Culture: Anthropological Perspectives*. Columbia University Press, New York.

Index

Note: Page numbers in **bold** type refer to **figures**
Page numbers in *italic* type refer to *tables*